STUDENT'S SOLUTIONS MANUAL

TONI C. GARCIA
Arizona State University

ELEMENTARY STATISTICS
EIGHTH EDITION

Neil A. Weiss
Arizona State University

Addison-Wesley
is an imprint of

The author and publisher of this book have used their best efforts in preparing this book. These efforts include the development, research, and testing of the theories and programs to determine their effectiveness. The author and publisher make no warranty of any kind, expressed or implied, with regard to these programs or the documentation contained in this book. The author and publisher shall not be liable in any event for incidental or consequential damages in connection with, or arising out of, the furnishing, performance, or use of these programs.

Reproduced by Pearson Addison-Wesley from electronic files supplied by the author.

Copyright © 2012, 2008, 2005 Pearson Education, Inc.
Publishing as Pearson Addison-Wesley, 75 Arlington Street, Boston, MA 02116.

All rights reserved. No part of this publication may be reproduced, stored in a retrieval system, or transmitted, in any form or by any means, electronic, mechanical, photocopying, recording, or otherwise, without the prior written permission of the publisher. Printed in the United States of America.

ISBN-13: 978-0-321-69141-5
ISBN-10: 0-321-69141-5

1 2 3 4 5 6 OPM 14 13 12 11 10

Addison-Wesley
is an imprint of

www.pearsonhighered.com

Contents

Copyright © 2012 Pearson Education, Inc. Publishing as Addison-Wesley

CHAPTER 1 ANSWERS

Exercises 1.1

1.1 (a) The *population* is the collection of all individuals or items under consideration in a statistical study.

(b) A *sample* is that part of the population from which information is obtained.

1.3 Descriptive methods are used for organizing and summarizing information and include graphs, charts, tables, averages, measures of variation, and percentiles.

1.5 (a) An *observational study* is a study in which researchers simply observe characteristics and take measurements.

(b) A *designed experiment* is a study in which researchers impose treatments and controls and *then* observe characteristics and take measurements.

1.7 This study is inferential. Data from a sample of Americans are used to make an estimate of (or an inference about) average TV viewing time for all Americans.

1.9 This study is descriptive. It is a summary of the assessment results for level of performance of all senior geography majors in 2003 and 2004 at one institution.

1.11 This study is descriptive. It is a summary of the annual final closing values of the Dow Jones Industrial Average at the end of December for the years 2000-2008.

1.13 (a) This study is inferential. It would have been impossible to survey all U.S. adults about their opinions on Darwinism. Therefore, the data must have come from a sample. Then inferences were made about the opinions of all U.S. adults.

(b) The population consists of all U.S. adults. The sample consists only of those U.S. adults who took part in the survey.

1.15 (a) The statement is descriptive since it only tells what was said by the respondents of the survey.

(b) Then the statement would be inferential since the data has been used to provide an estimate of what <u>all</u> Americans believe.

1.17 Designed experiment. The researchers did not simply observe the two groups of children, but instead randomly assigned one group to receive the Salk vaccine and the other to get a placebo.

1.19 Observational study. The researchers had no control over who became an elite distance runner and who did not. They simply observed the skinfold thickness of a sample of people in each group.

1.21 Designed experiment. The researchers did not simply observe the three groups of patients, but instead randomly assigned some patients to receive optimal pharmacologic therapy, some to receive optimal pharmacologic therapy and a pacemaker, and some to receive optimal pharmacologic therapy and a pacemaker-defibrillator combination.

1.23 (a) This statement is inferential since it is a statement about all Americans based on a poll. We can be reasonably sure that this is the case since the time and cost of questioning every single American on this issue would be prohibitive. Furthermore, by the time everyone could be questioned, many would have changed their minds.

(b) To make it clear that this is a descriptive statement, the new statement could be, "Of 1032 American adults surveyed, 73% favored a law that would require every gun sold in the United States to be test-fired first, so law enforcement would have its fingerprint in case it were ever used in a crime." To rephrase it as an inferential

Copyright © 2012 Pearson Education, Inc. Publishing as Addison-Wesley.

statement, use "Based on a sample of 1032 American adults, it is estimated that 73% of American adults favor a law that would require every gun sold in the United States to be test-fired first, so law enforcement would have its fingerprint in case it were ever used in a crime."

1.25 (a) The figure 42800 is an inferential statistic since it is indicated in the statement that it is a projection (probably based on incomplete data for the year 2004). The data may be incomplete in part because in April, 2005, there might still be deaths to occur that are the result of traffic accidents in 2004.

(b) The figure 42643 is a descriptive statistic since it reflects the actual number of traffic deaths for the year 2003.

Exercises 1.2

1.27 A census is generally time consuming, costly, frequently impractical, and sometimes impossible.

1.29 The sample should be representative so that it reflects as closely as possible the relevant characteristics of the population under consideration.

1.31 Dentists form a high-income group whose incomes are not representative of the incomes of Seattle residents in general.

1.33 (a) Probability sampling consists of using a randomizing device such as tossing a coin or consulting a random number table to decide which members of the population will constitute the sample.

(b) No. It is possible for the randomizing device to randomly produce a sample that is not representative.

(c) Probability sampling eliminates unintentional selection bias, permits the researcher to control the chance of obtaining a non-representative sample, and guarantees that the techniques of inferential statistics can be applied.

1.35 Simple random sampling.

1.37 (a) GLS, GLA, GLT, GSA, GST, GAT, LSA, LST, LAT, SAT.

(b) There are 10 samples, each of size three. Each sample has a one in 10 chance of being selected. Thus, the probability that a sample of three officials is the first sample on the list presented in part (a) is 1/10. The same is true for the second sample and for the tenth sample.

1.39 (a)

E,M,P,L	E,M,L,B	E,P,A,B	M,P,A,B
E,M,P,A	E,M,A,B	E,L,A,B	M,L,A,B
E,M,P,B	E,P,L,A	M,P,L,A	P,L,A,B
E,M,L,A	E,P,L,B	M,P,L,B	

(b) One procedure for taking a random sample of four representatives from the six is to write the initials of the representatives on six separate pieces of paper, place the six slips of paper into a box, and then, while blindfolded, pick four of the slips of paper. Or, number the representatives 1-6, and use a table of random numbers or a random-number generator to select four different numbers between 1 and 6.

(c) 1/15; 1/15

1.41 (a)

C,W,H	C,W,V	C,W,A	C,H,V	C,H,A
C,V,A	W,H,V	W,H,A	W,V,A	H,V,A

(b) 1/10; 1/10

1.43 (a) I am using Table I to obtain a list of 10 random numbers between 1 and

Copyright © 2012 Pearson Education, Inc. Publishing as Addison-Wesley.

500 as follows.

I start at the three digit number in line number 14 and column numbers 10-12, which is the number 452.

I now go down the table and record the three-digit numbers appearing directly beneath 452. Since I want numbers between 1 and 500 only, I throw out numbers between 501 and 999, inclusive. I also discard the number 000.

After 452, I skip 667, 964, 593, 534, and record 016.

Now that I've reached the bottom of the table, I move directly rightward to the adjacent column of three-digit numbers and go up.

I record 343, 242, skip 748, 755, record 428, skip 852, 794, 596, record 378, skip 890, record 163, skip 892, 847, 815, 729, 911, 745, record 182, 293, and 422.

I've finished recording the 10 random numbers. In summary, these are:

452	016	343	242	428
378	163	182	293	422

(b) We can use Excel to generate random numbers. If we enter the expression =RAND() in a cell, Excel returns a random number between 0 and 1. If we multiply this number by 500, we will get a number between 0 and 500, and if we take the integer part of that result, we will get one of the integers 0 through 499. If we add 1 to this result, we will get a random number between 1 and 500. For example, in cell A1, we enter the expression =INT(500*RAND())+1. One of integers 1 through 500 will result. We copy the content of A1 to the clipboard and paste it into cells A2 through A30. Then we choose the first 10 unique three-digit numbers for our sample. Our result is 489, 451, 61, 114, 389, 381, 364, 166, 221, 437, 266, 46, 422, 388, 401, 387, 276, 248, 21, 198. The first 10 unique three-digit numbers are 489, 451, 61, 114, 389, 381, 364, 166, 221, and 437. Your result may be different from ours. [Note: Each time the ENTER key is depressed, the random numbers will be recalculated, so make sure that you do not press the ENTER key until you have recorded all of your random numbers.]

1.45 (a) The possible samples of size one are   G   L   S   A   T

    (b) There is no difference between obtaining a sample of size one and selecting one official at random.

1.47 No. Only adults with access to the Internet would have been able to respond to the survey. Thus, not every adult had an equal chance of being chosen for the sample.

Exercises 1.3

1.49 (a) Answers will vary, but here is the procedure: (1) Divide the population size, 500, by the sample size, 10, and round down to the nearest whole number if necessary; this gives 50. (2) Use a table of random numbers (or a similar device) to select a number between 1 and 50, call it k. (3) List every 50th number, starting with k, until 10 numbers are obtained; thus, the first number on the required list of 10 numbers is k, the second is $k+50$, the third is $k+100$, and so forth (e.g., if $k=6$, then the numbers on the list are 6, 56, 106, ...).

    (b) Systematic random sampling is easier.

    (c) The answer depends on the purpose of the sampling. If the purpose of sampling is not related to the size of the sales outside the U.S., systematic sampling will work. However, since the listing is a ranking by amount of sales, if k is low (say 2), then the sample will contain firms that, on the average, have higher sales outside the U.S. than the

Copyright © 2012 Pearson Education, Inc. Publishing as Addison-Wesley.

population as a whole. If the k is high, (say 49) then the sample will contain firms that, on the average, have lower sales than the population as a whole. In either of those cases, the sample would not be representative of the population in regard to the amount of sales outside the U.S.

1.51 (a) Number the suites from 1 to 48, use a table of random numbers to randomly select three of the 48 suites, and take as the sample the 24 dormitory residents living in the three suites obtained.

(b) Probably not, since friends are more likely to have similar opinions than are strangers.

(c) There are 384 students in total. Freshmen make up 1/3 of them. Sophomores make up 7/24 of them, Juniors 1/4, and Seniors 1/8. Multiplying each of these fractions by 24 yields the proportional allocation, which dictates that the number of freshmen, sophomores, juniors, and seniors selected should be, respectively, 8, 7, 6, and 3. Thus a stratified sample of 24 dormitory residents can be obtained as follows: Number the freshmen dormitory residents from 1 to 128 and use a table of random numbers to randomly select 8 of the 128 freshman dormitory residents; number the sophomore dormitory residents from 1 to 112 and use a table of random numbers to randomly select 7 of the 112 sophomore dormitory residents; and so forth.

1.53 Stratified Sampling. The entire population is naturally divided into subpopulations, one from each lake, and random sampling is done from each lake. The stratified sampling is not with proportional allocation since that would require knowing how many fish were in each lake.

1.55 (a) This is a poll taken by calling randomly selected U.S. adults. Thus, the sampling design appears to be simple random sampling, although it is possible that a more complex design was used to ensure that various political, religious, educational, or other types of groups were proportionately represented in the sample.

(b) The sample size for the second question was 78% of 1010 or 788.

(c) The sample size for the third question was 28% of 788 or 221.

1.57 (a) It is also true for systematic random sampling if the population size divided by the sample size results in an <u>integer</u> for m. The chance for each member to be selected is then still equal to the sample size divided by the population size. For example, suppose the population size is N=10 and the sample size is n=2. The chance that each member in simple random sampling to be selected is 2/10 = 1/5. In systematic random sampling for the same example, m=5. The possible samples of size two are 1 and 6, 2 and 7, 3 and 8, 4 and 9, and 5 and 10. Therefore, the chance that a member is selected is equal to the chance of one of those five samples being selected, which is the same as simple random sampling of 1/5.

(b) It is not true for systematic random sampling if the population size divided by the sample size <u>does not result in an integer</u> for m. For example, suppose the population size is N=15 and the sample size is n=2. After dividing the population size by the sample size and rounding down to the nearest whole number, we get m=7. You would select every 7th member after a random starting place k, between 1 and 7, is determined. If k=1, you would select the first and eighth member. If k=7, you would select the seventh and fourteenth member. In this situation, the last member (fifteenth) can never be selected. Therefore, the last member of the sample does not have the same chance of being selected as any other member in the population.

1.59 From the information about the sample, we can conclude that the population of interest consists of all adults in the continental U.S. The sample size was 2010 except that for questions about politics, only registered voters

Copyright © 2012 Pearson Education, Inc. Publishing as Addison-Wesley.

were considered part of the sample. The sample size for those questions was 1637.

The overall procedure for drawing the sample was multistage (actually, three stages were used) sampling: the first stage was to randomly select 520 geographic points in the continental U.S.; then proportional sampling was used to randomly sample a number of households with telephones from each of the 520 regions in proportion to its population; finally, once each household was selected, a randomizing procedure was used to ensure that the correct numbers of adult male and female respondents were included in the sample.

The last paragraph indicates the confidence that the poll-takers had in the results of the survey, that is, that there is a 95% chance that the sample results will not differ by more than 2.2 percentage points in either direction from the true percentage that would have been obtained by surveying all adults in the actual population, or by more than 2.5 percentage points in either direction from the true percentage that would have been obtained by surveying all registered voters in the population. The last sentence says that smaller samples have a larger margin of error,@ an explanation for the difference in the maximum percentage points of error for all adults and for registered voters.

Exercises 1.4

1.61 (a) Experimental units are the individuals or items on which the experiment is performed.

(b) When the experimental units are humans, we call them subjects.

1.63 (a) There were three treatments.

(b) The first group, the one receiving only the pharmacologic therapy, would be considered the control group.

(c) There were three treatment groups. The first received only pharmacologic therapy, the second received pharmacologic therapy plus a pacemaker, and the third received pharmacologic therapy plus a pacemaker-defibrillator combination.

(d) The first group (control) contained 1/5 of the 1520 patients or 304. The other two groups each contained 2/5 of the 1520 patients or 608.

(e) Each patient could be randomly assigned a number from 1 to 1520. Any patient assigned a number between 1 and 304 would be assigned to the control group; any patient assigned to the next 608 numbers (305 to 912) would be assigned to receive the pharmacologic therapy plus a pacemaker; and any patient assigned a number between 913 and 1520 would receive pharmacologic therapy plus a pacemaker-defibrillator combination. Each random number would be used only once to ensure that the resulting treatment groups were of the intended sizes.

1.65 (a) Experimental units: batches of the product being sold

(b) Response variable: the number of units of the product sold

(c) Factors: two factors - display type and pricing scheme

(d) Levels of each factor: three types of display of the product and three pricing schemes

(e) Treatments: the nine different combinations of display type and price resulting from testing each of the three pricing schemes with each of the three display types

1.67 (a) Experimental units: female lions

(b) Response variable: whether or not the female lion approached the male lion dummy

Copyright © 2012 Pearson Education, Inc. Publishing as Addison-Wesley.

      (c) Factors: length and color of the mane on the male lion dummy

      (d) Levels of each factor: two different mane lengths and two different mane colors

      (e) Treatments: the four combinations of mane length and color

1.69 Double-blinding guards against bias, both in the evaluations and in the responses. In the Salk vaccine experiment, double-blinding prevented a doctor's evaluation from being influenced by knowing which treatment (vaccine or placebo) a patient received; it also prevented a patient's response to the treatment from being influenced by knowing which treatment he or she received.

Review Problems for Chapter 1

1. Student exercise.

2. Descriptive statistics are used to display and summarize the data to be used in an inferential study. Preliminary descriptive analysis of a sample often reveals features of the data that lead to the choice or reconsideration of the choice of the appropriate inferential analysis procedure.

3. Descriptive study. The scores are merely reported.

4. Descriptive study. The paragraph describes the results of 1,020 respondents surveyed.

5. Inferential study. The results of a sample are used to make inferences about the age distribution of all British backpackers in South Africa.

6. (a) Since 18% reported that they had abused Vicodin, this figure applies to the sample and is therefore descriptive.

      (b) The 4.3 million youths abusing Vicodin is clearly inferential since it applies to the entire population, not just the sample.

7. (a) An *observational study* is a study in which researchers simply observe characteristics and take measurements.

      (b) A *designed experiment* is a study in which researchers impose treatments and controls and *then* observe characteristics and take measurements.

8. This is an observational study. To be a designed experiment, the researchers would have to have the ability to assign some children at random to live in persistent poverty during the first 5 years of life or to not suffer any poverty during that period. Clearly that is not possible.

9. This is a designed experiment since the researcher is imposing a treatment and then observing the results.

10. A literature search should be made before planning and conducting a study.

11. (a) A representative sample is one that reflects as closely as possible the relevant characteristics of the population under consideration.

      (b) Probability sampling involves the use of a randomizing device such as tossing a coin or die, using a random number table, or using computer software that generates random numbers to determine which members of the population will make up the sample.

      (c) A sample is a simple random sample if all possible samples of a given size are equally likely to be the actual sample selected.

12. Because Yale is a very expensive school, incomes of parents of Yale students will not be representative of the incomes of all college students' parents.

13. (a) This method does not involve probability sampling. No randomizing device is being used and people who do not visit the campus cafeteria have no chance of being included in the sample.

Copyright © 2012 Pearson Education, Inc. Publishing as Addison-Wesley.

(b) The dart throwing is a randomizing device that makes all samples of size 20 equally likely. This is probability sampling.

14. (a) H,P,S   H,P,A   H,P,E   H,S,A   H,S,E

   H,A,E   P,S,A   P,S,E   P,A,E   S,A,E

   (b) Since each of the 10 samples of size three is equally likely, there is a 1/10 chance that the sample chosen is the first sample in the list, 1/10 chance that it is the second sample in the list, and 1/10 chance that it is the tenth sample in the list.

   (c) (i) Make five slips of paper with each airline on one slip. Draw three slips at random. (ii) Make 10 slips of paper, each having one of the combinations in part (a). Draw one slip at random. (iii) Number the five airlines from 1 to 5. Use a random number table or random number generator to obtain three distinct random numbers between 1 and 5 inclusive.

   (d) Your method and result may differ from ours. We rolled a die (ignoring 6's and duplicates) and got 2, 5, 2, 6, 4. So our sample consists of Pinnacle Airlines, Atlantic Southeast Airlines, and Alaska Airlines.

15. (a) Table I can be employed to obtain a sample of 15 random numbers between 1 and 100 as follows. First, I pick a random starting point by closing my eyes and putting my finger down on the table.

   My finger falls on three digits located at the intersection of a line with three columns. (Notice that the first column of digits is labeled "00" rather than "01".) This is my starting point.

   I now go down the table and record all three-digit numbers appearing directly beneath the first three-digit number that are between 001 and 100 inclusive. I throw out numbers between 101 and 999, inclusive. I also discard the number 0000. When the bottom of the column is reached, I move over to the next sequence of three digits and work my way back up the table. Continue in this manner. When 10 distinct three-digit numbers have been recorded, the sample is complete.

   (b) Starting in row 10, columns 7-9, we skip 484, 797, record 082, skip 586, 653, 452, 552, 155, record 008, skip 765, move to the right and record 016, skip 534, 593, 964, 667, 452, 432, 594, 950, 670, record 001, skip 581, 577, 408, 948, 807, 862, 407, record 047, skip 977, move to the right, skip 422 and all of the rest of the numbers in that column, move to the right, skip 732, 192, record 094, skip 615 and all of the rest of the numbers in that column, move to the right, record 097, skip 673, record 074, skip 469, 822, record 052, skip 397, 468, 741, 566, 470, record 076, 098, skip 883, 378, 154, 102, record 003, skip 802, 841, move to the right, skip 243, 198, 411, record 089, skip 701, 305, 638, 654, record 041, skip 753, 790, record 063.

   The final list of numbers is 082, 008, 016, 001, 047, 094, 097, 074, 052, 076, 098, 003, 089, 041, 063.

   (c) Using Excel, we enter the expression =INT(100*RAND())+1 in cell A1, copy the content of A1 to the clipboard and paste it into cells A2 to A20. Then we use the first 15 unique numbers as our random sample. Our results were the numbers 46, 99, 90, 31, 75, 98, 79, 14, 44, 13, 66, 49, 37, 87, 73, 26, 61, 71, 72, 2. Thus our sample consists of the first 15 numbers 46, 99, 90, 31, 75, 98, 79, 14, 44, 13, 66, 49, 37, 87, 73. Your sample may be different.

16. (a) Systematic random sampling is done by first dividing the population size by the sample size and rounding the result down to the next integer, say m. Then we select one random number, say k, between 1 and m inclusive. That number will be the first member of the sample. The remaining members of sample will be those numbered k+m, k+2m, k+3m, ... until a sample of size n has been chosen. Systematic sampling will

Copyright © 2012 Pearson Education, Inc. Publishing as Addison-Wesley.

yield results similar to simple random sampling as long as there is nothing systematic about the way the members of the population were assigned their numbers.

(b) In cluster sampling, clusters of the population (such as blocks, precincts, wards, etc.) are chosen at random from all such possible clusters. Then every member of the population lying within the chosen clusters is sampled. This method of sampling is particularly convenient when members of the population are widely scattered and is most appropriate when the members of each cluster are representative of the entire population. Cluster sampling can save both time and expense in doing the survey, but can yield misleading results if individual clusters are made up of subjects with very similar views on the topic being surveyed.

(c) In stratified random sampling with proportional allocation, the population is first divided into subpopulations, called strata, and simple random sampling is done within each stratum. Proportional allocation means that the size of the sample from each stratum is proportional to the size of the population in that stratum. This type of sampling may improve the accuracy of the survey by ensuring that those in each stratum are more proportionately represented than would be the case with cluster sampling or even simple random sampling. Ideally, the members of each stratum should be homogeneous relative to the characteristic under consideration. If they are not homogeneous within each stratum, simple random sampling would work just as well.

17. (a) Answers will vary, but here is the procedure: (1) Divide the population size, 100, by the sample size 15, and round down to the nearest whole number; this gives 6. (2) Use a table of random numbers (or a similar device) to select a number between 1 and 6, call it k. (3) List every 6th number, starting with k, until 15 numbers are obtained; thus the first number on the required list of 15 numbers is k, the second is $k+6$, the third is $k+12$, and so forth (e.g., if $k=4$, then the numbers on the list are 4, 10, 16, ...).

(b) Yes, unless for some reason there is some kind of trend or a cyclical pattern in the listing of the athletes.

18. (a) The number of full professors should be (205/820) x 40 = 10. Similarly, proportional allocation dictates that 16 associate professors, 12 assistant professors, and 2 instructors be selected.

(b) The procedure is as follows: Number the full professors from 1 to 205, and use Table I to randomly select 10 of the 205 full professors; number the associate professors from 1 to 328, and use Table I to randomly select 16 of the 328 associate professors; and so on.

19. The statement under the vote is a disclaimer as to the validity of the survey. Since the vote reflects only the responses of volunteers who chose to vote, it can not be regarded as representative of the public in general, some of whom do not use the Internet, nor as representative of Internet users since the sample was not chosen at random from either group.

20. (a) This is a designed experiment.

(b) The treatment group consists of the 158 patients who took AVONEX. The control group consists of the 143 patients who were given a placebo. The treatments were the AVONEX and the placebo.

21. The three basic principles of experimental design are control, randomization, and replication. Control refers to methods for controlling factors other than those of primary interest. Randomization means randomly dividing the subjects into groups in order to avoid unintentional selection bias in constituting the groups. Replication means using enough experimental units or subjects so that groups resemble each other closely and so that there is a good chance of detecting differences among the

Copyright © 2012 Pearson Education, Inc. Publishing as Addison-Wesley.

treatments when such differences actually exist.

22. (a) Experimental units: tomato plants

(b) Response variable: yield of tomatoes

(c) Factor(s): tomato variety and density of plants

(d) Levels of each factor: These are not given, but tomato varieties tested would be the levels of variety and the different densities of plants would be the levels of density.

(e) Treatments: Each treatment would be one of the combinations of a variety planted at a given plant density.

23. (a) Experimental Units: The children

(b) Response variable: Whether or not the child was able to open the bottle

(c) Factors: The container designs

(d) Levels of each factor: Three (types of containers)

(e) Treatments: The container designs

24. This is a completely randomized design. All of the experimental units (batches of doughnuts) were assigned at random to the four treatments (four different fats).

25. (a) This is a completely randomized design since the 24 cars were randomly assigned to the 4 brands of gasoline.

(b) This is a randomized block design. The four different gasoline brands are randomly assigned to the four cars in each of the six car model groups. The blocks are the six groups of four identical cars each.

(c) If the purpose is to learn about the mileage rating of one particular car model with each of the four gasoline brands, then the completely randomized design is appropriate. But if the purpose is to learn about the performance of the gasoline across a variety of cars (and this seems more reasonable), then the randomized block design is more appropriate and will allow the researcher to determine the effect of car model as well as of gasoline type on the mileage obtained.

26. The explanation informs the reader that only random sampling of a population can be used to draw valid inferences about the population as a whole.

27. The data in this study were clearly not collected via a controlled experiment in which some participants were forced to do crossword puzzles, practice musical instruments, play board games, or read while others were not allowed to do any of those activities. Therefore, any data relative to these activities and dementia arose as a result of observing whether or not the subjects in the study carried out any of those activities and whether or no they had some form of dementia. Since this would be an observational study, no statement of cause and effect can rightfully be made. It cannot be claimed as result of the study that "Crosswords Reduce Risk of Dementia."

28. The researchers did not impose or manipulate any of the conditions of this study. They didn't decide who had cancer, who didn't have cancer, who had hepatitis B, or who had hepatitis C. This study was an observational study and not a controlled experiment. Observational studies can only reveal an association, not causation. Therefore, the statement in quotes is valid. If the researchers wanted to establish causation, they would need a designed experiment.

Copyright © 2012 Pearson Education, Inc. Publishing as Addison-Wesley.

CHAPTER 2 ANSWERS

Exercises 2.1

2.1   (a) Hair color, model of car, and brand of popcorn are qualitative variables.

      (b) Number of eggs in a nest, number of cases of flu, and number of employees are discrete, quantitative variables.

      (c) Temperature, weight, and time are quantitative continuous variables.

2.3   (a) Qualitative data result from observing and recording values of a qualitative variable, such as, color or shape.

      (b) Discrete, quantitative data are values of a discrete quantitative variable. Values usually result from counting something.

      (c) Continuous, quantitative data are values of a continuous variable. Values are usually the result of measuring something such as temperature that can take on any value in a given interval.

2.5   Of qualitative and quantitative (discrete and continuous) types of data, only qualitative yields nonnumerical data.

2.7   (a) The second column consists of *quantitative, discrete* data. This column provides the ranks of the cities with the highest temperatures.

      (b) The third column consists of *quantitative, continuous* data since temperatures can take on any value from the interval of numbers found on the temperature scale. This column provides the highest temperature in each of the listed cities.

      (c) The information that Phoenix is in Arizona is *qualitative* data since it is nonnumeric.

2.9   (a) The first column consists of *quantitative, discrete* data. This column provides the ranks of the top ten countries with the highest number of Wi-Fi locations, as of October 28, 2009. These are whole numbers.

      (b) The countries listed in the second column are *qualitative* data since they are nonnumerical.

      (c) The third column consists of *quantitative, discrete* data. This column provides the number of Wi-Fi locations in each of the countries. These are whole numbers.

2.11 The first column contains *quantitative, discrete* data in the form of ranks. These are whole numbers. The second and third columns contain *qualitative* data in the form of names. The last column contains the number of viewers of the programs. Total number of viewers is a whole number and therefore *quantitative, discrete* data.

2.13 The first column contains *quantitative, discrete* data in the form of ranks. These are whole numbers. The second and fourth columns are nonnumerical and are therefore *qualitative* data. The third and fifth columns are measures of time and weight, both of which are *quantitative, continuous* data.

Exercises 2.2

2.15 A frequency distribution of qualitative data is a table that lists the distinct values of data and their frequencies. It is useful to organize the data and make it easier to understand.

2.17 (a) True. Having identical frequency distributions implies that the total number of observations and the numbers of observations in each class are identical. Thus, the relative frequencies will also be identical.

Copyright © 2012 Pearson Education, Inc. Publishing as Addison-Wesley.

(b)   False. Having identical relative frequency distributions means that the ratio of the count in each class to the total is the same for both frequency distributions. However, one distribution may have twice (or some other multiple) the total number of observations as the other. For example, two distributions with counts of 5, 4, 1 and 10, 8, 2 would be different, but would have the same relative frequency distribution.

(c)   If the two data sets have the same number of observations, either a frequency distribution or a relative-frequency distribution is suitable. If, however, the two data sets have different numbers of observations, using relative-frequency distributions is more appropriate because the total of each set of relative frequencies is 1, putting both distributions on the same basis for comparison.

2.19 (a)-(b)

The classes are the NCAA wrestling champions and are presented in column 1. The frequency distribution of the champions is presented in column 2. Dividing each frequency by the total number of champions, which is 25, results in each class's relative frequency. The relative frequency distribution is presented in column 3.

Champion	Frequency	Relative Frequency
Iowa	13	0.52
Iowa St.	1	0.04
Minnesota	3	0.12
Arizona St.	1	0.04
Oklahoma St.	7	0.28
	25	1.00

(b)   We multiply each of the relative frequencies by 360 degrees to obtain the portion of the pie represented by each team. The result is

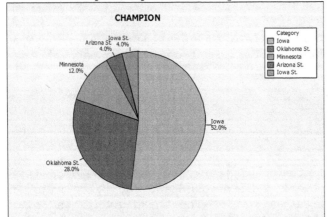

(c)   We use the bar chart to show the relative frequency with which each TEAM occurs. The result is

Copyright © 2012 Pearson Education, Inc. Publishing as Addison-Wesley.

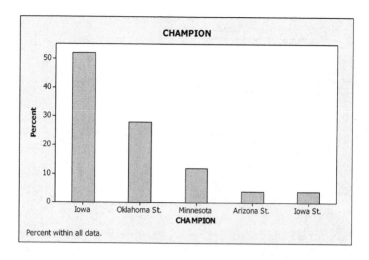

2.21 (a)-(b)

The classes are the class levels and are presented in column 1. The frequency distribution of the class levels is presented in column 2. Dividing each frequency by the total number of students in the introductory statistics class, which is 40, results in each class's relative frequency. The relative frequency distribution is presented in column 3.

Class Level	Frequency	Relative Frequency
Fr	6	0.150
So	15	0.375
Jr	12	0.300
Sr	7	0.175
	40	1.000

(c) We multiply each of the relative frequencies by 360 degrees to obtain the portion of the pie represented by each class level. The result is

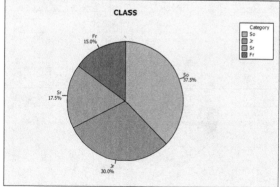

(d) We use the bar chart to show the relative frequency with which each CLASS level occurs. The result is

Copyright © 2012 Pearson Education, Inc. Publishing as Addison-Wesley.

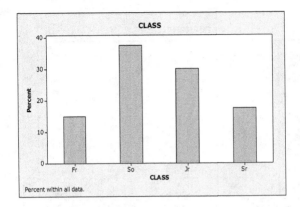

2.23 (a) - (b)

The classes are the days and are presented in column 1. The frequency distribution of the days is presented in column 2. Dividing each frequency by the total number road rage incidents, which is 69, results in each class's relative frequency. The relative frequency distribution is presented in column 3.

Class Level	Frequency	Relative Frequency
Su	5	0.0725
M	5	0.0725
Tu	11	0.1594
W	12	0.1739
Th	11	0.1594
F	18	0.2609
Sa	7	0.1014
	69	1.0000

(c) We multiply each of the relative frequencies by 360 degrees to obtain the portion of the pie represented by each day. The result is

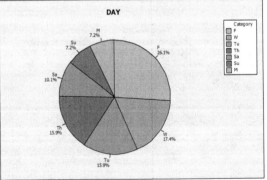

(d) We use the bar chart to show the relative frequency with which each DAY occurs. The result is

Copyright © 2012 Pearson Education, Inc. Publishing as Addison-Wesley.

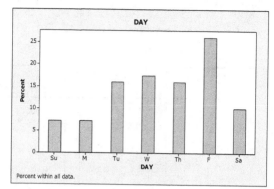

2.25 (a) We first find the relative frequencies by dividing each of the frequencies by the total sample size of 509.

Color	Frequency	Relative
Brown	152	0.2986
Yellow	114	0.2240
Red	106	0.2083
Orange	51	0.1002
Green	43	0.0845
Blue	43	0.0845
	509	1.0000

(b) We multiply each of the relative frequencies by 360 degrees to obtain the portion of the pie represented by each color of M&M. The result is

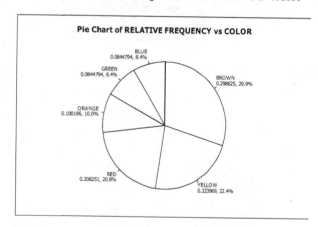

(c) We use the bar chart to show the relative frequency with which each color occurs. The result is

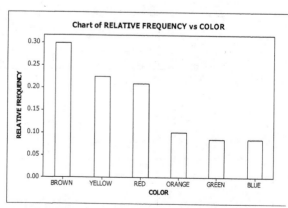

Copyright © 2012 Pearson Education, Inc. Publishing as Addison-Wesley.

2.27 (a) We first find the relative frequencies by dividing each of the frequencies by the total sample size of 98,993.

Rank	Frequency	Relative Frequency
Professor	24,418	0.2467
Associate professor	21,732	0.2195
Assistant professor	40,379	0.4079
Instructor	10,960	0.1107
Other	1,504	0.0152
	98,993	1.0000

(b) We multiply each of the relative frequencies by 360 degrees to obtain the portion of the pie represented by each rank. The result is

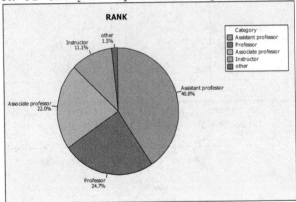

(c) We use the bar chart to show the relative frequency with which each rank occurs. The result is

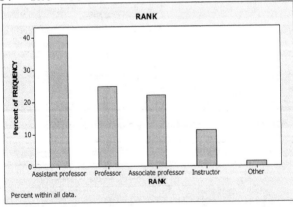

2.29 (a) We first find the relative frequencies by dividing each of the frequencies by the total sample size of 200.

Copyright © 2012 Pearson Education, Inc. Publishing as Addison-Wesley.

Color	Frequency	Relative Frequency
Red	88	0.44
Black	102	0.51
Green	10	0.05
	200	1.00

(b) We multiply each of the relative frequencies by 360 degrees to obtain the portion of the pie represented by each color. The result is

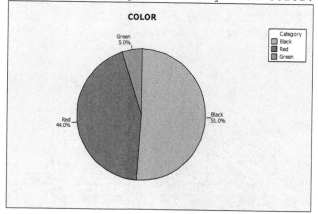

(c) We use the bar chart to show the relative frequency with which each color occurs. The result is

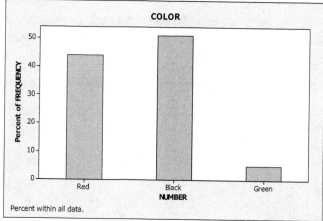

2.31 (a) Using Minitab, retrieve the data from the Weiss-Stats-CD. Column 1 contains the type of the hospital. From the tool bar, select **Stat ▶**

Tables ▶ Tally Individual Variables, double-click on TYPE in the first box so that TYPE appears in the **Variables** box, put a check mark next to Counts and Percents under Display, and click **OK.** The result is

TYPE	Count	Percent
NPC	2919	50.79
IOC	889	15.47
SLC	1119	19.47
FGH	221	3.85
NLT	129	2.24
NFP	451	7.85
HUI	19	0.33
N=	5747	

Copyright © 2012 Pearson Education, Inc. Publishing as Addison-Wesley.

(b) The relative frequencies were calculated in part(a) by putting a check mark next to Percents. 50.79% of the hospitals were Nongovernment not-for-profit community hospitals, 15.47% were Investor-owned (for-profit) community hospitals, 19.47% were State and local government community hospitals, 3.85% were Federal government hospitals, 2.24% were Nonfederal long term care hospitals, 7.85% were Nonfederal psychiatric hospitals, and 0.33% were Hospital units of institutions.

(c) Using Minitab, select **Graph** ▸ **Pie Chart**, check Chart counts of unique values, double-click on TYPE in the first box so that TYPE appears in the Categorical Variables box. Click Pie Options, check decreasing volume, click OK. Click Labels, enter TYPE in for the title, click Slice Labels, check Category Name, Percent, and Draw a line from label to slice, Click OK twice. The result is

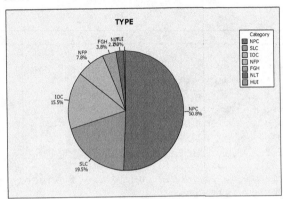

(d) Using Minitab, select **Graph** ▸ **Bar Chart**, select Counts of unique values, select Simple option, click OK. Double-click on TYPE in the first box so that TYPE appears in the Categorical Variables box. Select Chart Options, check decreasing Y, check show Y as a percent, click OK. Select Labels, enter in TYPE as the title. Click OK twice. The result is

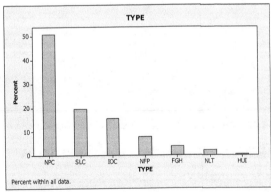

2.33 (a) Using Minitab, retrieve the data from the Weiss-Stats-CD. Column 2 contains the preference for how the members want to receive the ballots and column 3 contains the highest degree obtained by the members. From the tool bar, select **Stat** ▸ **Tables** ▸ **Tally Individual Variables**, double-click on PREFERENCE and DEGREE in the first box so that both PREFERENCE and DEGREE appear in the **Variables** box, put a check mark next to Counts and Percents under Display, and click **OK**. The results are

Copyright © 2012 Pearson Education, Inc. Publishing as Addison-Wesley.

PREFERENCE	Count	Percent	DEGREE	Count	Percent
Both	112	19.79	MA	167	29.51
Email	239	42.23	Other	11	1.94
Mail	86	15.19	PhD	388	68.55
N/A	129	22.79	N=	566	
N=	566				

(b) The relative frequencies were calculated in part(a) by putting a check mark next to Percents. For the PREFERENCE variable; 19.79% of the members prefer to receive the ballot by both e-mail and mail, 42.23% prefer e-mail, 15.19% prefer mail, and 22.79% didn't list a preference. For the Degree variable; 29.51% obtained a Master's degree, 68.55% obtained a PhD, and 1.94% received a different degree.

(c) Using Minitab, select **Graph** ▶ **Pie Chart**, check Chart counts of unique values, double-click on PREFERENCE and DEGREE in the first box so that PREFERENCE and DEGREE appear in the Categorical Variables box. Click Pie Options, check decreasing volume, click OK. Click Multiple Graphs, check On the Same Graphs, Click OK. Click Labels, click Slice Labels, check Category Name, Percent, and Draw a line from label to slice, Click OK twice. The results are

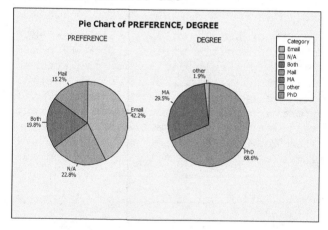

(d) Using Minitab, select **Graph** ▶ **Bar Chart**, select Counts of unique values, select Simple option, click OK. Double-click on PREFERENCE and DEGREE in the first box so that PREFERENCE and DEGREE appear in the Categorical Variables box. Select Chart Options, check decreasing Y, check show Y as a percent, click OK. Click OK twice. The results are

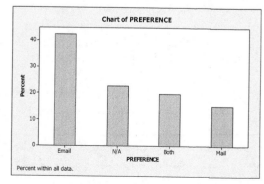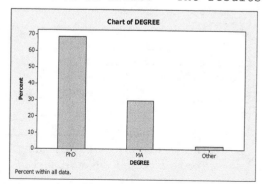

Exercises 2.3

2.35 For class limits, marks, cutpoints and midpoints to make sense, data must be numerical. They do not make sense for qualitative data classes because such data are nonnumerical.

Copyright © 2012 Pearson Education, Inc. Publishing as Addison-Wesley.

2.37   In the first method for depicting classes called cutpoint grouping, we used the notation **a – under b** to mean values that are greater than or equal to **a** and up to, but not including **b**, such as 30 – under 40 to mean a range of values greater than or equal to 30, but strictly less than 40. In the alternate method called limit grouping, we used the notation **a–b** to indicate a class that extends from **a** to **b**, including both. For example, 30–39 is a class that includes both 30 and 39. The alternate method is especially appropriate when all of the data values are integers. If the data include values like 39.7 or 39.93, the first method is more advantageous since the cutpoints remain integers; whereas, in the alternate method, the upper limits for each class would have to be expressed in decimal form such as 39.9 or 39.99.

2.39   For limit grouping, we find the class mark, which is the average of the lower and upper class limit. For cutpoint grouping, we find the class midpoint, which is the average of the two cutpoints.

2.41   An advantage of the frequency histogram over a frequency distribution is that it is possible to get an overall view of the data more easily. A disadvantage of the frequency histogram is that it may not be possible to determine exact frequencies for the classes when the number of observations is large.

2.43   If the classes consist of single values, stem-and-leaf diagrams and frequency histograms are equally useful. If only one diagram is needed and the classes consist of more than one value, the stem-and-leaf diagram allows one to retrieve all of the original data values whereas the frequency histogram does not. If two or more sets of data of different sizes are to be compared, the relative frequency histogram is advantageous because all of the diagrams to be compared will have the same total relative frequency of 1.00. Finally, stem-and-leaf diagrams are not very useful with very large data sets and may present problems with data having many digits in each number.

2.45   You can reconstruct the stem-and-leaf diagram using two lines per stem. For example, instead of listing all of the values from 10 to 19 on a '1' stem, you can make two '1' stems. On the first, you record the values from 10 to 14 and on the second, the values from 15 to 19. If there are still two few stems, you can reconstruct the diagram using five lines per stem, recording 10 and 11 on the first line, 12 and 13 on the second, and so on.

2.47   For the ages of householders, given as a whole number, limit grouping is probably the best because the data are given as whole numbers and there are probably too many distinct observations to list them as single-value grouping.

2.49   For the number of automobiles per family, single-value grouping is probably the best because the data is discrete with relatively few distinct observations.

2.51   For carapace length for a sample of giant tarantulas, cutpoint grouping is probably the best because the data is continuous and the data was recorded to the nearest hundredth of a millimeter.

2.53   (a) Since the data values range from 1 to 7, we construct a table with classes based on a single value. The resulting table follows.

Copyright © 2012 Pearson Education, Inc. Publishing as Addison-Wesley.

Number of Persons	Frequency
1	7
2	13
3	9
4	5
5	4
6	1
7	1
	40

(b) To get the relative frequencies, divide each frequency by the sample size of 40.

Number of Persons	Relative Frequency
1	0.175
2	0.325
3	0.225
4	0.125
5	0.100
6	0.025
7	0.025
	1.000

(c) The frequency histogram in Figure (a) is constructed using the frequency distribution presented in part (a) of this exercise. Column 1 demonstrates that the data are grouped using classes based on a single value. These single values in column 1 are used to label the horizontal axis of the frequency histogram. Suitable candidates for vertical axis units in the frequency histogram are the integers within the range 0 through 13, since these are representative of the magnitude and spread of the frequencies presented in column 2. When classes are based on a single value, the middle of each histogram bar is placed directly over the single numerical value represented by the class. Also, the height of each bar in the frequency histogram matches the respective frequency in column 2.

Figure (a)   Figure (b)

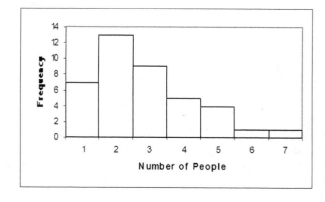

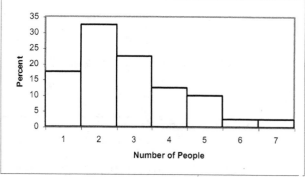

Copyright © 2012 Pearson Education, Inc. Publishing as Addison-Wesley.

(d)   The relative-frequency histogram in Figure (b) is constructed using the relative-frequency distribution presented in part (b) of this exercise. It has the same horizontal axis as the frequency histogram. We notice that the relative frequencies presented in column 2 range in size from 0.025 to 0.325. Thus, suitable candidates for vertical-axis units in the relative-frequency histogram are increments of 0.05 (or 5%), from zero to 0.35 (or 35%). The middle of each histogram bar is placed directly over the single numerical value represented by the class. Also, the height of each bar in the relative-frequency histogram matches the respective relative frequency in column 2.

2.55   (a)   Since the data values range from 1 to 10, we construct a table with classes based on a single value. The resulting table follows.

Number of Radios	Frequency
1	1
2	1
3	3
4	12
5	6
6	4
7	5
8	4
9	6
10	3
	45

(b)   To get the relative frequencies, divide each frequency by the sample size of 45.

Number of Radios	Relative Frequency
1	0.022
2	0.022
3	0.067
4	0.267
5	0.133
6	0.089
7	0.111
8	0.089
9	0.133
10	0.067
	1.000

(c)   The frequency histogram in Figure (a) is constructed using the frequency distribution presented in part (a) of this exercise. Column 1 demonstrates that the data are grouped using classes based on a single value. These single values in column 1 are used to label the horizontal axis of the frequency histogram. Suitable candidates for vertical axis units in the frequency histogram are the integers within

Copyright © 2012 Pearson Education, Inc. Publishing as Addison-Wesley.

the range 1 through 12, since these are representative of the magnitude and spread of the frequencies presented in column 2. When classes are based on a single value, the middle of each histogram bar is placed directly over the single numerical value represented by the class. Also, the height of each bar in the frequency histogram matches the respective frequency in column 2.

Figure (a)   Figure (b)

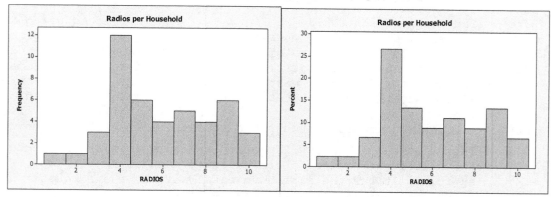

(d)   The relative-frequency histogram in Figure (b) is constructed using the relative-frequency distribution presented in part (b) of this exercise. It has the same horizontal axis as the frequency histogram. We notice that the relative frequencies presented in column 2 range in size from 0.022 to 0.267. Thus, suitable candidates for vertical-axis units in the relative-frequency histogram are increments of 0.05 (or 5%), from zero to 0.30 (or 30%). The middle of each histogram bar is placed directly over the single numerical value represented by the class. Also, the height of each bar in the relative-frequency histogram matches the respective relative frequency in column 2.

2.57 (a)   The first class to construct is 40-44. Since all classes are to be of equal width, and the second class begins with 45, we know that the width of all classes is 45 - 40 = 5. All of the classes are presented in column 1. The last class to construct is 60-64, since the largest single data value is 61. Having established the classes, we tally the age figures into their respective classes. These results are presented in column 2, which lists the frequencies.

Age	Frequency
40-44	4
45-49	3
50-54	4
55-59	8
60-64	2
	21

(b)   Dividing each frequency by the total number of observations, which is 21, results in each class's relative frequency. The relative frequencies for all classes are presented in column 2. The resulting table follows.

Copyright © 2012 Pearson Education, Inc. Publishing as Addison-Wesley.

Age	Relative Frequency
40-44	0.190
45-49	0.143
50-54	0.190
55-59	0.381
60-64	0.095
	1.000

(c)   The frequency histogram in Figure (a) is constructed using the frequency distribution presented in part (a) of this exercise. The lower class limits of column 1 are used to label the horizontal axis of the frequency histogram. Suitable candidates for vertical-axis units in the frequency histogram are the even integers 2 through 8, since these are representative of the magnitude and spread of the frequency presented in column 2. The height of each bar in the frequency histogram matches the respective frequency in column 2.

(d)   The relative-frequency histogram in Figure (b) is constructed using the relative-frequency distribution presented in part (b) of this exercise. It has the same horizontal axis as the frequency histogram. We notice that the relative frequencies presented in column 2 vary in size from 0.095 to 0.381. Thus, suitable candidates for vertical axis units in the relative-frequency histogram are increments of 0.10 (or 10%), from zero to 0.40 (or 40%). The height of each bar in the relative-frequency histogram matches the respective relative frequency in column 2.

Figure (a)   Figure (b)

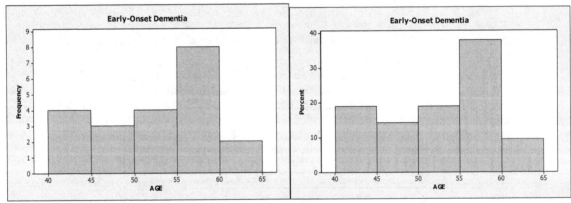

2.59 (a)   The first class to construct is 12-17. Since all classes are to be of equal width, and the second class begins with 18, we know that the width of all classes is 18 - 12 = 6. All of the classes are presented in column 1. The last class to construct is 60-65, since the largest single data value is 61. Having established the classes, we tally the anxiety questionnaire score figures into their respective classes. These results are presented in column 2, which lists the frequencies.

Copyright © 2012 Pearson Education, Inc. Publishing as Addison-Wesley.

Anxiety	Frequency
12-17	2
18-23	3
24-29	6
30-35	5
36-41	10
42-47	4
48-53	0
54-59	0
60-65	1
	31

(b)   Dividing each frequency by the total number of observations, which is 31, results in each class's relative frequency. The relative frequencies for all classes are presented in column 2. The resulting table follows.

Anxiety	Relative Frequency
12-17	0.065
18-23	0.097
24-29	0.194
30-35	0.161
36-41	0.323
42-47	0.129
48-53	0.000
54-59	0.000
60-65	0.032
	1.000

(c)   The frequency histogram in Figure (a) is constructed using the frequency distribution presented in part (a) of this exercise. The lower class limits of column 1 are used to label the horizontal axis of the frequency histogram. Suitable candidates for vertical-axis units in the frequency histogram are the even integers 0 through 10, since these are representative of the magnitude and spread of the frequency presented in column 2. The height of each bar in the frequency histogram matches the respective frequency in column 2.

(d)   The relative-frequency histogram in Figure (b) is constructed using the relative-frequency distribution presented in part (b) of this exercise. It has the same horizontal axis as the frequency histogram. We notice that the relative frequencies presented in column 2 vary in size from 0.000 to 0.323. Thus, suitable candidates for vertical axis units in the relative-frequency histogram are increments of 0.05 (or 5%), from zero to 0.35 (or 35%). The height of each bar in the relative-frequency histogram matches the respective relative frequency in column 2.

Copyright © 2012 Pearson Education, Inc. Publishing as Addison-Wesley.

Figure (a)   Figure (b)

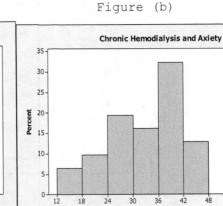

2.61 (a) The first class to construct is 52 – under 54. Since all classes are to be of equal width 2, the second class is 54 – under 56. All of the classes are presented in column 1. The last class to construct is 74 – under 76, since the largest single data value is 75.3. Having established the classes, we tally the cheetah speeds into their respective classes. These results are presented in column 2, which lists the frequencies.

Speed	Frequency
52 – under 54	2
54 – under 56	5
56 – under 58	6
58 – under 60	8
60 – under 62	7
62 – under 64	3
64 – under 66	2
66 – under 68	1
68 – under 70	0
70 – under 72	0
72 – under 74	0
74 – under 76	1
	35

(b) Dividing each frequency by the total number of observations, which is 35, results in each class's relative frequency. The relative frequencies for all classes are presented in column 2

Speed	Relative Frequency
52 – under 54	0.057
54 – under 56	0.143
56 – under 58	0.171
58 – under 60	0.229
60 – under 62	0.200
62 – under 64	0.086
64 – under 66	0.057
66 – under 68	0.029
68 – under 70	0.000
70 – under 72	0.000
72 – under 74	0.000
74 – under 76	0.029
	1.000

(c) The frequency histogram in Figure (a) is constructed using the frequency distribution obtained in part (a) of this exercise Column 1 demonstrates that the data are grouped using classes with class widths

Copyright © 2012 Pearson Education, Inc. Publishing as Addison-Wesley.

of 2. Suitable candidates for vertical axis units in the frequency histogram are the integers within the range 0 through 8, since these are representative of the magnitude and spread of the frequencies presented in column 2. Also, the height of each bar in the frequency histogram matches the respective frequency in column 2.

(d)   The relative-frequency histogram in Figure (b) is constructed using the relative-frequency distribution obtained in part (b) of this exercise. It has the same horizontal axis as the frequency histogram. We notice that the relative frequencies presented in column 3 range in size from 0.000 to 0.229. Thus, suitable candidates for vertical axis units in the relative-frequency histogram are increments of 0.05 (5%), from zero to 0.25 (25%). The height of each bar in the relative-frequency histogram matches the respective relative frequency in column 2.

Figure (a)   Figure (b)

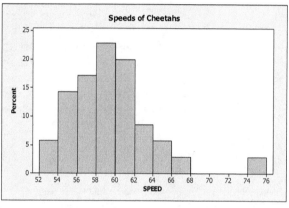

2.63   (a)   The first class to construct is 0 - under 1. Since all classes are to be of equal width 1, the second class is 1 - under 2. All of the classes are presented in column 1. The last class to construct is 7 - under 8, since the largest single data value is 7.6. Having established the classes, we tally the fuel tank capacities into their respective classes. These results are presented in column 2, which lists the frequencies.

Oxygen Distribution	Frequency
0 - under 1	1
1 - under 2	10
2 - under 3	5
3 - under 4	4
4 - under 5	0
5 - under 6	0
6 - under 7	1
7 - under 8	1
	22

(b)   Dividing each frequency by the total number of observations, which is 22, results in each class's relative frequency. The relative frequencies for all classes are presented in column 2

Copyright © 2012 Pearson Education, Inc. Publishing as Addison-Wesley.

Oxygen Distribution	Relative Frequency
0 – under 1	0.045
1 – under 2	0.455
2 – under 3	0.227
3 – under 4	0.182
4 – under 5	0.000
5 – under 6	0.000
6 – under 7	0.045
7 – under 8	0.045
	1.000

(c)   The frequency histogram in Figure (a) is constructed using the frequency distribution obtained in part (a) of this exercise Column 1 demonstrates that the data are grouped using classes with class widths of 2. Suitable candidates for vertical axis units in the frequency histogram are the integers within the range 0 through 10, since these are representative of the magnitude and spread of the frequencies presented in column 2. Also, the height of each bar in the frequency histogram matches the respective frequency in column 2.

(d)   The relative-frequency histogram in Figure (b) is constructed using the relative-frequency distribution obtained in part (b) of this exercise. It has the same horizontal axis as the frequency histogram. We notice that the relative frequencies presented in column 3 range in size from 0.000 to 0.455. Thus, suitable candidates for vertical axis units in the relative-frequency histogram are increments of 0.05 (5%), from zero to 0.50 (50%). The height of each bar in the relative-frequency histogram matches the respective relative frequency in column 2.

Figure (a)   Figure (b)

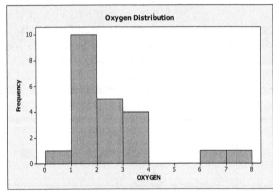   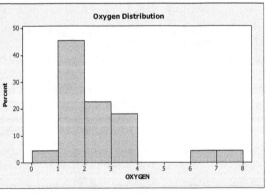

2.65   The horizontal axis of this dotplot displays a range of possible ages. To complete the dotplot, we go through the data set and record each age by placing a dot over the appropriate value on the horizontal axis.

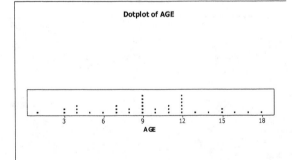

Copyright © 2012 Pearson Education, Inc. Publishing as Addison-Wesley.

2.67 (a)   The data values range from 7 to 18, so the scale must accommodate those values. We stack dots above each value on two different lines using the same scale for each line. The result is

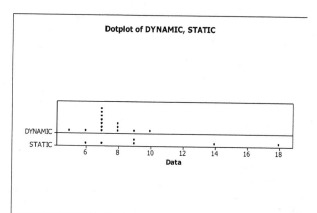

  (b)   The Dynamic system does seem to reduce acute postoperative days in the hospital on the average. The Dynamic data are centered at about 7 days, whereas the Static data are centered at about 11 days and are much more spread out than the Dynamic data.

2.69 Since each data value consists of 2 digits, each beginning with 1, 2, 3, or 4, we will construct the stem-and-leaf diagram with these four values as the stems. The result is

```
1|  238
2|  1678899
3|  34459
4|  04
```

2.71 (a)   Since each data value lies between 2 and 93, we will construct the stem-and-leaf diagram with one line per stem. The result is

```
0|  2234799
1|  11145566689
2|  023479
3|  004555
4|  19
5|  5
6|  9
7|  9
8|
9|  3
```

Copyright © 2012 Pearson Education, Inc. Publishing as Addison-Wesley.

  (b)   Using two lines per stem, the same data result in the following diagram:

```
0|  2234
0|  799
1|  1114
1|  5566689
2|  0234
2|  79
3|  004
3|  555
4|  1
4|  9
5|
5|  5
6|
6|  9
7|
7|  9
8|
8|
9|  3
```

  (c)   The stem with one line per stem is more useful. One gets the same impression regarding the shape of the distribution, but the two lines per stem version has numerous lines with no data, making it take up more space than necessary to interpret the data and giving it too many lines.

2.73 (a) Since we have two digit numbers, the last digit becomes the leaf and the first digit becomes the stem. For this data, we have stems of 6, 7, and 8. Splitting the data into five lines per stem, we put the leaves 0-1 in the first stem, 2-3 in the second stem, 4-5 in the third stem, 6-7 in the fourth stem, and 8-9 in the fifth stem. The result is

```
6|  99
7|  11
7|  222233
7|  44444444445555555555
7|  666667
7|  88
8|  1
```

  (b)   Using one or two lines per stem would have given us too few lines.

2.75 The graph indicates that:

  (a)   20% of the patients have cholesterol levels between 205 and 209, inclusive.

  (b)   20% are between 215 and 219; and 5% are between 220 and 224. Thus, 25% (i.e., 20% + 5%) have cholesterol levels of 215 or higher.

  (c)   35% of the patients have cholesterol levels between 210 and 214, inclusive. With 20 patients in total, the number having cholesterol levels between 210 and 214 is 7 (i.e., 0.35 x 20).

2.77 (a) Using Minitab, there is not a direct way to get a grouped frequency distribution. However, you can use an option in creating your histogram that will report the frequencies in each of the classes, essentially creating a grouped frequency distribution. Retrieve the data from the Weiss-Stats-CD. Column 2 contains the number of albums sold, in millions, for the top recording artists. From the tool bar, ,

Copyright © 2012 Pearson Education, Inc. Publishing as Addison-Wesley.

select **Graph ▶ Histogram**, choose **Simple** and click **OK**. Double click on
ALBUMS to enter ALBUMS in the **Graph variables** box. Click **Labels**, click
the **Data Labels** tab, then check **Use y-value labels**. Click **OK** twice.
The result is

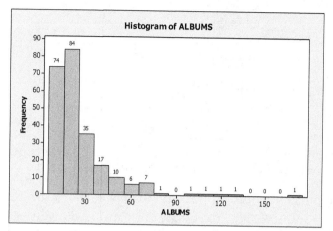

Above each bar is a label for each of the frequencies. Also, the
labeling on the horizontal axis is the midpoint for class, where each
class has a width of 10. The first class would be 5 - under 15, the
second class would be 15 - under 25, etc. To get the relative-
frequency distribution, follow the same steps as above, but also Click
the **Scale** button, click the **Y-scale Type**, check **Percent**, then click **OK**
twice. The result is

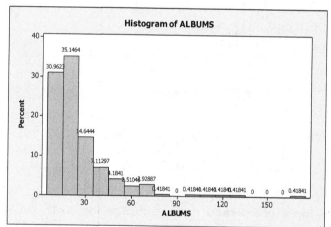

Above each bar is the percentage for that class, essentially creating a
relative-frequency distribution. You could also transfer these results
into a table.

(b) After entering the data from the WeissStats CD, in Minitab, select

Graph ▶ Histogram, choose **Simple** and click **OK**. Double click on ALBUMS
to enter ALBUMS in the **Graph variables** box and click **OK**. The result is

Copyright © 2012 Pearson Education, Inc. Publishing as Addison-Wesley.

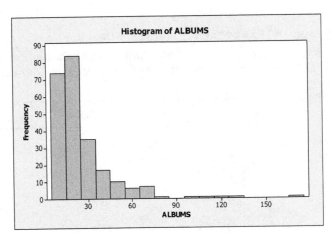

The graph shows that there are only a few artists who sell many units and many artists who sell relatively few units.

(c) To obtain the dotplot, select **Graph ▶ Dotplot**, select **Simple** in the **One Y** row, and click **OK**. Double click on ALBUMS to enter ALBUMS in the **Graph variables** box and click **OK**. The result is

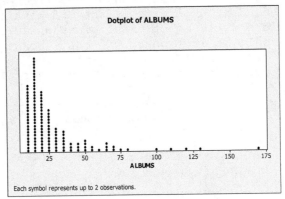

(d) The graphs are similar, but not identical. This is because Minitab grouped the data values slightly differently for the two graphs. The overall impression, however, remains the same.

2.79 (a) After entering the data from the WeissStats CD, in Minitab, select **Graph ▶ Stem-and-Leaf**, double click on RATE to enter RATE in the **Graph variables** box and enter a <u>10</u> in the **Increment** box, and click **OK**. The result is

```
Stem-and-leaf of RATE   N = 51
Leaf Unit = 1.0

    1    1  9
   14    2  0145678888999
  (13)   3  0014445667788
   24    4  012223345566677899
    6    5  00124
    1    6  2
```

(b) Repeat part (a), but this time enter a <u>5</u> in the **Increment** box. The result is

```
Stem-and-leaf of RATE   N = 51
Leaf Unit = 1.0
```

Copyright © 2012 Pearson Education, Inc. Publishing as Addison-Wesley.

```
  1   1   9
  4   2   014
 14   2   5678888999
 20   3   001444
 (7)  3   5667788
 24   4   01222334
 16   4   5566677899
  6   5   00124
  1   5
  1   6   2
```

(c) Repeat part (a) again, but this time enter a <u>2</u> in the **Increment** box. The result is

```
Stem-and-leaf of RATE   N  = 51
Leaf Unit = 1.0

  1   1   9
  3   2   01
  3   2
  5   2   45
  7   2   67
 14   2   8888999
 17   3   001
 17   3
 21   3   4445
 25   3   6677
 (2)  3   88
 24   4   01
 22   4   22233
 17   4   455
 14   4   66677
  9   4   899
  6   5   001
  3   5   2
  2   5   4
  1   5
  1   5
  1   6
  1   6   2
```

(d) The second graph is the most useful. The third one has more classes than necessary to comprehend the shape of the distribution and has a number of empty stems. Typically, we like to have five to fifteen classes and the first and second diagrams satisfy that condition, but the second one provides a better idea of the shape of the distribution.

2.81 (a) The classes are presented in column 1. With the classes established, we then tally the exam scores into their respective classes. These results are presented in column 2, which lists the frequencies. Dividing each frequency by the total number of exam scores, which is 20, results in each class's relative frequency. The relative frequencies for all classes are presented in column 3. The class mark of each class is the average of the lower and upper limits. The class marks for all classes are presented in column 4.

Copyright © 2012 Pearson Education, Inc. Publishing as Addison-Wesley.

Score	Frequency	Relative Frequency	Class Mark
30–39	2	0.10	34.5
40–49	0	0.00	44.5
50–59	0	0.00	54.5
60–69	3	0.15	64.5
70–79	3	0.15	74.5
80–89	8	0.40	84.5
90–100	4	0.20	95.0
	20	1.00	

(b)   The first six classes have width 10; the seventh class had width 11.

(c)   Answers will vary, but one choice is to keep the first six classes the same and make the next two classes 90-99 and 100-109. Another possibility is 31-40, 41-50, …, 91-100.

2.83   Answers will vary, but by following the steps we first decide on the approximate number of classes. Since there are 37 observations, we should have 7-14 classes. This exercise states we should have approximately eight classes. Step 2 says that we calculate an approximate class width as (278.8 – 129.2)/8 = 18.7. A convenient class width close to 18.7 would be a class width of 20. Step 3 says that we choose a number for the lower cutpoint which is less than or equal to our minimum observation of 129.2. Let's choose 120. Beginning with a lower cutpoint of 120 and width of 20, we have a first class of 120 – under 140, a second class of 140 – under 160, a third class of 160 – under 180, a fourth class of 180 - under 200, a fifth class of 200 – under 220, a sixth class of 220 – under 240, a seventh class of 240 – under 260, and an eighth class of 260 – under 280. This would be our last class since the largest observation is 278.8.

2.85   Consider columns 1 and 2 of the energy-consumption data given in Exercise 2.56 part (b). Compute the class mark for each class presented in column 1. Pair each class mark with its corresponding relative frequency found in column 2. Construct a horizontal axis, where the units are in terms of class marks and a vertical axis where the units are in terms of relative frequencies. For each class mark on the horizontal axis, plot a point whose height is equal to the relative frequency of the class. Then join the points with connecting lines. The result is a relative-frequency polygon.

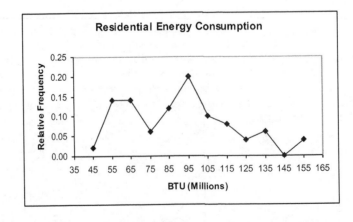

2.87   In single value grouping the horizontal axis would be labeled with the value of each class.

Copyright © 2012 Pearson Education, Inc. Publishing as Addison-Wesley.

2.89 (a)   Consider parts (a) and (b) of the Cheetah speed data given in Exercise 2.61. The classes are now reworked to present just the lower cutpoint of each class. The frequencies are reworked to sum the frequencies of all classes representing values less than the specified lower cutpoint. These successive sums are the cumulative frequencies. The relative frequencies are reworked to sum the relative frequencies of all classes representing values less than the specified cutpoints. These successive sums are the cumulative relative frequencies. (Note: The cumulative relative frequencies can also be found by dividing the each cumulative frequency by the total number of data values.)

Less than	Cumulative Frequency	Cumulative Relative Frequency
52	0	0.000
54	2	0.057
56	7	0.200
58	13	0.371
60	21	0.600
62	28	0.800
64	31	0.886
66	33	0.943
68	34	0.971
70	34	0.971
72	34	0.971
74	34	0.971
76	35	1.000

(b)   Pair each cutpoint with its corresponding cumulative relative frequency found in column 3. Construct a horizontal axis, where the units are in terms of the cutpoints and a vertical axis where the units are in terms of cumulative relative frequencies. For each cutpoint on the horizontal axis, plot a point whose height is equal to the cumulative relative frequency. Then join the points with connecting lines. The result, presented in the following figure, is an *ogive* using cumulative relative frequencies. (Note: A similar procedure could be followed using cumulative frequencies.)

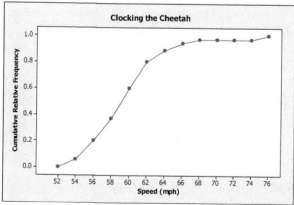

2.91 (a)   After rounding to the nearest 10 and then dropping the final zero, the stem-and-leaf diagram with five lines per stem is

```
 9|  1
 9|
 9|  5
 9|  6667
 9|  8888999999
10|  0000011
10|  223333
```

Copyright © 2012 Pearson Education, Inc. Publishing as Addison-Wesley.

```
        10|
        10|  6
```

(b) After truncating each observation to the 10s digit, the stem-and-leaf
     diagram with five lines per stem is

```
         9|  1
         9|
         9|  455
         9|  67777
         9|  88888999999
        10|  01111
        10|  2233
        10|
        10|  6
```

(c) The overall impression of the shape of the distribution is the same for
     the diagrams in parts (a) and (b) although there is a slight shift to lower
     values in part (b). This is due to truncating instead of rounding.

Section 2.4

2.93 (a) The distribution of a data set is a table, graph, or formula that
              provides the values of the observations and how often they occur.

          (b) Sample data are the values of a variable for a sample of the
               population.

          (c) Population data are the values of a variable for the entire population.

          (d) Census data are the same as population data, a complete listing of all
               data values for the entire population.

          (e) A sample distribution is the distribution of sample data.

          (f) A population distribution is the distribution of population data.

          (g) A distribution of a variable is the same as a population distribution.

2.95 A large simple random sample from a bell-shaped distribution would be
          expected to have roughly a bell-shaped distribution since more sample values
          should be obtained, on average, from the middle of the distribution.

2.97 Three distribution shapes that are symmetric are bell-shaped, triangular,
          and rectangular, shown in that order below. It should be noted that there
          are others as well.

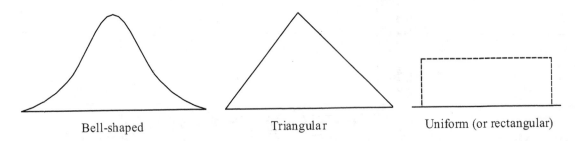

Bell-shaped   Triangular   Uniform (or rectangular)

2.99   (a) Except for the one data value between 74 and 76, this distribution is
                close to bell-shaped. That one value makes the distribution slightly
                right skewed.
           (b) The distribution is slightly right skewed.
2.101 (a) The distribution of burrow depths is left skewed.
           (b) The distribution is left skewed.
2.103 (a) The distribution of shell thickness is approximately bell-shaped.
           (b) The distribution is nearly symmetric.

Copyright © 2012 Pearson Education, Inc. Publishing as Addison-Wesley.

2.105 (a) The distribution of cholesterol levels appears to be slightly left skewed.

(b) This distribution is nearly symmetric, but is slightly left skewed. Given that the data originated from patients who had high cholesterol levels, one would not expect symmetry. Individuals with low cholesterol levels were not patients and were not included in the testing.

(b) This distribution is approximately symmetric.

2.107 (a) The distribution of length of stay is nearly reverse J-shaped. It is certainly right skewed.

(b) The distribution is right skewed.

2.109 (a) The distribution for Year 1 is right skewed and the distribution for Year 2 is reverse J shaped.

(b) Both distributions are right skewed.

(c) Both distributions are rights skewed, however the distribution for Year 1 has a longer right tail indicating more variation than the distribution for Year 2.

2.111 (a) After entering the data from the WeissStats CD, in Minitab, select **Graph ▶ Histogram**, select **Simple** and click **OK**. Double click on ALBUMS to enter ALBUMS in the **Graph variables** box and click **OK**. The result is

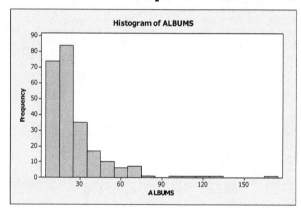

(b) The distribution of ALBUMS is definitely right skewed and comes very close to being reverse J-shaped. If the first class had a higher frequency than the second, we would call it reverse J shaped.

(c) The distribution is right skewed.

2.113 (a) In Exercise 2.79, we used Minitab to obtain a stem-and-leaf diagram using 2 lines per stem. That diagram is shown below

```
Stem-and-leaf of RATE   N  = 51
Leaf Unit = 1.0

    1   1  9
    4   2  014
   14   2  5678888999
   20   3  001444
   (7)  3  5667788
   24   4  01222334
   16   4  5566677899
    6   5  00124
    1   5
    1   6  2
```

(b) The distribution of crime rates is slightly right skewed. Without the

Copyright © 2012 Pearson Education, Inc. Publishing as Addison-Wesley.

largest observation of 62, it would be approximately bell-shaped.

(c) We classify the distribution as right skewed.

2.115 (a) After entering the data from the WeissStats CD, in Minitab, select

Graph ▶ Histogram, select **Simple** and click **OK**. Double click on LENGTH to enter LENGTH in the **Graph variables** box and click **OK**. The result is

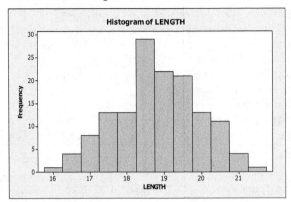

(b) The distribution of LENGTH is approximately bell shaped.

(c) The distribution is symmetric.

2.117 The precise answers to this exercise will vary from class to class or individual to individual. Thus your results will likely differ from our results shown below.

(a) We obtained 50 random digits from a table of random numbers. The digits were

4 5 4 6 8 9 9 7 7 2 2 2 9 3 0 3 4 0 0 8 8 4 4 5 3

9 2 4 8 9 6 3 0 1 1 0 9 2 8 1 3 9 2 5 8 1 8 9 2 2

(b) Since each digit is equally likely in the random number table, we expect that the distribution would look roughly rectangular.

(c) Using single value classes, the frequency distribution is given by the following table. The histogram is shown below.

Value	Frequency	Relative-Frequency
0	5	.10
1	4	.08
2	8	.16
3	5	.10
4	6	.12
5	3	.06
6	2	.04
7	2	.04
8	7	.14
9	8	.16

Copyright © 2012 Pearson Education, Inc. Publishing as Addison-Wesley.

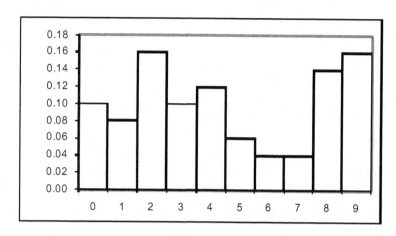

We did not expect to see this much variation.

(d) We would have expected a histogram that was a little more 'even', more like a rectangular distribution, but the relatively small sample size can result in considerable variation from what is expected.

(e) We should be able to get a more evenly distributed set of data if we choose a larger set of data.

(f) Class project.

2.119 (a) Your result will differ from, but be similar to, the one below which was obtained using Minitab. Choose **Calc ▶ Random Data ▶ Normal...**, type 3000 in the **Generate rows of data** text box, click in the **Store in column(s)** text box and type C1, and click **OK**.

(b) Then choose **Graph ▶ Histogram**, choose the **Simple** version, click **OK**, enter C1 in the **Graph variables** text box, click on the **Scale** button and then on the **Y-Scale Type** tab. Check the **Percent** box and click **OK** twice.

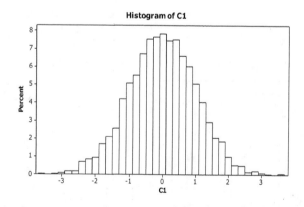

(c) The histogram in part (b) has the shape of a bell symmetric distribution. The sample of 3000 is representative of the population from which the sample was taken.

Section 2.5

2.121 (a) A truncated graph is one for which the vertical axis starts at a value other than its natural starting point, usually zero.

Copyright © 2012 Pearson Education, Inc. Publishing as Addison-Wesley.

    (b)   A legitimate motivation for truncating the axis of a graph is to place the emphasis on the ups and downs of the distribution rather than on the actual height of the graph.

    (c)   To truncate a graph and avoid the possibility of misinterpretation, one should start the axis at zero and put slashes in the axis to indicate that part of the axis is missing.

2.123 (a)   A large lower portion of the graph is eliminated. When this is done, differences between district and national averages appear greater than in the original figure.

    (b)   Even more of the graph is eliminated. Differences between district and national averages appear even greater than in part (a).

    (c)   The truncated graphs give the misleading impression that, in 2008, the district average is much greater relative to the national average than it actually is.

2.125 (a)   The problem with the bar chart is that it is truncated. That is, the vertical axis, which should start at $0 (in trillions), starts with $7.6 (in trillions) instead. The part of the graph from $0 (in trillions) to $7.6 (in trillions) has been cut off. This truncation causes the bars to be out of correct proportion and hence creates the misleading impression that the money supply is changing more than it actually is.

    (b)   A version of the bar chart with an untruncated and unmodified vertical axis is presented in Figure (a). Notice that the vertical axis starts at $0.00 (in trillions). Increments are in trillion dollars. In contrast to the original bar chart, this one illustrates that the changes in money supply from week to week are not that different. However, the "ups" and "downs" are not as easy to spot as in the original, truncated bar chart.

    (c)   A version of the bar chart in which the vertical axis is modified in an acceptable manner is presented in Figure (b). Notice that the special symbol "//" is used near the base of the vertical axis to signify that the vertical axis has been modified. Thus, with this version of the bar chart, not only are the "ups" and "downs" easy to spot but the reader is also aptly warned by the slashes that part of the vertical axis between $0.00 (in trillions) and $7.6 (in trillions) has been removed.

Figure (a)   Figure (b)

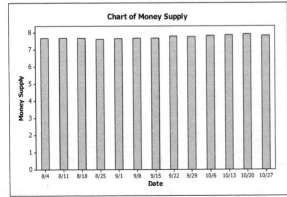
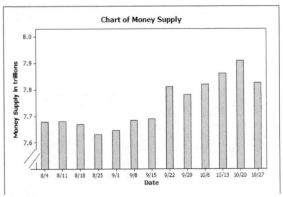

2.127 (b)   Without the vertical scale, it would appear that oil prices dropped about 75% from the first price to the last day shown on the graph.

    (c)   The actual drop was from about $82 per barrel to $58 per barrel, a drop of about $24 dollars per barrel. This is a drop of 24/82 = 0.29 or

Copyright © 2012 Pearson Education, Inc. Publishing as Addison-Wesley.

about 29%.

(d) The graph is potentially misleading because if the reader doesn't pay attention to the vertical scale, he or she may be led to conclude that the oil price was much more volatile than it actually was.

(e) The graph could be made less potentially misleading by either making the vertical scale range from zero to 85 or by starting at zero and putting a break in the vertical axis to call attention to the fact that part of the vertical axis is missing.

2.129 (a) The brochure shows a "new" ball with twice the radius of the "old" ball. The intent is to give the impression that the "new" ball lasts roughly twice as long as the "old" ball. However, if the "new" ball has twice the radius of the "old" ball, the "new" ball will have eight times the volume of the "old" ball (since the volume of a sphere is proportional to the cube of its radius, or the radius $2^3 = 8$). Thus, the scaling is improper because it gives the impression that the "new" ball lasts eight times as long as the "old" ball rather than merely two times as long.

Old Ball   New Ball

(b) One possible way for the manufacturer to illustrate the fact that the "new" ball lasts twice as long as the "old" ball is to present pictures of two balls, side by side, each of the same magnitude as the picture of the "old" ball and to label this set of two balls "new ball". This will illustrate the point that a purchaser will be getting twice as much for the money.

Old Ball   New Ball

Review Problems For Chapter 2

1. (a) A variable is a characteristic that varies from one person or thing to another.

(b) Variables are quantitative or qualitative.

(c) Quantitative variables can be discrete or continuous.

(d) Data are values of a variable.

(e) The data type is determined by the type of variable being observed.

2. A frequency distribution of qualitative data is a table that lists the distinct values of data and their frequencies. It is useful to organize the data and make it easier to understand. A relative-frequency distribution of qualitative data is a table that lists the distinct values of data and a ratio of the class frequency to the total number of observations, which is called the relative frequency.

Copyright © 2012 Pearson Education, Inc. Publishing as Addison-Wesley.

3.   For both quantitative and qualitative data, the frequency and relative-frequency distributions list the values of the distinct classes and their frequencies and relative frequencies. For single value grouping of quantitative data, it is the same as the distinct classes for qualitative data. For class limit and cutpoint grouping in quantitative data, we create groups that form distinct classes similar to qualitative data.

4.   The two main types of graphical displays for qualitative data are the bar chart and the pie chart.

5.   The bars do not abut in a bar chart because there is not any continuity between the categories. Also, this differentiates them from histograms.

6.   Answers will vary. One advantage of pie charts is that it shows the proportion of each class to the total. One advantage of bar charts is that it emphasizes each individual class in relation to each other. One disadvantage of pie charts is that if there are too many classes, the chart becomes confusing. Also, if a class is really small relative to the total, it is hard to see the class in a pie chart.

7.   Single value grouping is appropriate when the data are discrete with relatively few distinct observations.

8.   (a)   The second class would have lower and upper limits of 9 and 14. The class mark of this class would be the average of these limits and equal 11.5.

     (b)   The third class would have lower and upper limits of 15 and 20.

     (c)   The fourth class would have lower and upper limits of 21 and 26. This class would contain an observation of 23.

9.   (a)   If the width of the class is 5, then the class limits will be four whole numbers apart. Also, the average of the two class limits is the class mark of 8. Therefore, the upper and lower class limits must be two whole numbers above and below the number 8. The lower and upper class limits of the first class are 6 and 10.

     (b)   The second class will have lower and upper class limits of 11 and 15. Therefore, the class mark will be the average of these two limits and equal 13.

     (c)   The third class will have lower and upper class limits of 16 and 20.

     (d)   The fourth class has limits of 21 and 25, the fifth class has limits of 26 and 30. Therefore, the fifth class would contain an observation of 28.

10.   (a)   The common class width is the distance between consecutive cutpoints, which is 15 - 5 = 10.

      (b)   The midpoint of the second class is halfway between the cutpoints 15 and 25, and is therefore 20.

      (c)   The sequence of the lower cutpoints is 5, 15, 25, 35, 45, ... Therefore, the lower and upper cutpoints of the third class are 25 and 35.

      (d)   Since the third class has lower and upper cutpoints of 25 and 35, an observation of 32.4 would belong to this class.

11.   (a)   The midpoint is halfway between the cutpoints. Since the class width is 8, 10 is halfway between 6 and 14.

      (b)   The class width is also the distance between consecutive midpoints. Therefore, the second midpoint is at 10 + 8 = 18.

      (c)   The sequence of cutpoints is 6, 14, 22, 30, 38, ... Therefore the

Copyright © 2012 Pearson Education, Inc. Publishing as Addison-Wesley.

lower and upper cutpoints of the third class are 22 and 30.

(d)   An observation of 22 would go into the third class since that class contains data greater than or equal to 22 and strictly less than 30.

12.   (a)   If lower class limits are used to label the horizontal axis, the bars are placed between the lower class limit of one class and the lower class limit of the next class.

(b)   If lower cutpoints are used to label the horizontal axis, the bars are placed between the lower cutpoint of one class and the lower cutpoint of the next class.

(c)   If class marks are used to label the horizontal axis, the bars are placed directly above and centered over the class mark for that class.

(d)   If midpoints are used to label the horizontal axis, the bars are placed directly above and centered over the midpoint for that class.

13.   (a)   A single value frequency histogram for the prices of DVD players would be identical to the dotplot in example 2.16 because the classes would be 197 through 224 and the height for each bar in the frequency histogram would reflect the number of dots above each observation in the dotplot.

(b)   No. The frequency histogram with cutpoint or class limit grouping would combine some of the single values together which would change the frequencies and the heights of the bars corresponding to those classes.

14.   Bell-shaped   Right skewed   Reverse J shape   Uniform

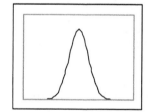 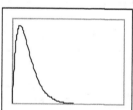 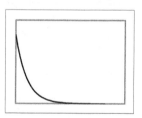 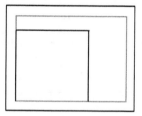

15.   (a)   Slightly skewed to the right. Assuming that the most typical heights are around 5'10", most heights below that figure would still be above 5'4", whereas heights above 5'10" extend to around 7'.

(b)   Skewed to the right. High incomes extend much further above the mean income than low incomes extend below the mean.

(c)   Skewed to the right. While most full-time college students are in the 17-22 age range, there are very few below 17 while there are many above 22.

(d)   Skewed to the right. The main reason for the skewness to the right is that those students with GPAs below fixed cutoff points have been suspended by the time they would have been seniors.

16.   (a)   The distribution of the large simple random sample will reflect the distribution of the population, so it would be left-skewed as well.

(b)   No. The randomness in the samples will almost certainly produce different sets of observations resulting in shapes that are not identical.

(c)   Yes. We would expect both of the simple random samples to reflect the shape of the population and be left-skewed.

17.   (a)   The first column ranks the hydroelectric plants. Thus, it consists of

Copyright © 2012 Pearson Education, Inc. Publishing as Addison-Wesley.

quantitative, discrete data.

(b) The fourth column provides measurements of capacity. Thus, it consists of *quantitative, continuous* data.

(c) The third column provides nonnumerical information. Thus, it consists of *qualitative* data.

18. (a) The first class to construct is 40-44. Since all classes are to be of equal width, and the second class begins with 45, we know that the width of all classes is 45 - 40 = 5. All of the classes are presented in column 1 of the grouped-data table in the figure below. The last class to construct does not go beyond 65-69, since the largest single data value is 69.

Age at Inauguration	Frequency	Relative Frequency	Class Mark
40-44	2	0.045	42
45-49	7	0.159	47
50-54	13	0.295	52
55-59	12	0.273	57
60-64	7	0.159	62
65-69	3	0.068	67
	44	1.000	

(b) By averaging the lower and upper limits for each class, we arrive at the class mark for each class. The class marks for all classes are presented in column 4.

(c) Having established the classes, we tally the ages into their respective classes. These results are presented in column 2, which lists the frequencies. Dividing each frequency by the total number of observations, which is 44, results in each class's relative frequency. The relative frequencies for all classes are presented in column 3.

(d) The frequency histogram presented below is constructed using the frequency distribution presented above; i.e., columns 1 and 2. Notice that the lower cutpoints of column 1 are used to label the horizontal axis of the frequency histogram. Suitable candidates for vertical-axis units in the frequency histogram are the even integers within the range 2 through 14, since these are representative of the magnitude and spread of the frequencies presented in column 2. The height of each bar in the frequency histogram matches the respective frequency in column 2.

Copyright © 2012 Pearson Education, Inc. Publishing as Addison-Wesley.

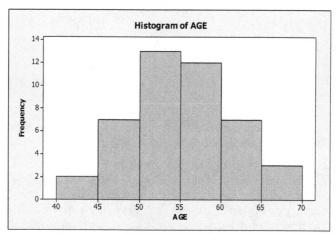

(e) The overall shape of the inauguration ages is somewhere between triangular and bell-shaped.

(f) The distribution is roughly symmetric.

19. The horizontal axis of this dotplot displays a range of possible ages for the 44 Presidents of the United States. To complete the dotplot, we go through the data set and record each age by placing a dot over the appropriate value on the horizontal axis.

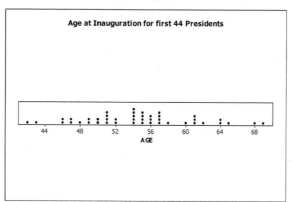

20. (a) Using *one* line per stem in constructing the ordered stem-and-leaf diagram means vertically listing the numbers comprising the stems *once*. The leaves are then placed with their respective stems in order. The ordered stem-and-leaf diagram using one line per stem is presented in the following figure.

    4| 236677899
    5| 0011112244444555566677778
    6| 0111244589

    (b) Using *two* lines per stem in constructing the ordered stem-and-leaf diagram means vertically listing the numbers comprising the stems *twice*. In turn, if the leaf is one of the digits 0 through 4, it is ordered and placed with the first of the two stem lines. If the leaf is one of the digits 5 through 9, it is ordered and placed with the second of the two stem lines. The ordered stem-and-leaf diagram using two lines per stem is presented in the following figure.

Copyright © 2012 Pearson Education, Inc. Publishing as Addison-Wesley.

```
4|  23
4|  6677899
5|  0011112244444
5|  555566677778
6|  0111244
6|  589
```

(c) The stem-and-leaf diagram in part (b) corresponds to the frequency distribution of Problem 18.

21. (a) The frequency and relative-frequency distribution presented below is constructed using classes based on a single value. Since each data value is one of the integers 0 through 6, inclusive, the classes will be 0 through 6, inclusive. These are presented in column 1. Having established the classes, we tally the number of busy tellers into their respective classes. These results are presented in column 2, which lists the frequencies. Dividing each frequency by the total number of observations, which is 25, results in each class's relative frequency. The relative frequencies for all classes are presented in column 3.

Number Busy	Frequency	Relative Frequency
0	1	0.04
1	2	0.08
2	2	0.08
3	4	0.16
4	5	0.20
5	7	0.28
6	4	0.16
	25	1.00

(b) The following relative-frequency histogram is constructed using the relative-frequency distribution presented in part (a); i.e., columns 1 and 3. Column 1 demonstrates that the data are grouped using classes based on a single value. These single values in column 1 are used to label the horizontal axis of the relative-frequency histogram. We notice that the relative frequencies presented in column 3 range in size from 0.04 to 0.28 (4% to 28%). Thus, suitable candidates for vertical axis units in the relative-frequency histogram are increments of 0.05, starting with zero and ending at 0.30. The middle of each histogram bar is placed directly over the single numerical value represented by the class. Also, the height of each bar in the relative-frequency histogram matches the respective relative frequency in column 3.

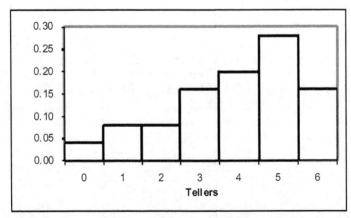

Copyright © 2012 Pearson Education, Inc. Publishing as Addison-Wesley.

(c) The overall shape of this distribution is left skewed.

(d) The distribution is left skewed.

(e)

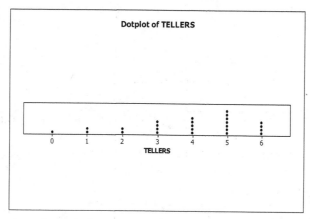

(f) Since both the histogram and the dotplot are based on single value grouping, they both convey exactly the same information.

22. (a) The classes will begin with the class 55 – under 60. The second class will be 60 – under 65. The classes will continue like this until the last class, which will be 90 – under 95, since the largest observation is 92.2. The classes can be found in column 1 of the frequency distribution in part (c).

(b) The midpoints of the classes are the averages of the lower and upper cutpoint for each class. For example, the first midpoint will be 57.5, the second midpoint will be 62.5. The midpoints will continue like this until the last midpoint, which will be 92.5. The midpoints can be found in column 4 of the frequency distribution in part (c).

(c) The first and second columns of the following table provide the frequency distribution. The first and third columns of the following table provide the relative-frequency distribution for percentages of on-time arrivals for the airlines.

Percent On-Time	Frequency	Relative Frequency	Midpoint
55 – under 60	2	0.105	57.5
60 – under 65	2	0.105	62.5
65 – under 70	5	0.263	67.5
70 – under 75	3	0.158	72.5
75 – under 80	5	0.263	77.5
80 – under 85	1	0.053	82.5
85 – under 90	0	0.000	87.5
90 – under 95	1	0.053	92.5
	19	1.000	

(d) The frequency histogram is constructed using columns 1 and 2 from the frequency distribution from part (c). The cutpoints are used to label the horizontal axis. The vertical axis is labeled with the frequencies that range from 0 to 5.

Copyright © 2012 Pearson Education, Inc. Publishing as Addison-Wesley.

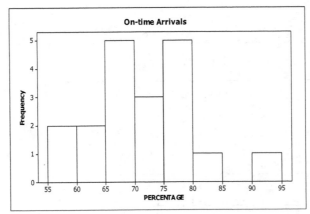

(e) After rounding each observation to the nearest whole number, the stem-and-leaf diagram with two lines per stem for the rounded percentages is

```
5|  99
6|  3
6|  567789
7|  34
7|  566788
8|  1
8|
9|  2
```

(f) After obtaining the greatest integer in each observation, the stem-and-leaf diagram with two lines per stem is

```
5|  89
6|  34
6|  57778
7|  244
7|  66777
8|  0
8|
9|  2
```

(g) The stem-and-leaf diagram in part (f) corresponds to the frequency distribution in part (d) because the observations weren't rounded in constructing the frequency distribution.

23. (a) The dotplot of the ages of the oldest player on each major league baseball team is

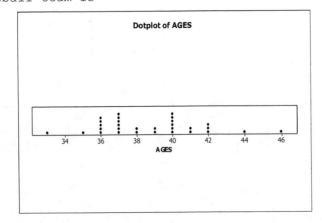

(b) The overall shape of the distribution of ages is bimodal.

Copyright © 2012 Pearson Education, Inc. Publishing as Addison-Wesley.

(c) The distribution is roughly symmetric.

24. (a)-(b) The two pie-charts are shown below. In each case, the proportion of
the circle for each category is found by multiplying 360 degrees by the
category frequency and dividing by the total frequency (941 for
Buybacks and 369 for Homicides).

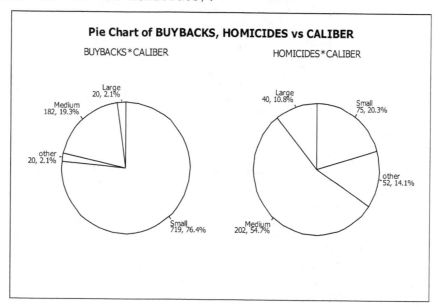

(c) The most striking characteristic of the two charts is that most of the
buybacks are of small caliber guns while most of the homicides are
committed with medium caliber guns. It should also be noted that small
and medium caliber guns are the two largest categories of both buybacks
and homicides, accounting for 95.7% of the buybacks and 75.0% of the
homicides.

25. (a) The population consists of the states in the United States. The
variable is the division.

(b) The frequency and relative frequency distribution for Region is shown
below.

Region	Frequency	Relative Frequency
East North Central	5	0.10
East South Central	4	0.08
Middle Atlantic	3	0.06
Mountain	8	0.16
New England	6	0.12
Pacific	5	0.10
South Atlantic	8	0.16
West North Central	7	0.14
West South Central	4	0.08
	50	1.000

(c) The pie chart for Region is shown below.

Copyright © 2012 Pearson Education, Inc. Publishing as Addison-Wesley.

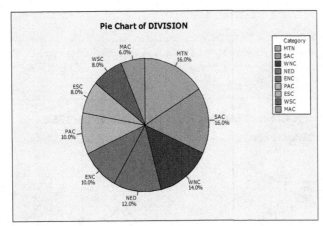

(d)   The bar chart for Region is shown below.

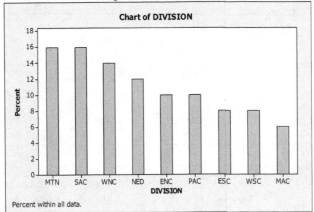

(e)   The distribution of Region seems to be almost uniformly distributed
       between the categories. The smallest region is the Middle Atlantic at
       a frequency of 3, the largest region is the Mountain and South
       Atlantic at a frequency of 8.

26.   (a)   The first class to construct is 1 - under 3. Since all classes are
              to be of equal width, we know that the width of all classes is 3 - 1 =
              2. All of the classes are presented in column 1 of Figure (a) below.
              The last class to construct is 13 - under 15, since the largest
              single data value is 14164.53. Having established the classes, we
              tally the highs into their respective classes. These results are
              presented in column 2, which lists the frequencies. Dividing each
              frequency by the total number of observations, which is 25, results in
              each class's relative frequency. The relative frequencies for all
              classes are presented in column 3.

Copyright © 2012 Pearson Education, Inc. Publishing as Addison-Wesley.

High close (in thousands)	Freq.	Relative Frequency
1 – under 3	7	0.28
3 – under 5	4	0.16
5 – under 7	2	0.08
7 – under 9	1	0.04
9 – under 11	5	0.20
11 – under 13	4	0.16
13 – under 15	2	0.08
	25	1.000

(b)   The following relative-frequency histogram is constructed using the relative-frequency distribution presented above; i.e., columns 1 and 3. The lower cutpoints of column 1 are at the left edges of each rectangle in the relative-frequency histogram. We notice that the relative frequencies presented in column 3 range in size from 0.00 to 0.28. Thus, suitable candidates for vertical axis units in the relative-frequency histogram are increments of 5%, from 0.00 (0%) to 0.30 (30%). The height of each bar in the relative-frequency histogram matches the respective relative frequency in column 3.

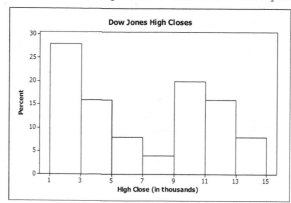

27.   Answers will vary, but here is one possibility:

28.   (a)   The break in the <u>third bar is to emphasize that</u> the bar as shown is not as tall as it should be.

(b)   The bar for the space available in coal mines is at a height of about 30 billion tonnes. To accurately represent the space available in saline aquifers (10,000 billion tonnes), the third bar would have to be over 300 times as high as the first bar and more than 10 times as high as the second bar. If the first two bars were kept at the sizes shown, there wouldn't be enough room on the page for the third bar to

Copyright © 2012 Pearson Education, Inc. Publishing as Addison-Wesley.

be shown at its correct height. If a reasonable height were chosen for the third bar, the first bar wouldn't be visible. The only apparent solution is to present the third bar as a broken bar.

29. (a) Covering up the numbers on the vertical axis totally obscures the percentages.

(b) Having followed the directions in part (a), we might conclude that the percentage of women in the labor force for 2000 is about three and one-half times that for 1960.

(c) Not covering up the vertical axis, we find that the percentage of women in the labor force for 2000 is about 1.8 times that for 1960.

(d) The graph is potentially misleading because it is truncated. Notice that vertical axis units begin at 30 rather than at zero.

(e) To make the graph less potentially misleading, we can start it at zero instead of 30.

30. (a) Using Minitab, retrieve the data from the Weiss-Stats-CD. Column 2 contains the eye color and column 3 contains the hair color for the students. From the tool bar, select **Stat ▶ Tables ▶ Tally Individual Variables**, double-click on EYES and HAIR in the first box so that both EYES and HAIR appear in the **Variables** box, put a check mark next to Counts and Percents under Display, and click **OK**. The results are

EYES	Count	Percent	HAIR	Count	Percent
Blue	215	36.32	Black	108	18.24
Brown	220	37.16	Blonde	127	21.45
Green	64	10.81	Brown	286	48.31
Hazel	93	15.71	Red	71	11.99
N=	592		N=	592	

(b) Using Minitab, select **Graph ▶ Pie Chart**, check Chart counts of unique values, double-click on EYES and HAIR in the first box so that EYES and HAIR appear in the Categorical Variables box. Click Pie Options, check decreasing volume, click OK. Click Multiple Graphs, check On the Same Graphs, Click OK. Click Labels, click Slice Labels, check Category Name, Percent, and Draw a line from label to slice, Click OK twice. The results are

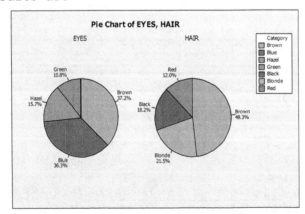

(c) Using Minitab, select **Graph ▶ Bar Chart**, select Counts of unique values, select Simple option, click OK. Double-click on EYES and HAIR in the first box so that EYES and HAIR appear in the Categorical Variables box. Select Chart Options, check decreasing Y, check show Y as a percent, click OK. Click OK twice. The results are

Copyright © 2012 Pearson Education, Inc. Publishing as Addison-Wesley.

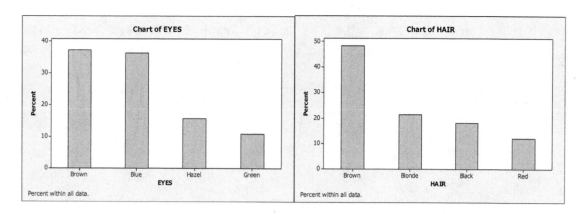

31. (a) The population consists of the states of the U.S. and the variable under consideration is the value of the exports of each state.

(b) Using Minitab, we enter the data from the WeissStats CD, choose **Graph ▶ Histogram**, click on **Simple** and click **OK**. Then double click on VALUE to enter it in the **Graph variables** box and click **OK**. The result is

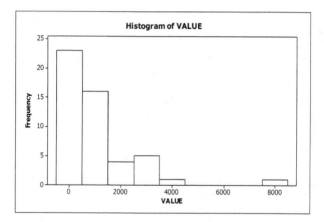

(c) For the dotplot, we choose **Graph ▶ Dotplot**, click on **Simple** from the **One Y** row and click **OK**. Then double click on VALUE to enter it in the **Graph variables** box and click **OK**. The result is

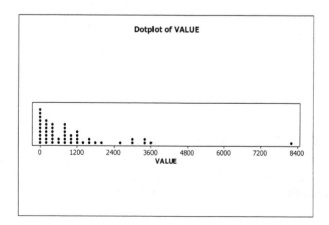

Copyright © 2012 Pearson Education, Inc. Publishing as Addison-Wesley.

(d) For the stem-and-leaf plot, we choose **Graph ▶ Stem-and-Leaf**, double click on VALUE to enter it in the **Graph variables** box and click **OK**. The result is

```
Stem-and-leaf of VALUE  N  = 50
Leaf Unit = 100

   23   0  00000000001112222344444
  (10)  0  5677888899
   17   1  012224
   11   1  5679
    7   2
    7   2  69
    5   3  034
    2   3  6
    1   4
    1   4
    1   5
    1   5
    1   6
    1   6
    1   7
    1   7
    1   8  2
```

(e) The overall shape of the distribution is reverse J shaped.

(f) The distribution is right skewed.

32. (a) The population consists of countries of the world, and the variable under consideration is the expected life in years for people in those countries.

(b) Using Minitab, we enter the data from the WeissStats CD, choose **Graph ▶ Histogram**, click on **Simple** and click **OK**. Then double click on LIFE EXP to enter it in the **Graph variables** box and click **OK**. The result is

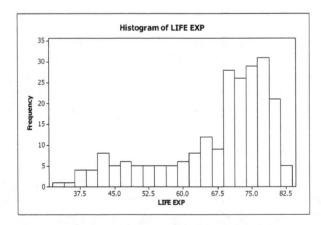

(c) For the dotplot, we choose **Graph ▶ Dotplot**, click on **Simple** from the **One Y** row and click **OK**. Then double click on LIFE EXP to enter it in the **Graph variables** box and click **OK**. The result is

Copyright © 2012 Pearson Education, Inc. Publishing as Addison-Wesley.

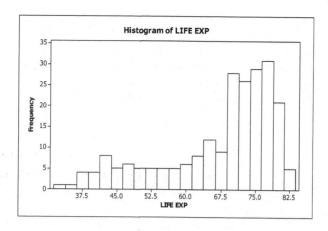

(d) For the stem-and-leaf plot, we choose **Graph ▶ Stem-and-Leaf**, double click on LIFE EXP to enter it in the **Graph variables** box and click **OK**. The result is

```
Stem-and-leaf of LIFE EXP   N  = 224
Leaf Unit = 1.0

   1    3  2
   9    3  56788999
  21    4  112222233444
  31    4  5567788889
  42    5  01111333444
  50    5  56667779
  70    6  00111112233333444444
  99    6  5555666677788888899999999999
 (52)   7  0000000001111111111111122222222233333333333444444444
  73    7  55555555555555666666666677777777777888888888888888889999999999
  11    8  00000011113
```

(e) The overall shape of the distribution is left skewed.
(f) This distribution is classified as left skewed.

33. (a) The population consists of cities in the U.S., and the variables under consideration are their annual average maximum and minimum temperatures.

(b) Using Minitab, we enter the data from the WeissStats CD, choose **Graph ▶ Histogram**, click on **Simple** and click **OK**. Double click on HIGH to enter it in the **Graph variables** box, and double click on LOW to enter it in the **Graph variables** box. Now click on the **Multiple graphs** button and click to **Show Graph Variables on separate graphs** and also check both boxes under **Same Scales for Graphs**, and click **OK** twice. The result is

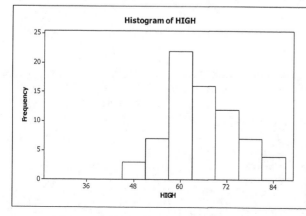
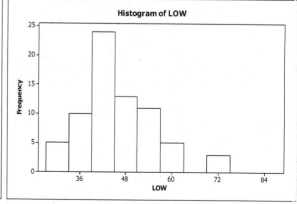

Copyright © 2012 Pearson Education, Inc. Publishing as Addison-Wesley.

(c) For the dotplot, we choose **Graph ▶ Dotplot**, click on **Simple** from the **Multiple Y's** row and click **OK**. Then double click on HIGH and then LOW to enter them in the **Graph variables** box and click **OK**. The result is

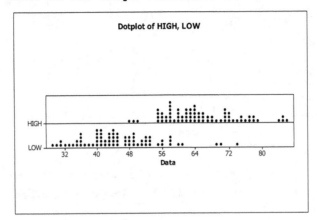

(d) For the stem-and-leaf diagram, we choose **Graph ▶ Stem-and-Leaf**, double click on HIGH and then on LOW to enter then in the **Graph variables** box and click **OK**. The result is

```
Stem-and-leaf of HIGH  N  = 71
Leaf Unit = 1.0

   3    4   789
   6    5   444
  21    5   555677777888999
 (17)   6   00001122222333444
  33    6   55556677779
  22    7   00001122234
  11    7   5577889
   4    8   444
   1    8   5


Stem-and-leaf of LOW  N  = 71
Leaf Unit = 1.0


   1    2   9
   7    3   001234
  19    3   555556779999
 (20)   4   00011111223333334444
  32    4   5567777888899
  19    5   11122223
  11    5   5667889
   4    6   1
   3    6   9
   2    7   04
```

(e) Both variables have distributions that are slightly right skewed.
(f) Both distributions are close to symmetric, but are slightly right skewed. It would take only a few changes in the data to make the distributions approximately symmetric.

Copyright © 2012 Pearson Education, Inc. Publishing as Addison-Wesley.

CHAPTER 3 ANSWERS

Exercises 3.1

3.1 The purpose of a measure of center is to indicate where the center or most typical value of a data set lies.

3.3 Of the mean, median, and mode, only the mode is appropriate for use with qualitative data.

3.5 (a) The mean is the sum of the values (45) divided by n (9). The result is 5. The median is the middle value in the ordered list and is also 5.

    (b) The mean is the sum of the values (135) divided by n (9). The result is 15. The median is the middle value in the ordered list and is 5, as previously. The median is more typical of most of the data than is the mean and works better here.

    (c) The mean does not have the property of being resistant to the influence of extreme observations.

3.7 The median is more appropriate as a measure of central tendency than the mean because, unlike the mean, the median is not affected strongly by the relatively few homes that have extremely large or small areas.

3.9 (a) The mean is calculated first by summing all 3 data values presented, which is 9, and then dividing this sum by 3. Thus the mean is 3.

    (b) The median requires ordering the data from the smallest to the largest values: 0, 4, 5. Since the number of data values is odd, the median will be the middle data value in the ordered list. The middle value is the second ordered observation. Thus, the median is 4.

    (c) The mode is the most frequently occurring data value or values. In these data, no value occurs more than once, so there is no mode.

3.11 (a) The mean is calculated first by summing all 4 data values presented, which is 11, and then dividing this sum by 4. Thus the mean is 2.75.

    (b) The median requires ordering the data from the smallest to the largest values: 1, 2, 4, 4. Since the number of data values is even, the median will be the mean of the two middle data values in the ordered list. The two middle values are the second and third ordered observation. Thus, the median is 3.

    (c) The mode is the most frequently occurring data value or values. In these data, the mode is 4 because it occurs twice.

3.13 (a) The mean is calculated first by summing all 5 data values presented, which is 25, and then dividing this sum by 5. Thus the mean is 5.

    (b) The median requires ordering the data from the smallest to the largest values: 1, 3, 4, 8, 9. Since the number of data values is odd, the median will be the middle data value in the ordered list. The middle value is the third ordered observation. Thus, the median is 4.

    (c) The mode is the most frequently occurring data value or values. In these data, no value occurs more than once, so there is no mode.

3.15 The mean number of values is calculated first by summing all seven values presented, which results in 51 days, and then dividing this by 7. Carrying this result to one more decimal place than the original data, the mean is 7.3 days.

Calculating the median requires ordering the data from the smallest to the largest values: 5 5 6 **6** 7 11 11. Since the number of data values is odd, the median will be the observation exactly in the middle of this ordering, i.e., in the fourth ordered position. Thus, the median number of days is 6.

The mode is the most frequently occurring data value or values. For these

Copyright © 2012 Pearson Education, Inc. Publishing as Addison-Wesley.

data, there are three modes because 5, 6, and 11 each occur twice.

3.17 The mean number of tornado touchdowns is calculated by first summing all 12 data values presented, which results in 941, and then dividing this sum by 12. Thus, the mean is 78.4. This figure is already rounded to one more decimal place than the original data.

Calculating the median requires ordering the data from the smallest to the largest values: 2, 3, 47, 57, 62, **68, 86**, 97, 98, 99, 118, 204. Since the number of data values is even, the median will be the mean of the two middle data values in the ordered list. The two middle values are the sixth and seventh ordered observations, or 68 and 86. Thus, the median age is found as (68 + 86)/2 = 77.0.

The mode is the most frequently occurring data value or values. In these data, no value occurs more than once, so there is no mode.

3.19 The mean wealth is calculated first by summing all 10 data values presented, which results in 285.1, and then dividing this sum by 10. Thus, the mean is $25.51 billion. This figure is already rounded to one more decimal place than the original data.

Calculating the median requires ordering the data from the smallest to the largest values: 19.0, 19.0, 20.0, 23.2, 23.2, 23.3, 23.4, 27.0, 50.0, 57.0. Since the number of data values is even, the median will be the mean of the two middle data values in the ordered list. The two middle values are the fifth and sixth ordered observations. Thus, the median wealth is (23.2 + 23.3)/2 = $23.25 billion dollars.

The mode is the most frequently occurring data value or values. In these data, $19.0 and $23.2 billion dollars are the modes because they both occur twice.

3.21 The mean mpg is calculated first by summing all 14 data values presented, which results in 197, and then dividing this sum by 14. Thus the mean is 14.07. Rounding to one decimal place yields 14.1 mpg.

Calculating the median requires ordering the data from the smallest to the largest values: 11, 12, 13, 13, 14, 14, **14, 14**, 14, 15, 15, 15, 16, 17. Since the number of data values is even, the median will be the mean of the two middle data values in the ordered list. The two middle values are the seventh and eighth ordered observations, which are both 14. Thus, the median is 14 mpg.

The mode is the most frequently occurring data value or values. In these data, the mode is 14 since 14 occurs 5 times, more than any other value.

3.23 (a) The mean number of cremation burials is calculated first by summing all 17 data values presented, which results in 4977, and then dividing this sum by 17. Thus, the mean is 292.8. This figure is already rounded to one more decimal place than the original data.

Calculating the median requires ordering the data from the smallest to the largest values: 21, 34, 35, 46, 46, 48, 51, 64, **83**, 86, 119, 258, 265, 385, 429, 523, 2484. Since the number of data values is odd, the median will be the middle data value in the ordered list. The middle value is the ninth ordered observation. Thus, the median is 83 cremation burials.

The mode is the most frequently occurring data value or values. In these data, the mode is 46 since 46 occurs two times, more than any other value.

(b) The median works best. The mean is highly affected by the largest value and the mode is well to the left of the distribution's center.

Copyright © 2012 Pearson Education, Inc. Publishing as Addison-Wesley.

Section 3.1, Measures of Center 59

3.25 (a) The mean number of accidents is calculated first by summing all 7 data values presented, which results in 619, and then dividing this sum by 7. Thus, the mean is 88.4. This figure is already rounded to one more decimal place than the original data.

Calculating the median requires ordering the data from the smallest to the largest values: 69, 76, 85, **88**, 98, 100, 103. Since the number of data values is odd, the median will be the middle data value in the ordered list. The middle value is the fourth ordered observation. Thus the median is 88.

The mode is the most frequently occurring data value or values. In these data, there is no mode since no number occurs more than once.

(b) The mean number of accidents is calculated first by summing all 7 data values presented, which results in 492, and then dividing this sum by 7. Thus, the mean is 70.3. This figure is already rounded to one more decimal place than the original data.

Calculating the median requires ordering the data from the smallest to the largest values: 53, 56, 58, **59**, 70, 94, 102. Since the number of data values is even, the median will be the middle data value in the ordered list. The middle value is the fourth ordered observation. Thus the median is 59.

(c) For built-up roads, the modal day is Friday since Friday would appear 103 times in the list, more than any other day. For non-built-up roads the modal day is Sunday since Sunday would appear 102 times in the list, more than any other day.

(d) Perhaps on Sunday, motorcycles are being used more for pleasure and sightseeing, possibly on unfamiliar roads. This explanation seems reasonable since Saturday is the day with the second most number of accidents on non-built-up roads. On the weekdays, which include the modal day Friday, motorcycles are perhaps being used more for transportation to and from work. This traffic is more likely to be on built-up roads.

3.27 For a given population, the population mean is not a variable; it is a parameter, a constant. The sample mean will vary from sample to sample and thus is a variable.

3.29 (a) n = number of data values = 4

(b) $\sum x_i = 12 + 8 + 9 + 17 = 46$

(c) $\bar{x} = \sum x_i/n = 46/4 = 11.5$

3.31 (a) n = number of data values = 10

(b) $\sum x_i = 23.3$

(c) $\bar{x} = \sum x_i/n = 23.3/10 = 2.33$, which is given to two decimal places.

3.33 (a) n = number of data values = 9

(b) $\sum x_i = 617$

(c) $\bar{x} = \sum x_i /n = 617/9 = 68.5556$, which, rounded to one decimal place, is 68.6 years

3.35 (a) The mode is defined as the data value or values that occur most frequently. In this exercise, the data values are the NCAA wrestling champion universities. During the period 1984 - 2008, Iowa was champion thirteen times, Oklahoma State seven times, Minnesota three

Copyright © 2012 Pearson Education, Inc. Publishing as Addison-Wesley.

times, and Arizona State and Iowa State each once. Since Iowa is the data value that occurs most frequently, it is the mode.

(b) Again, the data values are the five champions. There is no way to compute a mean or median for such nonnumerical data. Thus, it would not be appropriate to use either the mean or median here. The mode is the only measure of center that can be used for qualitative data.

3.37 (a) The mode is defined as the data value or values that occur most frequently. In this exercise, the data values are the four law schools of the justices of the U.S Supreme Court. As of October 29, 2008, Harvard was the law school for five of the justices, Yale for two justices, and Columbia and Northwestern each for one justice. Since Harvard is the data value that occurs most frequently, it is the mode.

(b) Again, the data values are the law schools. There is no way to compute a mean or median for such nonnumerical data. Thus, it would not be appropriate to use either the mean or median here. The mode is the only measure of center that can be used for qualitative data.

3.39 (a) The mode is defined as the data value or values that occur most frequently. In this exercise, the data values are the political view of the freshmen. During the year, Moderate was the most frequently occurring political view for the freshmen with a frequency of 246. Since moderate is the data value that occurs most frequently, it is the mode.

(b) Again, the data values are the three different political views. There is no way to compute a mean or median for such nonnumerical data. Thus, it would not be appropriate to use either the mean or median here. The mode is the only measure of center that can be used for qualitative data.

3.41 (a) The mode is defined as the data value or values that occur most frequently. In this exercise, the data values are the colors that the Roulette ball landed on. During 200 trials, black was the most frequently occurring color with a frequency of 102. Since black is the data value that occurs most frequently, it is the mode.

(b) Again, the data values are the three different colors. There is no way to compute a mean or median for such nonnumerical data. Thus, it would not be appropriate to use either the mean or median here. The mode is the only measure of center that can be used for qualitative data.

3.43 Using Minitab, retrieve the data from the WeissStats CD. Since the data are qualitative, the only appropriate measure of center is the mode. Select

Stat ▶ Tables ▶ Tally Individual Variables, then enter TYPE in the **Variables** box and click **OK**. The result is

TYPE	Count
NPC	2919
IOC	889
SLC	1119
FGH	221
NLT	129
NFP	451
HUI	19
N=	5747

The mode is the most occurring value. We see that NPC (Nongovernment not-for-profit community hospitals) occurred 2919 times, more than any other value. Therefore, NPC is the mode.

3.45 Using Minitab, retrieve the data from the WeissStats CD. Since both of the

Copyright © 2012 Pearson Education, Inc. Publishing as Addison-Wesley.

variables are qualitative, the only appropriate measure of center for preference and highest degree obtained is the mode. Select **Stat ▶ Tables ▶ Tally Individual Variables**, then enter <u>PREFERENCE</u> and <u>DEGREE</u> in the **Variables** box and click **OK**. The result is

PREFERENCE	Count		DEGREE	Count
Both	112		MA	167
Email	239		Other	11
Mail	86		PhD	388
N/A	129		N=	566
N=	566			

The mode is the most occurring value. For PREFERENCE, we see that email occurred 239 times, more than any other value. Therefore, email is the mode for PREFERENCE. For DEGREE, we see that PhD occurred 388 times, more than any other value. Therefore, PhD is the mode for DEGREE.

3.47 Using Minitab, retrieve the data from the WeissStats CD. Since the data are quantitative, any of the three measures of center could be used. However, since the data are rounded to the nearest 100,000 units, it is unlikely that any two artists sold exactly the same number of units. Most likely, there is no true mode.

To obtain the mean and median, select **Stat ▶ Basic Statistics ▶ Display Descriptive Statistics**, then enter <u>ALBUMS</u> in the **Variables** box and click **OK**. The result is

Variable	N	N*	Mean	SE Mean	StDev	Minimum	Q1	Median	Q3
ALBUMS	239	0	25.90	1.35	20.87	10.25	13.50	18.75	28.75

Variable	Maximum
ALBUMS	170.00

The data appear to be for artists who have sold 10 million or more units (top recording artists). From this display, we see that the mean is 25.90 million and the median is 18.75 million. In this case, the median is the most useful statistic since it tells us that half of the top recording artists sold 18.75 million units or fewer. The mean is heavily influenced by a few artists who sold many more units than the others. The median is more representative of the data as a whole.

3.49 Using Minitab, retrieve the data from the WeissStats CD. Since the data are quantitative, any of the three measures of center could be used. To obtain the mode, select **Stat ▶ Tables ▶ Tally Individual Variables**, then enter <u>RATE</u> in the **Variables** box and click **OK**. . The result is

RATE	Count
19	1
20	1
21	1
24	1
25	1
26	1
27	1
28	4
29	3
30	2
31	1
34	3
35	1
36	2
37	2
38	2

Copyright © 2012 Pearson Education, Inc. Publishing as Addison-Wesley.

```
40    1
41    1
42    3
43    2
44    1
45    2
46    3
47    2
48    1
49    2
50    2
51    1
52    1
54    1
62    1
N=    51
```

The mode is the most occurring value, which is 28 (occurring 4 times). Given that all of the data values arise from one to four times, the mode doesn't tell us much.

To obtain the mean and median, select **Stat ▶ Basic Statistics ▶ Display Descriptive Statistics**, then enter RATE in the **Variables** box and click **OK**. The result is

Variable	N	N*	Mean	SE Mean	StDev	Minimum	Q1	Median	Q3	Maximum
RATE	51	0	37.94	1.39	9.92	19.00	29.00	38.00	46.00	62.00

From this display, we see that the mean is 37.94 and the median is 38.00. In this case, the median is the most useful statistic since it tells us that half of the states had crime rates of 38 per 1000 population. The mean is not a true national crime rate since all states are counted equally in computing this mean, whereas states with greater populations should have more effect on the mean than states with smaller populations.

3.51 (a) In Minitab, open the worksheet from the WeissStats CD. This will enter the data values in columns C1 and C2 under the names INTEGRATED and

STANDARD. Select **Stat ▶ Basic Statistics ▶ Display Descriptive Statistics**, enter INTEGRATED and STANDARD in the **Variables** text box and click **OK**. The output appears in the Session Window as

Variable	N	N*	Mean	SE Mean	StDev	Minimum	Q1	Median	Q3
INTEGRATED	227	0	24.916	0.300	4.513	11.000	22.000	25.000	28.000
STANDARD	192	0	23.000	0.521	7.221	6.000	18.000	23.000	28.750

Variable	Maximum
INTEGRATED	35.000
STANDARD	40.000

From the output, we see that the mean score on client satisfaction for the integrated treatment is 24.916 and the median score is 25.000. The mean score on client satisfaction for the standard treatment is 23.000 and the median score is 23.000.

(b) It would appear from this data that the integrated treatment group had a slightly higher mean and median score than the standard treatment group. However, we can also see from the data that the standard treatment group had a higher maximum score than the integrated group.

3.53 (a) The mean of the ratings is 2.5. This is derived by summing the 14 individual ratings, which yields 35, and dividing by 14.

Copyright © 2012 Pearson Education, Inc. Publishing as Addison-Wesley.

(b) The median of the data is the average of the two observations appearing in the middle positions after ordering the data from the smallest to the largest values: 1, 1, 2, 2, 2, 2, **2**, **2**, 3, 3, 3, 4, 4, 4.   The median is the average of the seventh and eighth ordered observations, or (2 + 2)/2 = 2.

(c) In this case, the median provides a better descriptive summary of the data than does the mean. Notice that six of the fourteen values are 2s and these are more in line with the median 2 than with the mean 2.5. Furthermore, the data are not actually measurements, but reflect the opinions of the participants, each of whom must decide what a '2' or a '3' really means.

3.55 The expression $(\sum x_i)^2$ represents the square of the sum of the data, whereas $\sum x_i^2$ represents the sum of the squares of the data. To show that these two quantities are generally unequal, consider three observations on x as presented in column 1 of the following table. Also, consider each observation's square, as presented in column 2.

X	x^2
1	1
3	9
5	25
9	35

From the bottom row, we see that $(\sum x_i)^2 = 9^2 = 81$ and $\sum x_i^2 = 35$.   Thus, $(\sum x_i)^2 \neq \sum x_i^2$.

Exercises 3.2

3.57 The purpose of a measure of variation is to show the amount of variation or spread in a data set. Any value of central tendency does not adequately characterize the elements of a data set. Measures of central tendency merely describe the center of the observations, but do not show how observations differ from each other. A measure of variation is intended to capture the degree to which observations differ among themselves.

3.59 When we use the standard deviation as a measure of variation, the reference point is the mean since all of the differences used in computing the standard deviation are between the data values and the mean.

3.61 (a) The mean of this data set is 5. Thus

$$\sum (x_i - \overline{x})^2 = (-4)^2 + (-3)^2 + (-2)^2 + (-1)^2 + 0^2 + 1^2 + 2^2 + 3^2 + 4^2 = 60$$

Dividing by n - 1 = 8, s^2 = 60/8 = 7.5. The standard deviation, s, is the square root of 7.5, or 2.739.

(b) The mean of this data set is 15. Thus

$$\sum (x_i - \overline{x})^2 = (-14)^2 + (-13)^2 + (-12)^2 + (-11)^2 + 10^2 + (-9)^2 + (-8)^2 + (-7)^2 + 84^2 = 7980$$

Dividing by n - 1 = 8, s^2 = 7980/8 = 997.5. The standard deviation, s, is the square root of 997.5, or 31.583.

(c) Changing the 9 to 99 greatly increases the standard deviation, illustrating that it lacks the property of resistance to the influence of extreme data values.

3.63 (a) The range is 54 - 9 = 45.

Copyright © 2012 Pearson Education, Inc. Publishing as Addison-Wesley.

(b) The defining formula for s is $s = \sqrt{\dfrac{\sum (x_i - \bar{x})^2}{n-1}}$.

The first three columns of the following table present the calculations that are needed to compute s by the defining formula:

x	$x-\bar{x}$	$(x-\bar{x})^2$	x^2
21	-15	225	441
54	18	324	2916
9	-27	729	81
45	9	81	2025
51	15	225	2601
180	0	1584	8064

With n = 5 and using the bottom figure of column 1, we find that $\bar{x}$ = $\sum x_i/n$ = 180/5 = 36. This value is needed to construct the differences presented in column 2. Column 3 squares these differences. The figure presented at the bottom of column 3 is the computation needed for the numerator of the defining formula for s. Thus,

$$s = \sqrt{\frac{1584}{5-1}} = 19.900$$

(c) The computing formula for s is

$$s = \sqrt{\frac{\sum x_i^2 - \left(\sum x_i\right)^2 / n}{n-1}}$$

The computing formula for s with figures used from the bottom of Columns 1 and 4 of the previous table yields

$$s = \sqrt{\frac{8064 - 180^2/5}{5-1}} = 19.900$$

(d) The computing formula was a time-saver since only one column of intermediate numbers needed to be calculated, whereas the defining formula required three additional columns.

3.65 (a) The range is 5 - 0 = 5.

(b) The defining formula for s is $s = \sqrt{\dfrac{\sum (x_i - \bar{x})^2}{n-1}}$.

The following table presents the calculations that are needed to compute s by the defining formula:

X	$x-\bar{x}$	$(x-\bar{x})^2$
4	1.0	1.0
0	-3.0	9.0
5	2.0	4.0
9	0.0	14.0

With n = 3 and using the bottom figure of column 1, we find that $\bar{x}$ = $\sum x_i/n$ = 9/3 = 3.0. This value is needed to construct the differences presented in column 2. Column 3 squares these differences. The figure presented at the bottom of column 3 is the computation needed for the numerator of the defining formula for s. Thus,

Copyright © 2012 Pearson Education, Inc. Publishing as Addison-Wesley.

$$s = \sqrt{\frac{14.0}{3-1}} = 2.6$$

3.67 (a)   The range is 4 - 1 = 3.

(b)   The defining formula for s is $s = \sqrt{\dfrac{\sum (x_i - \bar{x})^2}{n-1}}$.

The following table presents the calculations that are needed to compute s by the defining formula:

X	x−x̄	(x−x̄)²
1	-1.75	3.06
2	-0.75	0.56
4	1.25	1.56
4	1.25	1.56
11	0.00	6.74

With n = 4 and using the bottom figure of column 1, we find that $\bar{x} = \sum x_i / n = 11/4 = 2.75$. This value is needed to construct the differences presented in column 2. Column 3 squares these differences. The figure presented at the bottom of column 3 is the computation needed for the numerator of the defining formula for s. Thus,

$$s = \sqrt{\frac{6.74}{4-1}} = 1.5$$

3.69 (a)   The range is 9 - 1 = 8.

(b)   The defining formula for s is $s = \sqrt{\dfrac{\sum (x_i - \bar{x})^2}{n-1}}$.

The following table presents the calculations that are needed to compute s by the defining formula:

X	x−x̄	(x−x̄)²
1	-4.0	16.0
9	4.0	16.0
8	3.0	9.0
4	-1.0	1.0
3	-2.0	4.0
25	0.00	46.0

With n = 5 and using the bottom figure of column 1, we find that $\bar{x} = \sum x_i / n = 25/5 = 5.0$. This value is needed to construct the differences presented in column 2. Column 3 squares these differences. The figure presented at the bottom of column 3 is the computation needed for the numerator of the defining formula for s. Thus,

$$s = \sqrt{\frac{46.00}{5-1}} = 3.4$$

3.71 The range is 11 - 5 = 6.

Copyright © 2012 Pearson Education, Inc. Publishing as Addison-Wesley.

The defining formula for s is $s = \sqrt{\dfrac{\sum (x_i - \overline{x})^2}{n-1}}$.

The following table presents the calculations that are needed to compute s by the defining formula:

x	x−x̄	(x−x̄)²
6	−1.3	1.69
7	−0.3	0.09
11	3.7	13.69
6	−1.3	1.69
5	−2.3	5.29
5	−2.3	5.29
11	3.7	13.69
51	−0.1	41.43

With n = 7 and using the bottom figure of column 1, we find that $\overline{x} = \sum x_i/n$ = 51/7 = 7.3. This value is needed to construct the differences presented in column 2. Column 3 squares these differences. The figure presented at the bottom of column 3 is the computation needed for the numerator of the defining formula for s. Thus,

$$s = \sqrt{\frac{41.43}{7-1}} = 2.63$$

3.73 The range of a data set is defined to be the difference between the largest and smallest data values in the data set. The largest number of tornado touchdowns is 204. The smallest number is 2. The range is 204 − 2 = 202 tornadoes.

The defining formula for s is $s = \sqrt{\dfrac{\sum (x_i - \overline{x})^2}{n-1}}$.

The first three columns of the following table present the calculations that are needed to compute s by the defining formula:

x	x−x̄	(x−x̄)²	x²
3	−75.4	5685.16	9
2	−76.4	5836.96	4
47	−31.4	985.96	2204
118	39.6	1568.16	13924
204	125.6	15775.36	41616
97	18.6	345.96	9409
68	−10.4	108.16	4624
86	7.6	57.76	7396
62	−16.4	268.96	3844
57	−21.4	457.96	3249
98	19.6	384.16	9604
99	20.6	424.36	9801
941	0.2	31898.92	105689

Copyright © 2012 Pearson Education, Inc. Publishing as Addison-Wesley.

$$s = \sqrt{\frac{31898.92}{12-1}} = 53.85$$

The computing formula for s is

$$s = \sqrt{\frac{\sum x_i^2 - (\sum x_i)^2 / n}{n-1}}$$

Columns 1 and 4 of the previous table present the calculations needed to compute s by the computing formula. Using the figures presented at the bottom of each of these columns and substituting into the formula itself, we get:

$$s = \sqrt{\frac{105689 - 941^2 / 12}{12-1}} = 53.85$$

This is, of course, the same result that we obtained by the defining formula.

$$s = \sqrt{\frac{0.10}{9-1}} = 0.11$$

3.75 The range of a data set is defined to be the difference between the largest and smallest data values in the data set. The highest value is 57 billion and the lowest value is 19 billion. The range is 57 - 19 = 38 billion.

The defining formula for s is $s = \sqrt{\frac{\sum (x_i - \bar{x})^2}{n-1}}$.

The following table presents the calculations that are needed to compute s by the defining formula:

x	x−$\bar{x}$	(x−$\bar{x}$)^2
57.0	28.49	811.68
50.0	21.49	461.82
27.0	-1.51	2.28
23.4	-5.11	26.11
23.3	-5.21	27.14
23.2	-5.31	28.20
23.2	-5.31	28.20
20.0	-8.51	72.42
19.0	-9.51	90.44
19.0	-9.51	90.44
285.1	0.00	1638.73

With n = 10 and using the bottom figure of column 1, we find that $\bar{x}$ = $\sum x_i / n$ = 285.1/10 = 28.51. This value is needed to construct the differences in column 2. Column 3 squares the differences presented in column 2. The figure presented at the bottom of column 3 is the computation needed for the numerator of the defining formula s. Thus,

$$s = \sqrt{\frac{1638.73}{10-1}} = 13.49$$

Copyright © 2012 Pearson Education, Inc. Publishing as Addison-Wesley.

3.77 The range of a data set is defined to be the difference between the largest and smallest data values in the data set. The highest overall mpg is 17 and the lowest overall mpg is 11. The range is 17 - 11 = 6.

The defining formula for s is $s = \sqrt{\dfrac{\sum(x_i - \bar{x})^2}{n-1}}$.

The following table presents the calculations that are needed to compute s by the defining formula:

x	x-x̄	(x-x̄)²
14	-0.071	0.00504
12	-2.071	4.28904
13	-1.071	1.14704
15	0.929	0.86304
14	-0.071	0.00504
15	0.929	0.86304
13	-1.071	1.14704
17	2.929	8.57904
14	-0.071	0.00504
14	-0.071	0.00504
14	-0.071	0.00504
15	0.929	0.86304
11	-3.071	9.43104
16	1.929	3.72104
197	0.000	30.9286

With n = 14 and using the bottom figure of column 1, we find that $\bar{x} = \sum x_i/n$ = 197/14 = 14.071. This value is needed to construct the differences in column 2. Column 3 squares the differences presented in column 2. The figure presented at the bottom of column 3 is the computation needed for the numerator of the defining formula s. Thus,

$$s = \sqrt{\frac{30.9286}{14-1}} = 1.542$$

Rounding to one decimal place, s = 1.5.

3.79 (a) We will use the computing formula to find s. The following table presents the calculations that are needed to compute s by the computing formula:

Copyright © 2012 Pearson Education, Inc. Publishing as Addison-Wesley.

x	x^2
83	6889
46	2116
64	4096
385	148225
46	2116
21	441
48	2304
86	7396
523	273529
429	184041
35	1225
51	2601
34	1156
258	66564
265	70225
119	14161
2484	6170256
4977	6957341

The computing formula for *s* is

$$s = \sqrt{\frac{\sum x_i^2 - (\sum x_i)^2 / n}{n-1}}$$

The previous table presents the calculations needed to compute *s* by the computing formula. Using the figures presented at the bottom of the columns and substituting into the formula itself, we get:

$$s = \sqrt{\frac{6957341 - 4977^2 / 17}{17-1}} = 586.3$$

(b) No. The standard deviation is based on deviations from the mean. However, these data are right skewed and contain a very large outlier (2484) that greatly increases the value of the mean. The deviation of 2484 from the mean accounts for about 87% of the total of the squared deviations from the mean, so the influence of that data value on both the mean and the standard deviation is very great.

3.81 (a) We would guess that the non-built up roads would have the greater variation due to more variation in the recreational use of those roads at various times of the week.

(b) The range for built-up roads is 103 − 69 = 34, and the range for non-built-up roads is 102 − 53 = 49.

For built-up roads, n = 17, $\sum x_i = 619$, $\sum x_i^2 = 55719$, and

Copyright © 2012 Pearson Education, Inc. Publishing as Addison-Wesley.

$$s = \sqrt{\frac{55719 - 619^2 / 7}{7 - 1}} = 12.79$$

For non-built-up-roads, $n = 7$, $\sum x_i = 492$, $\sum x_i^2 = 36930$, and

$$s = \sqrt{\frac{36930 - 492^2 / 7}{7 - 1}} = 19.79$$

Thus both measures of variation confirm that our guess is correct. The range and standard deviation are greater for non-built-up roads than for built-up roads.

3.83 The data in this set are qualitative (the hospital type for U.S. registered hospitals). Being nonnumerical, it is not possible to compute a range or standard deviation.

3.85 The data in this set are qualitative (ballot preference and highest degree obtained). Being nonnumerical, it is not possible to compute a range or standard deviation.

3.87 Using Minitab, retrieve the data from the WeissStats CD. . Select **Stat ▶**

Basic Statistics ▶ Display Descriptive Statistics, enter ALBUMS in the **Variables** text box. Now click on the **Statistics** button and check the range to add it to the list of statistics to be computed. Click **OK** twice. The output appears in the Session Window as

Variable	N	N*	Mean	SE Mean	StDev	Minimum	Q1	Median	Q3
ALBUMS	239	0	25.90	1.35	20.87	10.25	13.50	18.75	28.75

Variable	Maximum	Range
ALBUMS	170.00	159.75

The range (the difference between the maximum and the minimum data values) is 159.75 and the standard deviation is 20.87. Roughly speaking, on the average, the mean scores differ from the overall mean score by 20.87. For these data, which are right skewed and contain a few very large values, the standard deviation may not be a very good measure of variation due to the large influence of those large values. Typically, the range will be about 4 to 6 times the standard deviation. In this case, it is about 8 times the standard deviation, indicating that there may be outliers or highly influential data values.

3.89 Using Minitab, retrieve the data from the WeissStats CD. . Select **Stat ▶**

Basic Statistics ▶ Display Descriptive Statistics, enter RATE in the **Variables** text box. Now click on the **Statistics** button and check the range to add it to the list of statistics to be computed. Click **OK** twice. The output appears in the Session Window as

Variable	N	N*	Mean	SE Mean	StDev	Minimum	Q1	Median	Q3	Maximum
RATE	51	0	37.94	1.39	9.92	19.00	29.00	38.00	46.00	62.00

Variable	Range
RATE	43.00

The range (the difference between the maximum and the minimum data values) is 43 and the standard deviation is 9.92. Roughly speaking, on the average, the crime rates differ from the overall crime rate by 9.92. One should keep in mind, however, that the overall mean computed by Minitab assumes that the

Copyright © 2012 Pearson Education, Inc. Publishing as Addison-Wesley.

crime rates for each state count equally. In fact, the true overall mean crime rate would take into account that there are different numbers of students in each state.

3.91 (a) Using Minitab, retrieve the data from the WeissStats CD. . Select

Stat ▶ Basic Statistics ▶ Display Descriptive Statistics, enter INTEGRATED and STANDARD in the **Variables** text box. Now click on the **Statistics** button and check the range to add it to the list of statistics to be computed. Click **OK** twice. The output appears in the Session Window as

Variable	N	N*	Mean	SE Mean	StDev	Minimum	Q1	Median	Q3
INTEGRATED	227	0	24.916	0.300	4.513	11.000	22.000	25.000	28.000
STANDARD	192	0	23.000	0.521	7.221	6.000	18.000	23.000	28.750

Variable	Maximum	Range
INTEGRATED	35.000	24.000
STANDARD	40.000	34.000

The range of the questionnaire scores for the integrated treatment is 24. The range of the questionnaire scores for the standard treatment is 34. The standard deviation of the questionnaire scores for the integrated treatment is 4.513 and 7.221 for the standard treatment.

(b) Both the range and the standard deviation are greater for the standard treatment. This indicates that the standard treatment had more variation in its observations than the integrated treatments observations.

3.93 (a) We will compute s for each data set using the computing formula. The computing formula requires the following column manipulations.

Data Set I

x	x^2
0	0
0	0
10	100
12	144
14	196
14	196
14	196
15	225
15	225
15	225
16	256
17	289
23	529
24	576
189	3,157

Data Set II

x	x^2
10	100
12	144
14	196
14	196
14	196
15	225
15	225
15	225
16	256
17	289
142	2,052

For each data set, the calculations for s are presented in column 2 of the following table.

Copyright © 2012 Pearson Education, Inc. Publishing as Addison-Wesley.

Data Set	s	Range
I	$s = \sqrt{\dfrac{3157 - 189^2/14}{14-1}} = 6.8$	24
II	$s = \sqrt{\dfrac{2052 - 142^2/10}{10-1}} = 0.20.$	7

(b) The range for Data Set I is 24 - 0 = 24. The range for Data Set II is 17 - 10 = 7. These are recorded in column 3 of the previous table.

(c) Outliers increase the variation in a data set; in other words, removing the outliers from a data set results in a decrease in the variation.

3.95 (a) We create a table with columns headed by x, f, xf, $(x-\overline{x})$, $(x-\overline{x})^2$, and $(x-\overline{x})^2 f$ to estimate the sample mean and standard deviation of the days-to-maturity data.

Class Mark x	Frequency f	xf	$x-\overline{x}$	$(x-\overline{x})^2$	$f(x-\overline{x})^2$
34.5	3	103.5	-33.5	1122.25	3366.75
44.5	1	44.5	-23.5	552.25	552.25
54.5	8	436.0	-13.5	182.25	1458.00
64.5	10	645.0	-3.5	12.25	122.50
74.5	7	521.5	6.5	42.25	295.75
84.5	7	591.5	16.5	272.25	1905.75
94.5	4	378.0	26.5	702.25	2809.00
	40	2720.0			10510.00

The mean $\overline{x} = 2720/40 = 68$. This is subtracted from each of the class marks in column 1 to get the entries in column 4. The grouped data formula yields the following estimate of the standard deviation:

$$s = \sqrt{\frac{\sum(x-\overline{x})^2 f}{n-1}} = \sqrt{\frac{10510}{40-1}} = 16.4$$

(b) There is a small discrepancy between the estimated mean and estimated standard deviation and the actual values. These discrepancies occur because in the grouped data formulas, every actual data value in a given class is replaced by the class mark even though most values in the class are not equal to the class mark. For example, every data value in the first class is represented by the class mark 34.5, but the three data values are 36, 38, and 39, all of them larger than the class mark. This introduces a small error in each difference between an observation and the mean. It is unlikely that all of these small errors will cancel each other out causing both the mean and the standard deviation to exhibit small errors.

3.97 (a) In Figure 3.5, nine of the ten data points (90%) lie within two standard deviations to either side of the mean.

(b) In Figure 3.5, ten of the ten data points (100%) lie within three standard deviations to either side of the mean.

(c) In Figure 3.6, nine of the ten data points (90%) lie within two standard deviations to either side of the mean and ten of the ten data points (100%) lie within three standard deviations to either side of the mean.

Copyright © 2012 Pearson Education, Inc. Publishing as Addison-Wesley.

(d) Chebychev's Rule clarifies that the phrase "almost all" in Key Fact 3.2 means "at least 89%". Another reason for the importance of Chebychev's Rule is that is applies to all sets of data, regardless of the sample size or the shape of the distribution of the data, whether it be a sample or an entire population. This rule also makes it possible to make statements about a data set even if only the mean and standard deviation of the data are known.

3.99 (a) Since 30 of the 40 equates to 75%, k must be 2. Thus at least 30 of the books' costs lie within two standard deviations of the mean, or between $106.75 - 2(10.42) = and $106.75 + 2(10.42) or between $85.91 and $127.59.

(b) The difference between $138.01 and the mean $106.75 is $31.26 or 3($10.42). The difference between $106.75 and $75.49 is the same. Chebychev's Rule, using k = 3, says that at least 89% of the 40 books must have costs in this range. Since 0.89(40) = 35.6, at least 35.6 of the 40 sociology books cost between $75.49 and $138.01. Since the number of books must be a whole number, this really means that at least 36 of the 40 books' costs must fall in this range.

3.101 (a) Assuming that these weights have roughly a bell-shaped distribution, approximately 68% of the observations should lie within 4.16kg of the mean 45.30 or within the interval from 41.14 to 49.46. Approximately 95% of the observations should lie within 2(4.16)kg of the mean 45.30 or within the interval from 36.98 to 53.62. Approximately 99.7% of the observations should lie within 3(4.16)kg of the mean 45.30 or within the interval from 32.82 to 57.78.

(b) Forty-four of the 60 observations (73.3%) lie between 41.14 and 49.46.0. Fifty-three of the 60 observations (88.3%) lie between 36.98 and 53.62. All 60 of the observations (100%)lie between 32.82 and 57.78.

(c) The actual percentage within one standard deviation to either side of the mean is close but a little higher than expected from the Empirical Rule. The actual percentage within two standard deviations to either side of the mean is close but a little lower than expected from the Empirical Rule. The actual percentage within three standard deviations to either side of the mean is very close to the expected from the Empirical Rule.

(d) The histogram shown in Exercise 2.100 is a little right-skewed, but not far from bell-shaped. Thus one should expect that the Empirical Rule percentages would not be exact for these data.

(e) The differences between the actual percentages and the Empirical Rule percentages may be the result of this slight right skewness. In each case, the difference in percentages is the result of only three or four observations out of 60 (41 instead of 44 and 57 instead of 53). These could easily be the result of random variation in the sampling process. Thus, use of the Empirical rule is reasonable.

3.103 (a) The mean cost of the books was $106.75 and the standard deviation was $10.42. Thirty-eight out of 40 is 95%. If the costs have a bell-shaped distribution, we would expect the 38 costs to fall within two standard deviations ($20.84) to either side of the mean, or between $106.75 - $20.84 = $85.91 and $106.75 + $20.84 = $127.59. In exercise 3.99, we stated that at least 30 of the books fall within this range.

(b) The interval from $75.49 to 138.01 is $31.26 to either side of $106.75, or 3($10.42). We would expect that about 99.7% (or all 40) of the

Copyright © 2012 Pearson Education, Inc. Publishing as Addison-Wesley.

costs would fall in this range if the costs had a bell-shaped distribution. In exercise 3.99, we stated that at least 36 of the books fall within this range.

Section 3.3

3.105 The median and the interquartile range have the advantage over the mean and standard deviation that they are not sensitive to extreme values; they are said to be resistant.

3.107 No. An extreme observation may be an outlier, but it may also be an indication of skewness.

3.109 (a) The interquartile range is a descriptive measure of variation.

(b) It measures the spread of the middle two quarters (50%) of the data.

3.111 The adjacent values are just the minimum and maximum observations when there are no potential outliers.

3.113 (a) First arrange the 4 data values in increasing order and note the two middle values:

1 **2 3** 4

The median (second quartile) is the average of the two middle values, 2 and 3, and thus equals $(2 + 3)/2 = 2.5$.

The first quartile Q_1 is the median of the lower half of the ordered data set 1 2 and thus is the average of the first and second values, 1 and 2. $Q_1 = (1 + 2)/2 = 1.5$

The third quartile Q_3 is the median of the upper half of the ordered data set 3 4 and thus is the average of the first and second values, 3 and 4. $Q_3 = (3 + 4)/2 = 3.5$

(b) The interquartile range is $Q_3 - Q_1 = 3.5 - 1.5 = 2.0$.

(c) The five-number summary consists of the Minimum, Q_1, Median, Q_3, and the Maximum. Respectively, these are 1, 1.5, 2.5, 3.5, and 4.

3.115 (a) First arrange the 5 data values in increasing order:

1 2 3 4 5

Since this data set has on odd number of values (5), the median is the middle value (3rd) in the ordered list. The median, or second quartile, is 3.

The first quartile Q_1 is the median of the lower half of the ordered data set (including the median) 1 **2** 3 and thus is the second value. $Q_1 = 2$

The third quartile Q_3 is the median of the upper half of the ordered data set (including the median) 3 **4** 5 and thus is the second value. $Q_3 = 4$

(b) The interquartile range is $Q_3 - Q_1 = 4 - 2 = 2$.

(c) The five-number summary consists of the Minimum, Q_1, Median, Q_3, and the Maximum. Respectively, these are 1, 2, 3, 4, and 5.

3.117 (a) First arrange the 6 data values in increasing order and note the two middle values:

1 2 **3 4** 5 6

Copyright © 2012 Pearson Education, Inc. Publishing as Addison-Wesley.

The median (second quartile) is the average of the two middle values, 3 and 4, and thus equals $(3 + 4)/2 = 3.5$.

The first quartile Q_1 is the median of the lower half of the ordered data set 1 **2** 3 and thus is the second value. $Q_1 = 2$

The third quartile Q_3 is the median of the upper half of the ordered data set 4 **5** 6 and thus is second values. $Q_3 = 5$

(b) The interquartile range is $Q_3 - Q_1 = 5 - 2 = 3$.

(c) The five-number summary consists of the Minimum, Q_1, Median, Q_3, and the Maximum. Respectively, these are 1, 2, 3.5, 5, and 6.

3.119 (a) First arrange the 7 data values in increasing order:

1 2 3 4 5 6 7

Since this data set has on odd number of values (7), the median is the middle value (4th) in the ordered list. The median, or second quartile, is 4.

The first quartile Q_1 is the median of the lower half of the ordered data set (including the median) 1 **2** **3** 4 and thus is the mean of the second and third values. $Q_1 = (2 + 3)/2 = 2.5$

The third quartile Q_3 is the median of the upper half of the ordered data set (including the median) 4 **5** **6** 7 and thus is the mean of the second and third values. $Q_3 = (5 + 6)/2 = 5.5$

(b) The interquartile range is $Q_3 - Q_1 = 5.5 - 2.5 = 3$.

(c) The five-number summary consists of the Minimum, Q_1, Median, Q_3, and the Maximum. Respectively, these are 1, 2.5, 4, 5.5, and 7.

3.121 (a) First arrange the 20 data values in increasing order and note the two middle values:

45 48 64 70 73 74 74 78 78 **79** **79** 80 80 80 80 80 80 81 82 82

The median (second quartile) is the average of the two middle values, 79 and 79, and thus equals $(79 + 79)/2 = 79$.

The first quartile Q_1 is the median of the lower half of the ordered data set 45 48 64 70 **73** **74** 74 78 78 79 and thus is the average of the fifth and sixth values, 73 and 74. $Q_1 = (73 + 74)/2 = 73.5$

The third quartile Q_3 is the median of the upper half of the ordered data set 79 80 80 80 **80** **80** 80 81 82 82 and thus is the average of the fifth and sixth values, 80 and 80. $Q_3 = (80 + 80)/2 = 80.0$

Thus 25% of the data values are less than or equal to 73.5, the next 25% are between 73.5 and the median 79, the next 25% are between 79 and 80, and the upper 25% are greater than or equal to 80.

(b) The interquartile range is $Q_3 - Q_1 = 80.0 - 73.5 = 6.5$.

(c) The five-number summary consists of the Minimum, Q_1, Median, Q_3, and the Maximum. Respectively, these are 45, 73.5, 79.0, 80.0, and 82. In addition to the explanation of the quartiles in part (a), all of the values are greater than or equal to the Minimum, and all of the values are less than or equal to the Maximum.

(d) To identify any potential outliers, we first find the lower and upper limits. These are found by subtracting 1.5 x IQR from Q_1 and adding 1.5 x IQR to Q_3. The lower limit is $73.5 - 1.5(6.5) = 63.75$ and the upper limit is $80.0 + 1.5(6.5) = 89.75$. Potential outliers consist of any values that are less than 63.75 or greater than 89.75. We see that 45

Copyright © 2012 Pearson Education, Inc. Publishing as Addison-Wesley.

and 48 are both less than 63.75, so these two values are potential outliers.

(e) We will construct a modified boxplot so as to show the potential outliers. The boxplot consists of a rectangle with its ends at the first and third quartiles, a line through it at the median, and two "whiskers" that extend from the quartiles outward to the adjacent values. The adjacent values are the most extreme data values that do not lie outside the lower and upper limits. For these data, the adjacent values are 64 and 82. The potential outliers, 45 and 48, are plotted individually. The modified boxplot, prepared using Minitab, is

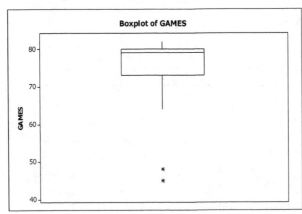

We see that the data is very left skewed, even without considering the two potential outliers. This is because, in most seasons, Gretzky played almost every game, but in a few seasons, he played considerably fewer than the maximum number of games, probably due to injuries.

3.123 (a) First arrange the data in increasing order:

1 1 3 3 4 4 5 6 6 7 **7** 9 9 10 12 12 13 15 18 23 55

Since this data set has on odd number of values (21), the median is the middle value (11th) in the ordered list. The median, or second quartile, is 7.

The first quartile Q_1 is the median of the lower half of the ordered data set (including the median) 1 1 3 3 4 **4** 5 6 6 7 7. Since this data set has an odd number of values (11), Q_1 is the sixth value, so $Q_1 = 4$.

The third quartile Q_3 is the median of the upper half of the ordered data set (including the median) 7 9 9 10 12 **12** 13 15 18 23 55. Since this data set has an odd number of values (11), Q_3 is the sixth value, so $Q_3 = 12$.

Thus 25% of the data values are less than or equal to 4, the next 25% are between 4 and the median 7, the next 25% are between 7 and 12, and the upper 25% are greater than or equal to 12.

(b) The interquartile range is $Q_3 - Q_1 = 12 - 4 = 8$.

(c) The five-number summary consists of the Minimum, Q_1, Median, Q_3, and the Maximum. Respectively, these are 1, 4, 7, 12, and 55. In addition to the explanation of the quartiles in part (a), all of the values are greater than or equal to the Minimum, and all of the values are less than or equal to the Maximum.

(d) To identify any potential outliers, we first find the lower and upper limits. These are found by subtracting 1.5 x IQR from Q_1 and adding 1.5

Copyright © 2012 Pearson Education, Inc. Publishing as Addison-Wesley.

x IQR to Q$_3$. Thus the lower limit is 4 - 1.5(8) = -8 and the upper limit is 12 + 1.5(8) = 24. Potential outliers consist of any values that are less than -8 or greater than 24. We see that 55 is the only potential outlier.

(e) We will construct a modified boxplot, which consists of a rectangle with its ends at the first and third quartiles, a line through it at the median, and two "whiskers" that extend from the quartiles outward to the adjacent values. The adjacent values are the most extreme data values that do not lie outside the lower and upper limits. For these data, the adjacent values are 1 and 23. The potential outlier, 55, is plotted individually. Thus the modified boxplot, prepared using Minitab, is

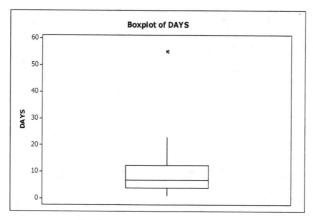

We see that the data are slightly right skewed if the potential outlier is not considered. The potential outlier is far from the other values, making it important to investigate the observation further.

3.125 (a) First arrange the data in increasing order:

57   66   88   96   **116   147**   147   154   154   175

Since this data set has on even number of values (10), the median is the mean of the middle two values (5th and 6th) in the ordered list. Thus the median, or second quartile, is (116 + 147)/2 = 131.5.

The first quartile Q$_1$ is the median of the lower half of the ordered data set 57   66   **88**   96   116. Since this data set has an odd number of values (5), Q$_1$ is the middle value (3rd) in the ordered list = 88.

The third quartile Q$_3$ is the median of the upper half of the ordered data set 147   147   **154**   154   175. Since this data set has an odd number of values (5), Q$_3$ is the middle value in the ordered list = 154.

Thus 25% of the data values are less than or equal to 88, the next 25% are between 88 and the median 131.5, the next 25% are between 131.5 and 154, and the upper 25% are greater than or equal to 154.

(b) The interquartile range is Q$_3$ - Q$_1$ = 154 - 88 = 66.

(c) The five-number summary consists of the Minimum, Q$_1$, Median, Q$_3$, and the Maximum. Respectively, these are 57, 88, 131.5, 154, and 175. In addition to the explanation of the quartiles in part (a), all of the values are greater than or equal to the Minimum, and all of the values are less than or equal to the Maximum.

(d) To identify any potential outliers, we first find the lower and upper limits. These are found by subtracting 1.5 x IQR from Q$_1$ and adding 1.5

Copyright © 2012 Pearson Education, Inc. Publishing as Addison-Wesley.

x IQR to Q_3. Thus the lower limit is $88 - 1.5(66) = -11$ and the upper limit is $154 + 1.5(66) = 253$. Potential outliers consist of any values that are less than -11 or greater than 253. We see that there are no potential outliers.

(e) We will construct a boxplot, which consists of a rectangle with its ends at the first and third quartiles, a line through it at the median, and two "whiskers" that extend from the quartiles outward to the minimum and maximum values. Thus the boxplot, prepared using Minitab, is

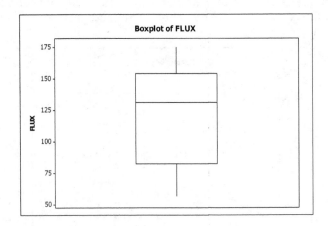

We see that the data are roughly symmetrical with no potential outliers.

3.127 (a) First arrange the data in increasing order:

21 70 125 195 389 649 656 664 682 1006 1300 1403 1433 **1800** 1982 2205 2515 3027 3634 4200 5299 5947 7886 8543 9310 11189 17341

Since this data set has on odd number of values (27), the median is the middle value (14th) in the ordered list. The median, or second quartile, is 1800.

The first quartile Q_1 is the median of the lower half of the ordered data set (including the median) 21 70 125 195 389 649 **656 664** 682 1006 1300 1403 1433 1800. Since this data set has an even number of values (14), Q_1 is the average of the seventh and eighth values, so $Q_1 = (656 + 664)/2 = 660.0$.

The third quartile Q_3 is the median of the upper half of the ordered data set (including the median) 1800 1982 2205 2515 3027 3634 **4200 5299** 5947 7886 8543 9310 11189 17341. Since this data set has an even number of values (14), Q_3 is the average of the seventh and eighth values, so $Q_3 = (4200 + 5299)/2 = 4749.5$.

Thus 25% of the data values are less than or equal to 660.0, the next 25% are between 660.0 and the median 1800, the next 25% are between 1800 and 1719.5, and the upper 25% are greater than or equal to 4749.5.

(b) The interquartile range is $Q_3 - Q_1 = 4749.5 - 660.0 = 4089.5$

(c) The five-number summary consists of the Minimum, Q_1, Median, Q_3, and the Maximum. Respectively, these are 21, 660, 1800, 4749.5, and 17341. In addition to the explanation of the quartiles in part (a), all of the values are greater than or equal to the Minimum, and all of the values are less than or equal to the Maximum.

Copyright © 2012 Pearson Education, Inc. Publishing as Addison-Wesley.

(d) To identify any potential outliers, we first find the lower and upper limits. These are found by subtracting 1.5 x IQR from Q_1 and adding 1.5 x IQR to Q_3. Thus the lower limit is $660.0 - 1.5(4089.5) = -5474.25$ and the upper limit is $4749.5 + 1.5(4089.5) = 10883.75$. Potential outliers consist of any values that are less than -5474.25 or greater than 10883.75. We see that 11189 and 17341 are potential outliers.

(e) We will construct a modified boxplot that consists of a rectangle with its ends at the first and third quartiles, a line through it at the median, and two "whiskers" that extend from the quartiles outward to the adjacent values. The adjacent values are the most extreme data values that do not lie outside the lower and upper limits. For these data, the adjacent values are 21 and 9310. The potential outliers, 11189 and 17341, are plotted individually. Thus the boxplot, prepared using Minitab, is

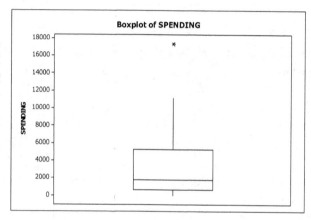

We have shown you the modified boxplot from Minitab to illustrate that software may produce a slightly different boxplot than we expected from our own computations. Minitab uses a different definition of the quartiles, which leads to a larger value of the IQR (4643). This, in turn, results in upper and lower limits that are farther from the median than ours are. Consequently, 11189 shows up as an adjacent value, not an outlier, in Minitab's modified boxplot. The overall picture of the data being right skewed with outlier(s) remains much the same.

3.129 (a) First, arrange the data in increasing order

6 7 8 8 8 8 9 **9** 9 10 10 10 10 10 10

Since there is an odd number of data values (15), the median is the middle (8th) value. Thus the median is 9.

The first quartile Q_1 is the median of the lower half of the ordered data set (including the median) 6 7 8 **8 8** 8 9 9. Since this data set has an even number of values (8), Q_1 is the average of the fourth and fifth values, so $Q_1 = (8 + 8)/2 = 8$.

The third quartile Q_3 is the median of the upper half of the ordered data set (including the median) 9 9 10 **10 10** 10 10 10. Since this data set has an even number of values (8), Q_3 is the average of the fourth and fifth values, so $Q_3 = (10 + 10)/2 = 10$.

(b) The quartiles are not particularly useful for this set of data due to the small range of the data and the large numbers of identical values at 8, 9, and 10. For example, Q_3 and the maximum are equal as a result

Copyright © 2012 Pearson Education, Inc. Publishing as Addison-Wesley.

of the highest six values all being 10.

3.131 The center of the distribution of weight loss, as measured by the median, is very similar for the two groups. However, group 1 has less variation in weight loss, both in the middle 50% and overall. Group 2 has a larger range than group 1. The distribution of weight loss for both groups is approximately symmetric.

3.133 The median hemoglobin level is much lower in the HB SS patients than in the other two types of patients (which were roughly the same). The least variation in hemoglobin levels occurred in the HB SC patients, while the greatest variation occurred in the HB ST patients All three distributions were roughly symmetric about their medians.

3.135 The central box is divided into equal parts by the median. The two whiskers will be the same length, although they may not necessarily be the same length as each part of the central box.

3.137 (a) Using Minitab, retrieve the data from the WeissStats CD. Then choose

Stat ▶ Basic Statistics ▶ Display Descriptive Statistics, enter PUPS

in the **Variables** text box, and click **OK**. The output appears in the Session Window as

Variable	N	N*	Mean	SE Mean	StDev	Minimum	Q1	Median	Q3	Maximum
PUPS	80	0	7.125	0.211	1.885	3.000	6.000	7.000	8.000	12.000

We see from the output that $Q_1 = 6.0$, the median = 7.0, and $Q_3 = 8.0$. Thus 25% of the values are less than or equal to 6.0, another 25% are between 6.0 and 7.0, the next 25% are between 7.0 and 8.0, and the top 25% are greater than 8.0.

(b) The interquartile range is $Q_3 - Q_1 = 8.0 - 6.0 = 2.0$. The middle 50% of the data values span an interval of 2.0.

(c) The five-number summary consists of the Minimum, Q_1, median, Q_3, and the Maximum. These are, respectively, 3, 6, 7, 8, and 12. In addition to the interpretation in part (a), all of the values are greater than or equal to the minimum, 3, and all are less than or equal to the maximum, 12.

(d) To identify any potential outliers, we first find the lower and upper limits. These are found by subtracting 1.5 x IQR from Q_1 and adding 1.5 x IQR to Q_3. Thus the lower limit is $6 - 1.5(2) = 3$, and the upper limit is $8 + 1.5(2) = 11$. Potential outliers consist of any values that are less than 3 or greater than 11. We see that 12 is a potential outlier. The maximum, 128, is greater than the upper limit, and it occurs once, so there is one potential outlier.

(e) To obtain a boxplot, we choose **Graph ▶ Boxplot**, select **Simple** in the **One Y** row and click **OK**. Enter PUPS in the **Graph Variables** text box and click **OK**. The result is

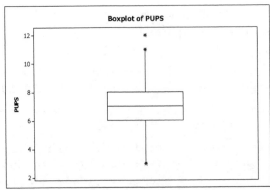

Copyright © 2012 Pearson Education, Inc. Publishing as Addison-Wesley.

With the exception of the one potential outlier, the distribution of the data is very symmetric, approximately bell-shaped.

3.139 (a) Using Minitab, retrieve the data from the WeissStats CD. Then choose

Stat ▶ Basic Statistics ▶ Display Descriptive Statistics, enter
<u>PERCENT</u> in the **Variables** text box, and click **OK**. The output appears in the Session Window as

Variable	N	N*	Mean	SE Mean	StDev	Minimum	Q1	Median	Q3
PERCENT	51	0	86.902	0.533	3.807	79.000	83.000	88.000	90.000

Variable	Maximum
PERCENT	93.000

We see from the output that $Q_1 = 83.0$, the median = 88.0, and $Q_3 = 90.0$. Thus 25% of the values are less than or equal to 83.0, another 25% are between 83.0 and 88.0, the next 25% are between 88.0 and 90.0, and the top 25% are greater than 90.0.

(b) The interquartile range is $Q_3 - Q_1 = 90.0 - 83.0 = 7.0$. The middle 50% of the data values span an interval of 7.0.

(c) The five-number summary consists of the Minimum, Q_1, median, Q_3, and the Maximum. These are, respectively, 79.0, 83.0, 88.0, 90.0, and 93.0. In addition to the interpretation in part (a), all of the values are greater than or equal to the minimum, 78.0, and all are less than or equal to the maximum, 93.0.

(d) To identify any potential outliers, we first find the lower and upper limits. These are found by subtracting 1.5 x IQR from Q_1 and adding 1.5 x IQR to Q_3. Thus the lower limit is $83.0 - 1.5(7.0) = 72.5$, and the upper limit is $90.0 + 1.5(7.0) = 100.5$. Potential outliers consist of any values that are less than 72.5 or greater than 100.5. We see that there are no potential outliers.

(e) To obtain a boxplot, we choose **Graph ▶ Boxplot**, select **Simple** in the **One Y** row and click **OK**. Enter <u>PERCENT</u> in the **Graph Variables** text box and click **OK**. The result is

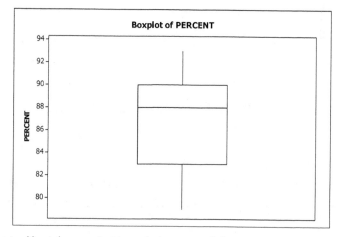

The distribution of the data roughly symmetrical, maybe slightly left skewed. The whiskers on the graph extend out to the minimum and maximum values.

3.141 (a) Using Minitab, retrieve the data from the WeissStats CD. Then choose

Stat ▶ Basic Statistics ▶ Display Descriptive Statistics, enter

Copyright © 2012 Pearson Education, Inc. Publishing as Addison-Wesley.

BODYTEMP in the **Variables** text box, and click **OK**. The output appears in the Session Window as

```
Variable   N   N*   Mean   SE Mean   StDev   Minimum      Q1   Median       Q3
BODYTEMP  93    0  98.124   0.0671   0.647    96.700   97.650   98.200   98.600

Variable  Maximum
BODYTEMP   99.400
```

We see from the output that $Q_1 = 97.65$, the median = 98.20, and $Q_3 = 98.60$. Thus 25% of the values are less than or equal to 97.65, another 25% are between 97.65 and 98.20, the next 25% are between 98.20 and 98.60, and the top 25% are greater than 98.60.

(b) The interquartile range is $Q_3 - Q_1 = 98.60 - 97.65 = 0.95$. The middle 50% of the data values span an interval of 0.95.

(c) The five-number summary consists of the Minimum, Q_1, median, Q_3, and the Maximum. These are, respectively, 96.70, 97.65, 98.20, 98.60, and 99.40. In addition to the interpretation in part (a), all of the values are greater than or equal to the minimum, 96.70, and all are less than or equal to the maximum, 99.40.

(d) To identify any potential outliers, we first find the lower and upper limits. These are found by subtracting 1.5 x IQR from Q_1 and adding 1.5 x IQR to Q_3. Thus the lower limit is $97.65 - 1.5(.095) = 96.225$, and the upper limit is $98.60 + 1.5(0.95) = 100.025$. Potential outliers consist of any values that are less than 96.225 or greater than 100.025. We see that there are no potential outliers.

(e) To obtain a boxplot, we choose **Graph ▶ Boxplot**, select **Simple** in the **One Y** row and click **OK**. Enter RATE in the **Graph Variables** text box and click **OK**. The result is

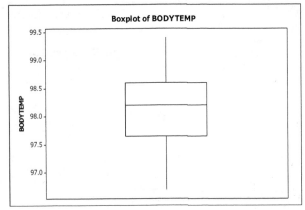

We see that the distribution of the data is roughly symmetrical with no outliers.

3.143 (a) Using Minitab, retrieve the data from the WeissStats CD. To obtain a boxplot choose **Graph ▶ Boxplot**, select **Simple** in the **Multiple Y's** row and click **OK**. Enter ITALIANS and ETRUSCANS in the **Graph Variables** text box and click **OK**. The result is

Copyright © 2012 Pearson Education, Inc. Publishing as Addison-Wesley.

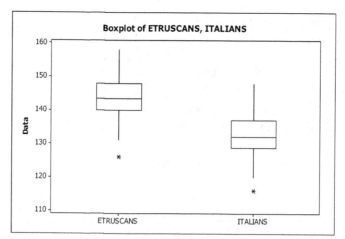

(b) The overall shape of the distributions for the Italian and Etruscan head breadths are very similar. They are both symmetric with a single lower potential outlier and similarly spread out quartiles. However, the distribution for the Etruscans is greater than the Italians. The median, quartiles, maximum, etc. are all larger for the Etruscans than the Italians.

3.145 (a) Using Minitab, retrieve the data from the WeissStats CD. Then choose

Graph ▶ Boxplot, select **Simple** in the **Multiple Y's** row and click **OK**.

Enter <u>STOMACH</u>, <u>BRONCHUS</u>, <u>COLON</u>, <u>OVARY</u>, and <u>BREAST</u> in the **Graph Variables** text box and click **OK**. The result is

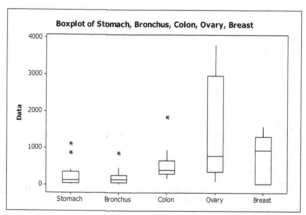

(b)   The median survival times were greatest for breast cancer patients, followed by those with cancer of the ovary, colon, bronchus, and stomach. The variation in survival times for colon, bronchus, and stomach cancer patients, as measured by the interquartile range, were considerably smaller than the variation experienced by the ovary and breast cancer patients. The first three also contained outliers in the data, whereas the last two did not. All of the data distributions except for breast cancer are right skewed, whereas the breast cancer survival times are slightly left skewed.

Section 3.4

3.147 The ultimate objective of dealing with sample data in inferential studies is to describe the entire population.

3.149 (a)   A standardized variable always has mean <u>0</u> and standard deviation <u>1</u>.

(b)   The z-score corresponding to an observed value of a variable tells you

Copyright © 2012 Pearson Education, Inc. Publishing as Addison-Wesley.

how many standard deviations the observation is from the mean and, by its sign, what direction it is from the mean.

(c) A positive z-score indicates that the observation is greater than the mean, whereas a negative z-score indicates that the observation is less than the mean.

3.151 The number 62.5 is a parameter since it is the mean of the entire population of players on the 2008 U.S. women's soccer team.

3.153 (a) The population mean is $\mu = \dfrac{\sum x_i}{N} = \dfrac{9}{3} = 3$.

(b) We will use the computing formula to find the standard deviation. For that, we need $\sum x_i^2 = 4^2 + 0^2 + 5^2 = 41$. The standard deviation is then

$$\sigma = \sqrt{\frac{\sum x_i^2}{N} - \mu^2} = \sqrt{\frac{41}{3} - 3^2} = 2.2 \ .$$

3.155 (a) The population mean is $\mu = \dfrac{\sum x_i}{N} = \dfrac{11}{4} = 2.75$.

(b) We will use the computing formula to find the standard deviation. For that, we need $\sum x_i^2 = 1^2 + 2^2 + 4^2 + 4^2 = 37$. The standard deviation is then

$$\sigma = \sqrt{\frac{\sum x_i^2}{N} - \mu^2} = \sqrt{\frac{37}{4} - 2.75^2} = 1.3 \ .$$

3.157 (a) The population mean is $\mu = \dfrac{\sum x_i}{N} = \dfrac{25}{5} = 5$.

(b) We will use the computing formula to find the standard deviation. For that, we need $\sum x_i^2 = 1^2 + 9^2 + 8^2 + 4^2 + 3^2 = 171$. The standard deviation is

then $\sigma = \sqrt{\dfrac{\sum x_i^2}{N} - \mu^2} = \sqrt{\dfrac{171}{5} - 5^2} = 3.0$.

3.159 (a) The population consists of all people living in the United States. The variable of interest is the age of each person.

(b) The mean is the total of the six ages divided by 6, i.e. 195/6 = 32.5. The median is the mean of the two middle ages when they are listed in increasing order as 7, 9, **29, 45**, 51, 54. Thus the median is (29 + 45)/2 = 37.0. These are statistics since they are derived from a sample. We write our results as $\bar{x}$ = 32.5 and M = 37.0.

(c) This mean and median are parameters since they are derived from the entire population. Thus we write μ = 35.8 and η = 35.3.

3.161 (a) The population mean is $\mu = \dfrac{\sum x_i}{N} = \dfrac{1430}{16} = 89.375$.

(b) We will use the computing formula to find the standard deviation. For that we need $\sum x_i^2 = 45^2 + 65^2 + ... + 50^2 = 149550$. The standard deviation is

Copyright © 2012 Pearson Education, Inc. Publishing as Addison-Wesley.

then $\sigma = \sqrt{\dfrac{\sum x_i^2}{N} - \mu^2} = \sqrt{\dfrac{149550}{16} - 89.375^2} = 36.9$.

(c)   The median is the mean of the two middle observations in the following ordered list of 16 observations: 40, 45, 60, 65, 65, 65, 65, **65, 80,** 80, 100, 125, 135, 145, 145, 150.   The mean of the two middle observations is (65 + 80)/2 = 72.5, so η = 72.5.

(d)   The mode is the value that occurs more often than any other value, so the mode is 65.

(e)   The first quartile is the median of the first eight values in the ordered list of part (c), so it is the mean of the fourth and fifth values.   Thus Q_1 = (65 + 65)/2 = 65.   The third quartile is the median of the last eight values in the ordered list of part (c), so Q_3 = (125 + 135)/2 = 130.   The IQR is the difference between Q_1 and Q_3.   Thus IQR = Q_3 - Q_1 = 130 - 65 = 65.

3.163 (a)   For Orlando, the population mean is $\mu = \dfrac{\sum x_i}{N} = \dfrac{1803}{5} = 360.6$.   For Las Vegas, the population mean is $\mu = \dfrac{\sum x_i}{N} = \dfrac{1083}{5} = 216.6$.

(b)   The range of values for Las Vegas is almost 300, while it is only a little under 150 for Orlando, so Orlando is likely to have the smaller standard deviation.

(c)   We will use the computing formula to find the standard deviation.   For Orlando, we need $\sum x_i^2 = 402^2 + 318^2 + ... + 403^2 = 666,995$.   The standard deviation is then $\sigma = \sqrt{\dfrac{\sum x_i^2}{N} - \mu^2} = \sqrt{\dfrac{666995}{5} - 360.6^2} = 58.0$.   For Las Vegas, we need $\sum x_i^2 = 81^2 + 123^2 ... + 354^2 = 287,631$.   The standard deviation is then $\sigma = \sqrt{\dfrac{\sum x_i^2}{N} - \mu^2} = \sqrt{\dfrac{287631}{5} - 216.6^2} = 103.0$.

(d)   Yes.   We expected the standard deviation to be smaller for Orlando than for Las Vegas, and that turned out to be the case.

3.165 (a)   The standardized version of x is z = (x - 32.9)/17.9.

(b)   The mean and standard deviation of z are 0 and 1 respectively.

(c)   For x = 81.3, the z score is (81.3 - 32.9)/17.9 = 2.70

For x = 20.8, the z score is (20.8 - 32.9)/17.9 = -0.68

(d)   The value 81.3 is 2.70 standard deviations above the mean 32.9.

The value 20.8 is 0.68 standard deviations below the mean 32.9.

(e)

$\bar{x}$-3s	$\bar{x}$-2s	$\bar{x}$-s	$\bar{x}$	$\bar{x}$+s	$\bar{x}$+2s	$\bar{x}$+3s
|	|	| • |	|	|	| • |	
-20.8	-2.9	15.0	32.9	50.8	68.7	86.6 x
-3	-2	-1	0	1	2	3 z

Copyright © 2012 Pearson Education, Inc. Publishing as Addison-Wesley.

3.167 (a) The standardized version of x is z = (x - 6.71)/0.67.

(b) For x = 5.2, the z score is (5.2 - 6.71)/0.67 = -2.25

For x = 8.1, the z score is (8.1 - 6.71)/0.67 = 2.07

The value 5.2 is 2.25 standard deviations below the mean 6.71.

The value 8.1 is 2.07 standard deviations above the mean 6.71.

3.169 (a) z = (21.4 - 25)/1.15 = -3.13.

(b) Yes. If the gas mileages have a distribution that is roughly bell-shaped, we would expect more than 99% of cars to get mileage within three standard deviations from the mean of 25 mpg. Your car's mpg is more than three standard deviations below the mean. Even if the distribution is not bell-shaped, Chebychev's Rule says that at least 89% of the mileages must be within three standard deviations.

3.171 The formulas used here to compute s and σ are

$$s = \sqrt{\frac{\sum (x_i - \bar{x})^2}{n-1}} \qquad \text{and} \qquad \sigma = \sqrt{\frac{\sum (x_i - \mu)^2}{N}}$$

For each data set, the relevant calculations are:

Data Set	Number of Observations	$\sum (x_i - \text{mean})^2$	s	σ
1	4	14.00	2.16	1.87
2	7	24.86	2.04	1.88
3	10	36.40	2.01	1.91

(a) The sample standard deviations are found in the fourth column of the previous table.

(b) The population standard deviations are found in the fifth column of the previous table.

(c) Comparing s and σ for a given data set, the two measures will tend to be closer together if the data set is large.

3.173 (a) Using Excel, we enter HEIGHT in A1 and the 18 data values in cells A2 through A19. The data values in inches are

178   173   165   163   173   168   168   170   173

173   175   165   165   168   163   175   168   165

To obtain the mean, in cell A20, we type =AVERAGE(A2:A19) and the result is 169.33 cm.

To obtain the population standard deviation directly, in cell A21 we type =STDEVP(A2:A19). This yields the value 4.45 cm.

You could also obtain the first result in Excel by selecting **DDXL ▶ Summaries**. Select **Summary of One Variable** from the Function type box, select cells A1 to A21 with the cursor, enter HEIGHT in the **Quantitative Variable** box by dragging it from the **Names and Columns** box and click **OK**. Unfortunately, this procedure produces a sample standard deviation instead of a population standard deviation, so you than have to convert s to σ using the formula derived in Exercise 3.172.

3.175 z = (205500 - 220258)/5237 = -2.82. Applying Chebychev's rule to that z-score with k = 82.82, we conclude that at least $100(1 - 1/2.82^2)\% = 87.4\%$ of

Copyright © 2012 Pearson Education, Inc. Publishing as Addison-Wesley.

the sale prices lie within 2.82 standard deviations to either side of the mean of $220,258. Therefore, the $205,500 price of the home you are contemplating buying is lower than at least 87.4% of the prices of comparable homes. It appears that this home is bargain.

3.176 (a) If the distributions of the math and verbal SAT scores for the graduating class are approximately the same shape, it would be reasonable to use z-scores to compare the standings of the student on the two tests relative to other students in the class.

(b) On the math SAT, the student's z-score is (740 − 528)/105 = 2.01, while on the verbal SAT, the student's z-score is (715 − 475)/98 = 2.45. Thus, relative to the other students in the class, the student did better on the verbal SAT.

Review Problems for Chapter 3

1. (a) Descriptive measures are numbers used to describe data sets.

   (b) Measures of center indicate where the center or most typical value of a data set lies.

   (c) Measures of variation indicate how much variation or spread the data set has.

2. The two most commonly used measures of center for quantitative data are the mean and the median. The mean uses all of the data, but can be influenced by the presence of a few outliers. The median is computed from the one or two center values in an ordered list of the data. It does not make use of all of the data, but it has the advantage that it is not influenced by the presence of a few outliers.

3. The only measure of center, among those we discussed, that is appropriate for qualitative data is the mode.

4. (a) standard deviation

   (b) interquartile range

5. (a) $\overline{x}$ (b) s   (c) μ (d) σ

6. (a) This is not necessarily true.

   (b) This is necessarily true.

7. Almost all of the observations in any data set lie within <u>3</u> standard deviations to either side of the mean.

8. (a) The components of the five-number summary are the minimum, Q_1, median, Q_3, and the maximum.

   (b) The median is a measure of the center. The interquartile range, which is found as $Q_3 - Q_1$, and the range, which is the difference between the maximum and the minimum, are measures of variation.

   (c) The boxplot is based on the five-number summary.

9. (a) An outlier is an observation that falls well outside the overall pattern of the data.

   (b) First compute the interquartile range IQR = $Q_3 - Q_1$. Then compute two limits as $Q_1 - 1.5$IQR and $Q_3 + 1.5$IQR. Observations that are lower than the lower limit or higher than the upper limit are potential outliers and require further study.

10. (a) A z-score for an observation x is obtained by subtracting the mean of

Copyright © 2012 Pearson Education, Inc. Publishing as Addison-Wesley.

the data set from x and then dividing the result by the standard deviation; i.e., $z = (x - \mu)/\sigma$ for population data or $z = (x - \overline{x})/s$ for sample data.

(b) A z-score indicates how many standard deviations an observation is below or above the mean of the data set.

(c) An observation with a z-score of 2.9 is likely to be greater than all or almost all of the other data values.

11. (a) The mean is calculated as $\overline{x} = \Sigma x_i/n$. For the sample of 20 guests, $\overline{x} = 47/20 = 2.35$ alcoholic drinks.

The median is found by ordering the observations from lowest to highest and finding the average of the two observations in the middle. The list is

0 0 1 1 1 1 1 2 2 **2 2** 2 3 3 4 4 4 4 5 5

The median is the average of the 10th and 11th values, and is therefore $(2 + 2)/2 = 2$.

The mode is the value that occurs the most times. In this data set, 1 and 2 both occur 5 times, and so the modes are 1 and 2.

(b) If the purpose is to help in estimating the cost of the party, the mean is the best since it takes into account the drinking practices of all of the guests. If the purpose is to describe the average guest, the median is more typical of the list of values. The mode is less useful for this data since there are two modes.

12. Many more marriages are characterized as being of short duration than other durations. Since the median is ordinarily preferred for data sets that have exceptional (very large or small) values, the median is more appropriate than the mean as a measure of central tendency for data on the duration of marriages.

13. Regarding death certificates, we are interested in the most frequent cause of death. Causes of death are qualitative. There is no way to compute a mean or median for such data. The mode is the only measure of central tendency that can be used for qualitative data.

14. The mean is $\overline{x} = \dfrac{\sum x_i}{n} = \dfrac{305.3}{10} = 30.53$; since n is even (10), the median is the mean of the two middle observations in the ordered list of the data values: 17.4 21.0 17.4 31.5 **32.0 33.0** 33.0 34.5 37.5 38.0. Thus the median is $(32 + 33)/2 = 32.5$. The mode is the most frequently occurring value. Since 33.0 occurs twice and no other value occurs more than once, 33.0 is the mode.

15. (a) $\overline{x} = \dfrac{\sum x_i}{n} = \dfrac{457}{10} = 45.7$ kilograms

(b) range $= 54 - 37 = 17$ kilograms

(c) $s = \sqrt{\dfrac{\sum x^2 - (\sum x)^2/n}{n-1}} = \sqrt{\dfrac{21109 - 457^2/10}{9}} = 5.0$ kg

Copyright © 2012 Pearson Education, Inc. Publishing as Addison-Wesley.

16. (a)

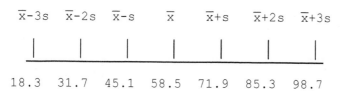

$$\bar{x}-3s \quad \bar{x}-2s \quad \bar{x}-s \quad \bar{x} \quad \bar{x}+s \quad \bar{x}+2s \quad \bar{x}+3s$$

18.3   31.7   45.1   58.5   71.9   85.3   98.7

(b)   Almost all of the ages of the 36 millionaires are between <u>18.3</u> and <u>98.7</u> years old.

17. (a)   The first quartile is the median of the lower half of the ordered list. Since there is a total of 36 ages, the first 18 are in the lower half and the median of these is the average of the middle two, the 9th and 10th values.   Thus $Q_1 = (48 + 48)/2 = 48$.

The second quartile is the median of the entire data set.   The number of pieces of data is 36, and so the position of the median is $(36 + 1)/2 = 18.5$, halfway between the eighteenth and nineteenth data values. Thus the median of the entire data set is $(59 + 60)/2 = 59.5$. That is, $Q_2 = 59.5$.

The third quartile is the median of the upper half of the ordered list. Since there is a total of 36 ages, the last 18 are in the upper half and the median of these is the average of the middle two, the 9th and 10th values.   Thus $Q_3 = (68 + 69)/2 = 68.5$.

Interpreting our results, we conclude that 25% of the ages are less than 48 years; 25% of the ages are between 48 and 59.5 years; 25% of the ages are between 59.5 and 68.5 years; and 25% of the ages are greater than 68.5 years.

(b)   The IQR = $Q_3 - Q_1 = 68.5 - 48 = 20.5$.   Thus, the middle 50% of the ages has a range of 20.5 years.

(c)   Min = 31, $Q_1 = 48$, $Q_2 = 59.5$, $Q_3 = 68.5$, Max = 79

(d)   The limits are given by:

Lower limit = $Q_1 - 1.5(IQR) = 48.0 - 1.5(20.5) = 17.25$

Upper limit = $Q_3 + 1.5(IQR) = 68.5 + 1.5(20.5) = 99.25$

(e)   There are no values below 17.25 or above 99.25, so there are no potential outliers.

(f)   A boxplot is constructed easily using the information in part (a).

(i)   low data value = 31 years

(ii)   high data value = 79 years

(iii) Q_1 = 48.0 years

(iv) Q_3 = 68.5 years

(v)   median = Q_2 = 59.5 years.

Copyright © 2012 Pearson Education, Inc. Publishing as Addison-Wesley.

These values are used to construct the boxplot as follows:

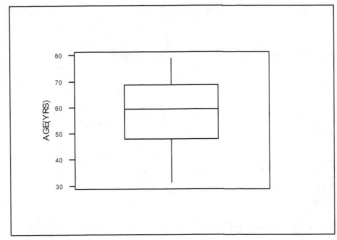

18. (a) The order list of the data is 0.7 1.0 1.1 1.1 1.2 1.5 1.8 1.8
        1.8 1.8 **1.9 2.0** 2.0 2.0 2.3 2.7 3.3 3.4 3.6 3.8 6.7 7.6.
        Since n is even (22), the median is mean of the two middle values, the
        11th and 12th. Thus the median is (1.9 + 2.0)/2 = 1.95.

    The first quartile is the median of the first 11 values in the list, so
    Q_1 = the 6th value = 1.5. The third quartile is the median of the last
    11 values in the list, so Q_3 = the 6th value from the upper end = 3.3.
    Thus, the five-number summary is Minimum = 0.7, Q_1 = 1.5, Median = 1.95,
    Q_3 = 3.3, and Maximum = 7.6.

    (b) The IQR = 3.3 – 1.95 = 1.35. The lower limit = Q1 – 1.5 x IQR = 1.5 –
        1.5(1.35) = –0.525, while the upper limit = Q3 + 1.5 x IQR = 3.3 +
        1.5(1.35) = 5.325. There are no values below the lower limit, but 6.7
        and 7.6 lie above the upper limit and are therefore potential outliers.

    (c) The boxplot consists of a rectangle with its ends at 1.5 and 3.3 and a
        line across the middle at 1.95. Whiskers extend from the ends of the
        box out to the adjacent values (the most extreme values on each end of
        the box that are not beyond the limits). Thus the lower whisker
        extends to 0.7, and the upper whisker extends to 3.8. The two
        potential outliers are plotted individually. The result is

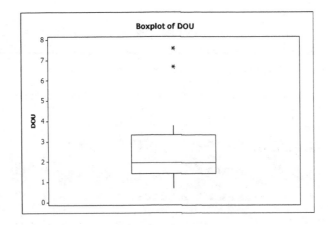

19. The median number of traffic fatalities in Wisconsin is about 770 for the
    years 1982-2003 while the median number in New Mexico is about 490. In
    fact, the maximum number in New Mexico (about 590) is less than the minimum
    number in Wisconsin (about 650) for those years. The IQR and the range of

Copyright © 2012 Pearson Education, Inc. Publishing as Addison-Wesley.

traffic fatalities are both slightly larger for Wisconsin than for New Mexico.

20. (a) $\mu = \dfrac{\sum x_i}{n} = \dfrac{167.6}{9} = 18.62$ thousand students.

(b) Squaring each data value and then summing, we find that

$\sum x_i^2 = 3570.7$. Then the population standard deviation is

$\sigma = \sqrt{\dfrac{\sum x_i^2}{N} - \mu^2} = \sqrt{\dfrac{3570.7}{9} - 18.62^2} = 7.07$ thousand students.

(c) z = (x - μ)/σ = (x - 18.62)/7.07.

(d) The mean of z is 0, and the standard deviation of z is 1. All standardized variables have mean 0 and standard deviation 1.

(e) Converting each x value (dotplot on the left) to its corresponding z-score results in the dotplot on the right.

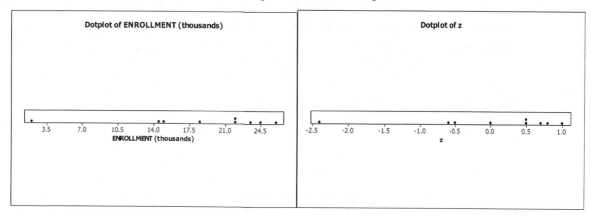

The relative positions of the points in the two plots are the same, but the scales are different.

(f)

x	z = (x - μ)/σ
25.9	(25.9 - 18.62)/7.07 = 1.03
15.0	(15.0 - 18.62)/7.07 = -0.51

The enrollment at Los Angeles is 1.03 standard deviations above the mean and the enrollment at Riverside is 0.51 standard deviation below the mean.

21. (a) The mean price given here is a sample mean. This is because it is the mean price per gallon for the sample of 900 gasoline service stations.

(b) The letter used to designate the mean of $1.811 is $\bar{x}$.

(c) The mean price given here is a statistic. It is a descriptive measure.

22. Using Minitab, retrieve the data from the WeissStats CD. Then choose **Stat**

▶ **Tables** ▶ **Tally Individual Variables**. Enter <u>REGION</u> and <u>DIVISION</u> in the **Variables** text box and click **OK**. The result is the table below.

Copyright © 2012 Pearson Education, Inc. Publishing as Addison-Wesley.

REGION	Count	DIVISION	Count
Midwest	12	East North Central	5
Northeast	9	East South Central	4
South	16	Middle Atlantic	3
West	13	Mountain	8
N=	50	New England	6
		Pacific	5
		South Atlantic	8
		West North Central	7
		West South Central	4
		N=	50

(a) We see from the first column of the above table that the South region occurs 16 times, more than any other region, so South is the mode of the regions.

(b) We see from the second column of the above table that the Mountain and the South Atlantic divisions each occur 8 times, more than any other region, so Mountain and South Atlantic are both modal divisions.

23.   (a) Using Minitab, retrieve the data from the WeissStats CD. Then choose

Stat ▶ Basic Statistics ▶ Display Descriptive Statistics. Click **Statistics** and check **Mode** and click **OK.** Enter <u>VALUE</u> in the **Variables** text box and click **OK.** The result is

Variable	N	N*	Mean	SE Mean	StDev	Minimum	Q1	Median	Q3	Maximum
VALUE	50	0	1563	281	1983	4	234	995	2118	11302

Variable	Mode	N for Mode
VALUE	*	0

We see that the mean is 1563 and the Median is 995. Examining the data, we find that no value occurs more than once, so there is no mode. If we wanted to estimate the total agricultural exports for the nation, the mean would be the best measure of center to use since we could just multiply the mean by 50 to get the total. If we wanted to know what value half of the states are below, then the median is the best measure of center to use. Another reason for using the median is that there are several outliers that lie above the median, including one that is very large (11301.7 for California). Since the mean is quite sensitive to outliers such as this one, the median is the better choice for describing these data.

(b) From the table above, Range = 11302 - 4 = 11298, and the sample standard deviation s = 1983.

(c) The five-number summary consists of the minimum, Q_1, the median, Q_3, and the maximum. These are, respectively, 4, 234, 995, 2118, and 11302. The interquartile range IQR = Q_3 - Q_1 = 2118 - 234= 1884.

(d) We first find the lower limit as Q_1 - 1.5 x IQR = 234 - 1.5(1884) = -2592. The upper limit is Q_3 + 1.5 x IQR = 2118 + 1.5(1884) = 4944. Any data values below the lower limit or above the upper limit are potential outliers. Thus 5198.6, 5246.8, and 11301.7 are potential outliers.

(e) To obtain a boxplot of the data, choose **Graph ▶ Boxplot**, select **Simple** from the **One Y** row and click **OK.** Enter <u>VALUE</u> in the **Variables** text box and click **OK.** The result is

Copyright © 2012 Pearson Education, Inc. Publishing as Addison-Wesley.

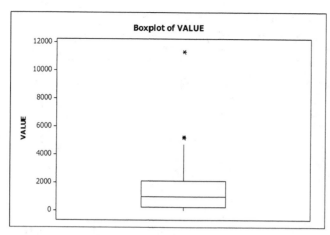

The data distribution is quite right skewed with a median of 995. The plot shows potential outliers on the high side of the data.

24.   (a) Using Minitab, retrieve the data from the WeissStats CD. Then choose

Stat ▶ Basic Statistics ▶ Display Descriptive Statistics. Enter 'LIFE EXP' in the **Variables** text box and click **OK**. The result is

Variable	N	N*	Mean	SE Mean	StDev	Minimum	Q1	Median	Q3
LIFE EXP	224	0	67.188	0.813	12.175	32.300	61.400	71.250	76.625
Variable	Maximum								
LIFE EXP	83.500								

We see that the mean is 67.188 and the median is 71.250. Finding the mode is easiest if we choose **Stat ▶ Tables ▶ Tally Individual Variables** and enter 'LIFE EXP' in the **Variables** text box. Examining the resulting table (not shown here), we find that three values (68.9, 75.4, and 75.5) each occur four times, more than any other values, so there are three modes. The mean is not appropriate for these data since each value is a mean life expectancy based on population sizes that are different for different countries. The median is the best measure of center to use since it tells us that half of the listed countries have life expectancies below this number. Another reason for using the median is that the distribution is left skewed with several potential outliers at the lower end, as will be seen in parts (d) and (e).

(b) From the table above, Range = 83.5 − 32.3 = 51.2, and the sample standard deviation s = 12.175. [Note: The sample standard deviation is based on a calculation that uses the sample mean. Since the sample mean being used does not account for the fact that the different countries have different populations, the true sample standard deviation may differ somewhat from this value.]

(c) The five-number summary consists of the minimum, Q_1, the median, Q_3, and the maximum. These are, respectively, 32.3, 61.4, 71.250, 76.625, and 83.500. The interquartile range IQR = $Q_3 − Q_1$ = 76.625 − 61.4= 15.225.

(d) We first find the lower limit as Q_1 − 1.5 x IQR = 61.4 − 1.5(15.225) = 38.5625. The upper limit is Q_3 + 1.5 x IQR = 76.625 + 1.5(15.225) = 99.4625. Any data values below the lower limit or above the upper limit are potential outliers. Thus 32.3, 35.3, 36.9, 37.0, and 38.0 are potential outliers.

(e) To obtain a boxplot of the data, choose **Graph ▶ Boxplot**, select **Simple** from the **One Y** row and click **OK**. Enter 'LIFE EXP' in the **Variables**

Copyright © 2012 Pearson Education, Inc. Publishing as Addison-Wesley.

text box and click **OK**.   The result is

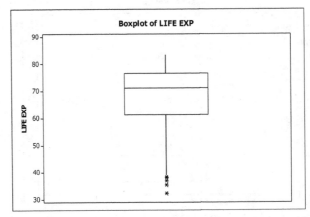

We see that the data are left skewed with the median at about 71.
There are several potential outliers at the lower end of the data.

25.   (a)   Using Minitab, retrieve the data from the WeissStats CD.   Then choose

Stat ▶ Basic Statistics ▶ Display Descriptive Statistics.   Enter <u>HIGH</u>

and <u>LOW</u> in the **Variables** text box and click **OK**.   The result is

Variable	N	N*	Mean	SE Mean	StDev	Minimum	Q1	Median	Q3	Maximum
HIGH	71	0	65.33	1.04	8.72	47.60	58.40	64.00	71.10	85.50
LOW	71	0	45.52	1.11	9.34	29.30	39.80	44.20	51.40	74.20

We see that the mean HIGH temperature is 65.33 and the median HIGH is
64.00.   The mean LOW temperature is 45.52 and the median LOW is 44.20.

Finding the mode is easiest if we choose **Stat ▶ Tables ▶ Tally**

Individual Variables and enter <u>HIGH</u> and <u>LOW</u> in the **Variables** text box.
Examining the resulting table (not shown here), we find for HIGH that
54.5, 55.9, 57.6, 59.8, 62.6, 63.6, 65.1, 67.4, 67.8, and 70.6 all
occur twice, more than any other values, so there are ten modes.   For
LOW, the values 35.8, 39.8, 41.2, 43.2, 44.8, 47.4, 48.6, and 52.5 all
occur twice, more than any other values, so there are eight modes.   The
median is the best measure of center to use since it tells us that half
of the listed cities have minimums (or maximums) below this number.
Another reason for using the median is that the distribution for LOW
has several potential outliers at the upper end, as will be seen in
parts (d) and (e).

(b)   From the table above, for HIGH, Range = 85.5 - 47.6 = 37.9, and the
sample standard deviation s = 8.72.   For LOW, Range = 74.20 - 29.30 =
44.90 and the sample standard deviation s = 9.34.

(c)   The five-number summary consists of the minimum, Q_1, the median, Q_3, and
the maximum.   For HIGH, these are, respectively, 47.60, 58.40, 64.00,
71.10, and 85.50.   The interquartile range for HIGH is IQR = Q_3 - Q_1 =
71.10 - 58.40= 12.70.   For LOW, the five-number summary is 29.30,
39.80, 44.20, 51.40, and 74.20.   IQR = Q_3 - Q_1 = 51.40 - 39.80 = 11.60.

(d)   For HIGH, we first find the lower limit as Q_1 - 1.5 x IQR = 58.40 -
1.5(12.70) = 39.35.   The upper limit is Q_3 + 1.5 x IQR = 71.10 +
1.5(12.70) = 90.15.   Any data values below the lower limit or above the
upper limit are potential outliers.   Examining the data, we find that
there are no potential outliers for HIGH.

For LOW, we first find the lower limit as Q_1 - 1.5 x IQR = 39.80 -
1.5(11.60) = 22.40.   The upper limit is Q_3 + 1.5 x IQR = 51.40 +
1.5(11.60) = 68.80.   Any data values below the lower limit or above the

Copyright © 2012 Pearson Education, Inc. Publishing as Addison-Wesley.

upper limit are potential outliers. Examining the data, we see that 69.1, 70.2, and 74.2 are potential outliers. If you are looking for someplace that is warm year around, these values are for Miami, Honolulu, and San Juan.

(e) To obtain a boxplot of the data, choose **Graph ▶ Boxplot**, select **Simple** from the **Multiple Y's** row and click **OK**. Enter <u>HIGH</u> and <u>LOW</u> in the **Variables** text box and click **OK**. The result is

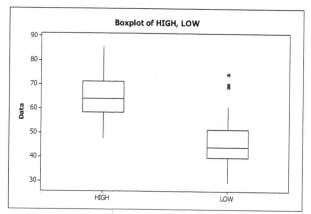

The distribution of HIGH temperatures is quite symmetrical, while the distribution of LOW temperatures is also quite symmetrical except for three outliers. The three cities with high LOW temperatures are all affected by warm southern ocean waters that keep the temperature quite moderate year around.

26. (a) To obtain a boxplot of the height data, choose **Graph ▶ Boxplot**, select **Simple** from the **Multiple Y's** row and click **OK**. Enter 'VEGETARIAN' and 'OMNIVORE' in the **Variables** text box and click **OK**. The result is

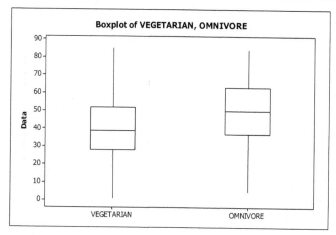

(b) The range from minimum to maximum seems to be similar for the vegetarian and omnivore daily protein intake. Also, the IQR seem to be similar in value for the two groups. The biggest difference is that the quartiles (including the median) for the omnivore daily protein intake is greater than that for the vegetarians. There were not any outliers in either group.

Copyright © 2012 Pearson Education, Inc. Publishing as Addison-Wesley.

Exercises 4.1

4.1 (a) $y = b_0 + b_1 x$ (b) Constants are b_0, b_1; variables are x,y.

    (c) The independent variable is x and the dependent variable is y.

4.3 (a) b_0 is the y-intercept; it is the value of y where the line crosses the y-axis.

    (b) b_1 is the slope; it indicates the change in the value of y for every 1 unit increase in the value of x.

4.5 (a) $y = 68.22 + 0.25x$ (b) $b_0 = 68.22$, $b_1 = 0.25$

    (c)   (d)

Miles x	Cost ($) Y
50	68.22 + 0.25(50) = 80.72
100	68.22 + 0.25(100)= 93.22
250	68.22 + 0.25(250)= 130.72

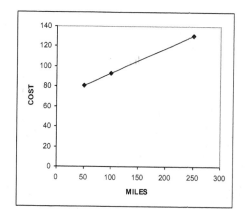

    (e) The visual cost estimate of driving the car 150 miles is a little more than $100. The exact cost is

    $y = 68.22 + 0.25(150) = 105.72$

4.7 (a) $b_0 = 32$, $b_1 = 1.8$

    (b)   (c)

X(°C)	y(°F)
-40	32 + 1.8(-40) = -40
0	32 + 1.8(0) = 32
20	32 + 1.8(20) = 68
100	32 + 1.8(100) = 212

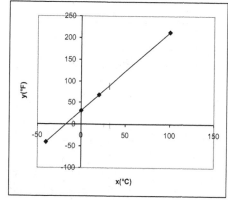

    (d) The visual Fahrenheit temperature estimate corresponding to a Celsius temperature of 28° is about 80°F. The exact temperature is y = 32 + 1.8(28) = 82.4°F.

4.9 (a) $b_0 = 68.22$, $b_1 = 0.25$

    (b) The y-intercept $b_0 = 68.22$ gives the y-value at which the straight line $y = 68.22 + 0.25x$ intersects the y-axis. The slope $b_1 = 0.25$ indicates that the y-value increases by 0.25 units for every increase in x of one unit.

    (c) The y-intercept $b_0 = 68.22$ is the cost (in dollars) for driving the car zero miles. The slope $b_1 = 0.25$ represents the fact that the cost per mile is $0.25; it is the amount the total cost increases for each additional mile driven.

4.11 (a) $b_0 = 32$, $b_1 = 1.8$

Copyright © 2012 Pearson Education, Inc. Publishing as Addison-Wesley.

(b) The y-intercept $b_0 = 32$ gives the y-value where the line $y = 32 + 1.8x$ intersects the y-axis. The slope $b_1 = 1.8$ indicates that the y-value increases by 1.8 units for every increase in x of one unit.

(c) The y-intercept $b_0 = 32$ is the Fahrenheit temperature corresponding to 0°C. The slope $b_1 = 1.8$ represents the fact that Fahrenheit temperature increases by 1.8° for every increase of the Celsius temperature of 1°.

4.13 (a) $b_0 = 3$, $b_1 = 4$

(b) slopes upward

(c)

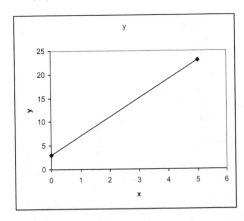

4.15 (a) $b_0 = 6$, $b_1 = -7$

(b) slopes downward

(c)

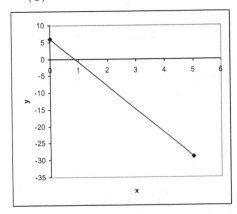

4.17 (a) $b_0 = -2$, $b_1 = 0.5$

(b) slopes upward

(c)

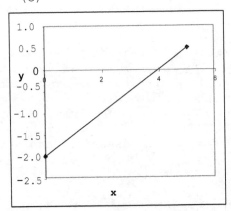

Copyright © 2012 Pearson Education, Inc. Publishing as Addison-Wesley.

4.19   (a) $b_0 = 2$, $b_1 = 0$
           (b)   horizontal
           (c)

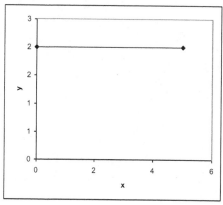

4.21   (a) $b_0 = 0$, $b_1 = 1.5$
           (b)   slopes upward
           (c)

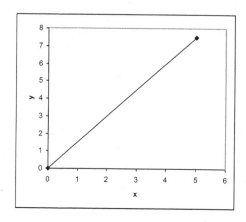

4.23   (a)   slopes upward
           (b) $y = 5 + 2x$
           (c)

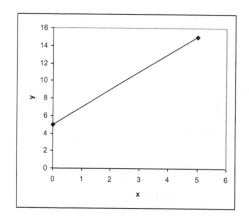

Copyright © 2012 Pearson Education, Inc. Publishing as Addison-Wesley.

4.25 (a) slopes downward

(b) y = -2 - 3x

(c)

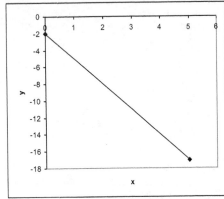

4.27 (a) slopes downward

(b) $y = -0.5x$

(c)

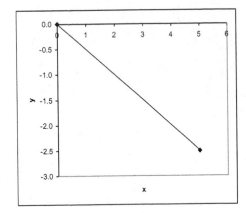

4.29 (a) horizontal

(b) y = 3

(c)

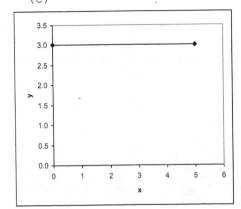

4.31 The line defined by the equation must pass through the two points (x,F) –
$(2,32$ and $(3,16)$. We need to solve for k and x_0. We have

Copyright © 2012 Pearson Education, Inc. Publishing as Addison-Wesley.

$$\begin{Bmatrix} 32 = -k\left(2 - x_0\right) \\ 16 = -k\left(3 - x_0\right) \end{Bmatrix} \quad or \quad \begin{Bmatrix} 32 = -2k + kx_0 \\ 16 = -3k + kx_0 \end{Bmatrix}$$

Subtracting the second equation from the first, we obtain 16 = k. Substituting 16 for k in the first equation, we find

$$32 = -2(16) + 16x_0 \ or \ 64 = 16\,x_0.,$$ from which we find x_0 = 4.

(a) Thus, the linear equation that relates the force exerted to the length compressed is F = -16(x - 4).

(b) The spring constant k = 16.

(c) The natural length of the spring is x_0 = 4 feet.

4.33 (a) If we can express a line in the form $y = b_0 + b_1x$, then there will be one and only one y-value corresponding to each *x*-value. However, that is not the case for a vertical line; one value of x results in an infinite number of y-values.

(b) The form of the equation of a vertical line is x = x_0, where x_0 is the *x*-coordinate of the vertical line.

(c) For a linear equation, the slope indicates how much the y-value on the line increases (or decreases) when the *x*-value increases by one unit. We cannot apply this concept to a vertical line, since the *x*-value is *not* permitted to change. Thus, a vertical line has no slope.

Exercises 4.2

4.35 (a) The criterion used to decide on the line that best fits a set of data points is called the least squares criterion.

(b) The criterion is that the line that best fits a set of data points is the one that has the smallest possible sum of the squares of the errors (errors are the differences between and actual and predicted y values).

4.37 (a) The dependent variable is called the response variable.

(b) The independent variable is called the predictor variable or the explanatory variable.

4.39 (a) In the context of regression, an <u>outlier</u> is a data point that lies far from the regression line, relative to the other data points.

(b) In regression analysis, an <u>influential observation</u> is a data point whose removal causes the regression equation to change considerably.

4.41 (a) Line A: $y = 3 - 0.6x$ Line B: $y = 4 - x$

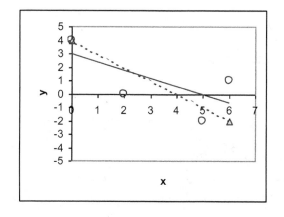

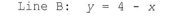
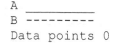

A _____
B - - - - - - - - -
Data points O

Copyright © 2012 Pearson Education, Inc. Publishing as Addison-Wesley.

(b)

Line A: $y = 3 - 0.6x$

x	y	$\hat{y}$	e	e^2
0	4	3.0	1.0	1.00
2	2	1.8	0.2	0.04
2	0	1.8	-1.8	3.24
5	-2	0.0	-2.0	4.00
6	1	-0.6	1.6	2.56
				10.84

Line B: $y = 4 - x$

x	y	$\hat{y}$	e	e^2
0	4	4	0	0
2	2	2	0	0
2	0	2	-2	4
5	-2	-1	-1	1
6	1	-2	3	9
				14

(c) According to the least-squares criterion, Line A fits the set of data points better than Line B. This is because the sum of squared errors, i.e., $\sum e^2$, is smaller for Line A than for Line B.

4.43 (a)

x	y	xy	x^2
2	1	2	4
4	3	12	16
6	4	14	20

Using Formula 4.1,

$$S_{xx} = 20 - 6^2/2 = 2, \ S_{xy} = 14 - (6)(4)/2 = 2, \ b_1 = S_{xy}/S_{xx} = 2/2 = 1$$

$$b_0 = \bar{y} - b_1 \bar{x} = 2 - 1(3) = -1$$

Thus, the regression line is $\hat{y} = -1 + x$.

Without using Formula 4.1, both points must be on the line $y = b_0 + b_1 x$,

so $\begin{Bmatrix} 1 = b_0 + b_1 2 \\ 3 = b_0 + b_1 4 \end{Bmatrix}$. Subtracting the first equation from the second, we

get $2 = b_1 2$, or $b_1 = 1$. Substituting 1 back in the first equation for b_1, $1 = b_0 + 1(2)$, so $b_0 = -1$, and the equation of the line is $y = -1 + x$.

(b)

x	y	xy	x^2
1	3	3	1
5	-3	-15	25
6	0	-12	26

Using Formula 4.1,

$$S_{xx} = 20 - 6^2/2 = 2, \ S_{xy} = -12 - (6)(0)/2 = -12, \ b_1 = S_{xy}/S_{xx} = -12/8 = -1.5$$

$$b_0 = \bar{y} - b_1 \bar{x} = 0 - (-1.5)(3) = 4.5$$

Thus, the regression line is $\hat{y} = 4.5 - 1.5x$.

Without using Formula 4.1, both points must be on the line $y = b_0 + b_1 x$,

so $\begin{Bmatrix} 3 = b_0 + b_1 1 \\ -3 = b_0 + b_1 5 \end{Bmatrix}$. Subtracting the first equation from the second, we

get $-6 = 4b_1$, or $b_1 = -1.5$. Substituting -1.5 back in the first

Copyright © 2012 Pearson Education, Inc. Publishing as Addison-Wesley.

equation for b_1, $3 = b_0 + (-1.5)(1)$, *so* $b_0 = 4.5$, and the equation of the line is y = 4.5 - 1.5x.

4.45

x	y	xy	x^2
3	-4	-12	9
1	0	0	1
2	-5	-10	4
6	-9	-22	14

(a)

$$S_{xx} = \sum x^2 - \left(\sum x\right)^2 / n = 14 - 6^2 / 3 = 2;$$

$$S_{xy} = \sum xy - \left(\sum x\right)\left(\sum y\right) / n = -22 - (6)(-9)/3 = -4$$

$$b_1 = S_{xy} / S_{xx} = -4/2 = -2; \quad b_0 = \bar{y} - b_1 \bar{x} = -3 - (-2)(2) = 1$$

$$\hat{y} = 1 - 2x$$

(b) Applying the regression equation for x = 1 and x = 3,

x $\hat{y}$

1 −1

3 −5

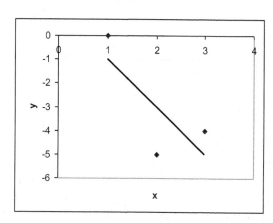

4.47

x	y	xy	x^2
3	4	12	9
4	5	20	16
1	0	0	1
2	-1	-2	4
10	8	30	30

(a)

$$S_{xx} = \sum x^2 - \left(\sum x\right)^2 / n = 30 - 10^2 / 4 = 5;$$

$$S_{xy} = \sum xy - \left(\sum x\right)\left(\sum y\right) / n = 30 - (10)(8)/4 = 10$$

$$b_1 = S_{xy} / S_{xx} = 10/5 = 2; \quad b_0 = \bar{y} - b_1 \bar{x} = 2 - (2)(2.5) = -3$$

$$\hat{y} = -3 + 2x$$

Copyright © 2012 Pearson Education, Inc. Publishing as Addison-Wesley.

(b) Applying the regression equation for x = 1 and x = 4

x $\hat{y}$

1 -1

4 5

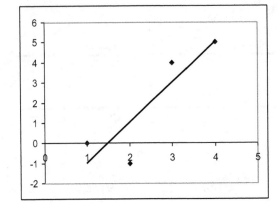

4.49

x	y	xy	x2
0	4	0	0
2	2	4	4
2	0	0	4
5	-2	-10	25
6	1	6	36
15	5	0	69

(a)

$$S_{xx} = \sum x^2 - \left(\sum x\right)^2 / n = 69 - 15^2 / 5 = 24;$$
$$S_{xy} = \sum xy - \left(\sum x\right)\left(\sum y\right) / n = 0 - (15)(5)/5 = -15$$
$$b_1 = S_{xy} / S_{xx} = -15/24 = -0.625; \quad b_0 = \bar{y} - b_1 \bar{x} = 1 - (-0.625)(3) = 2.875$$
$$\hat{y} = 2.875 - 0.625x$$

(b) Applying the regression equation for x = 0 and x = 6

x $\hat{y}$

0 2.875

6 -0.875

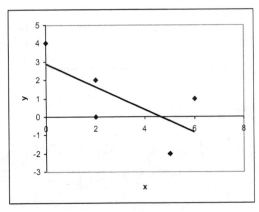

4.51 (a) Formulas for the slope (b_1) and intercept (b_0) of the regression equation are, respectively:

$$b_1 = \frac{\sum xy - \sum x \sum y / n}{\sum x^2 - (\sum x)^2 / n} \qquad b_0 = \frac{1}{n}\left(\sum y - b_1 \sum x\right)$$

Copyright © 2012 Pearson Education, Inc. Publishing as Addison-Wesley.

To compute b_0 and b_1, construct a table of values for x, y, xy, x^2, and their sums:

x	y	xy	x^2
6	290	1740	36
6	280	1680	36
6	295	1770	36
2	425	850	4
2	384	768	4
5	315	1575	25
4	355	1420	16
5	328	1640	25
1	425	425	1
4	325	1300	16
41	3422	13168	199

Thus: $b_1 = \dfrac{13168 - 41(3422)/10}{199 - 41^2/10} = -27.9029$ $b_0 = \dfrac{1}{10}(3422 - (-27.9029)41) = 456.6019$

and the regression equation is $\hat{y} = 456.6019 - 27.9029x$

(b) Begin by selecting two x-values within the range of the x-data. For the x-values 1 and 6, the calculated values for y (in $hundreds) are, respectively:

$$\hat{y} = 456.6019 - 27.9029(1) = 428.6990$$

$$\hat{y} = 456.6019 - 27.9029(6) = 289.1845$$

The regression equation can be graphed by plotting the pairs (1, 428.699) and (6, 289.185) and connecting these points with a line. This equation, the original set of 10 data points, and all of the predicted values are shown in the following graph:

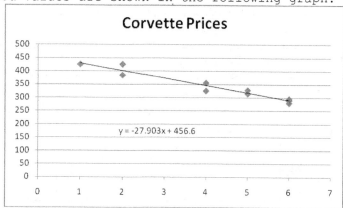

(c) Price tends to decrease as age increases.
(d) Corvettes depreciate an estimated $2,790.29 per year.
(e) The predictor variable is age. The response variable is price.
(f) There are no outliers or potential influential observations.
(g) Price of two-year old Corvette: $\hat{y} = 456.6019 - 27.9029(2) = 400.7961$

or $40,080.

Price of three-year old Corvette: $\hat{y} = 456.6019 - 27.9029(3) = 372.8932$

or $37,289.

Copyright © 2012 Pearson Education, Inc. Publishing as Addison-Wesley.

4.53 (a) Formulas for the slope (b_1) and intercept (b_0) of the regression equation are, respectively:

$$b_1 = \frac{\sum xy - \sum x \sum y / n}{\sum x^2 - (\sum x)^2 / n} \qquad b_0 = \frac{1}{n}\left(\sum y - b_1 \sum x\right)$$

To compute b_0 and b_1, construct a table of values for x, y, xy, x^2, and their sums:

x	y	xy	x^2
57	8.0	456.0	3249
85	22.0	1870.0	7225
57	10.5	598.5	3249
65	22.5	1462.5	4225
52	12.0	624.0	2704
67	11.5	770.5	4489
62	7.5	465.0	3844
80	13.0	1040.0	6400
77	16.5	1270.5	5929
53	21.0	1113.0	2809
68	12.0	816.0	4624
723	156.5	10486.0	48747

$$b_1 = \frac{10486 - 723(1565)/11}{48747 - 723^2/11} = 0.16285 \quad b_0 = \frac{1}{11}(1565 - 0.16285(723)) = 3.52369$$

and the regression equation is $\hat{y} = 3.52369 + 0.16285x$.

(b) Begin by selecting two x-values within the range of the x-data. For the x-values 52 and 85, the calculated values for y are, respectively:

$$\hat{y} = 3.52369 + 0.16285(52) = 11.992$$
$$\hat{y} = 3.52369 + 0.16285(85) = 17.366$$

The regression equation can be graphed by plotting the pairs (52,12.000) and (85, 17.379) and connecting these points with a line. This equation and the original set of 11 data points are presented as follows:

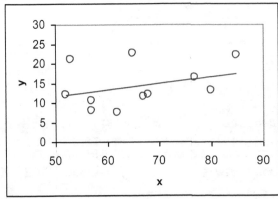

(c) The quantity of volatile compounds emitted tends to decrease as plant weight increases.

Copyright © 2012 Pearson Education, Inc. Publishing as Addison-Wesley.

(d)   The quantity of volatile compounds emitted increases by an estimated 0.163 hundred nanograms for each increase in plant weight of one gram.

(e)   The predictor variable is plant weight. The response variable is hundreds of nanograms of volatile compounds emitted by the plant.

(f)   There are no outliers or potential influential observations.

(g)   The quantity of volatile compounds emitted by a 75 gram plant:

$$\hat{y} = 3.52369 + 0.16285(75) = 15.737 \text{ hundred nanograms or 1574 nanograms.}$$

4.55 (a)   Formulas for the slope (b_1) and intercept (b_0) of the regression equation are, respectively:

$$b_1 = \frac{\sum xy - \sum x \sum y / n}{\sum x^2 - (\sum x)^2 / n} \qquad b_0 = \frac{1}{n}\left(\sum y - b_1 \sum x\right)$$

To compute b_0 and b_1, construct a table of values for x, y, xy, x^2, and their sums:

x	y	xy	x^2
10	92	920	100
15	81	1215	225
12	84	1008	144
20	74	1480	400
8	85	680	64
16	80	1280	256
14	84	1176	196
22	80	1760	484
117	660	9519	1869

$$b_1 = \frac{9519 - 117(660)/8}{1869 - 117^2/8} = -0.84561 \qquad b_0 = \frac{1}{8}(660 - (-0.84561)(117)) = 94.86698$$

and the regression equation is $\hat{y} = 94.86698 - 0.84561x$.

(b)   Begin by selecting two x-values within the range of the x-data. For the x-values 10 and 28, the calculated values for y are, respectively:

$$\hat{y} = 94.86698 - 0.84561(10) = 86.41$$

$$\hat{y} = 94.86698 - 0.84561(22) = 76.26$$

The regression equation can be graphed by plotting the pairs (10, 68.40) and (28, 278.16) and connecting these points with a line. This equation and the original set of 10 data points are presented as follows:

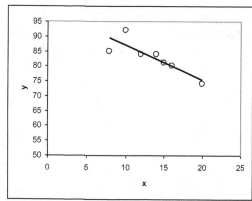

(c)   The exam scores tend to decrease as the number of study hours increases.

Copyright © 2012 Pearson Education, Inc. Publishing as Addison-Wesley.

(d)   The exam score decreases, on average, about 0.85 points for each hour of study time.

(e)   The predictor variable is study time, in hours. The response variable is the exam score.

(f)   There may be two outliers, (8,85) and (10,92). There are no potential influential observations.

(g)   For a student who studies 15 hours the predicted exam score is

$$\hat{y} = 94.86698 - 0.84561(15) = 82.18$$

4.57   The idea behind finding a regression line is based on the assumption that the data points are actually scattered about a line. Only the second data set appears to be scattered about a line. Thus, it is reasonable to determine a regression line only for the second set of data.

4.59   (a)   It is acceptable to use the regression equation to predict the price of a four-year-old Corvette since that age lies within the range of the ages in the sample data. It is not acceptable (and would be extrapolation) to use the regression equation to predict the price of a 10-year-old Corvette since that age falls outside the range of the ages in the sample data.

(b)   It is reasonable to use the regression equation to predict price for ages between one and six years old, inclusive.

4.61   One possibility is that there are students who understand the material well after only a short period of study. These students do not need to study as long as students who have a more difficult time comprehending the material and possibly never reach the same level of understanding. There are other possibilities. Perhaps some students are more organized and do their studying when they are wide awake while others put it off until they are tired and less efficient.

4.63   (a)

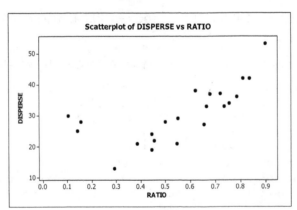

(b)   No. The data obviously do not fall in a line.

4.65   (a)   Using Minitab, with the data in columns named INAUGURATION and DEATH, choose **Graph ▶ Scatterplot...**, select the **Simple** version and click **OK**. Enter DEATH in Row 1 of the **Y variables** column and INAUGURATION in the **X variables** column, and click **OK**. The result is

Copyright © 2012 Pearson Education, Inc. Publishing as Addison-Wesley.

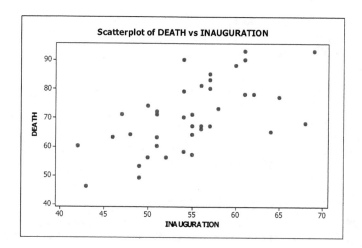

(b) A line appears to be reasonable, although there may be an outlier or two and an influential observation. There does not appear to be any curvature in the data.

(c) Choose **Stat ▶ Regression ▶ Regression**, enter <u>DEATH</u> in the **Response** text box, and <u>INAUGURATION</u> in the **Predictors** text box. Click **OK**. The results are

```
The regression equation is
DEATH = 5.4 + 1.18 INAUGURATION

Predictor        Coef  SE Coef      T      P
Constant         5.37    14.02   0.38  0.704
INAUGURATION   1.1848   0.2537   4.67  0.000

S = 9.70072   R-Sq = 37.1%   R-Sq(adj) = 35.4%

Analysis of Variance

Source           DF      SS      MS      F      P
Regression        1  2051.6  2051.6  21.80  0.000
Residual Error   37  3481.8    94.1
Total            38  5533.4

Unusual Observations

Obs  INAUGURATION  DEATH    Fit  SE Fit  Residual  St Resid
31           54.0  90.00  69.35    1.57     20.65     2.16R
39           69.0  93.00  87.12    3.90      5.88     0.66 X
```

R denotes an observation with a large standardized residual.
X denotes an observation whose X value gives it large leverage.

The regression equation (bold-faced in the output) indicates that the age at death is 5.4 years plus an increase of 1.18 years for every year later that inauguration occurs.

(d) There is one outlier, the point (54,90) and a potential influential observation (69,93)

(e) After removing the outlier, the regression equation becomes

DEATH = 4.1 + 1.20 INAUGURATION .

The effect of removing the outlier is to reduce the constant and increase the slope, both slightly.

(f) After replacing the outlier and removing the potential influential observation (69,93), the regression equation becomes

Copyright © 2012 Pearson Education, Inc. Publishing as Addison-Wesley.

```
DEATH = 8.9 + 1.12 INAUGURATION
```
Removing the influential observation increases the constant and reduces the slope, both significantly. In addition, part of the output is
```
Unusual Observations

Obs  INAUGURATION  DEATH    Fit   SE Fit  Residual  St Resid
 9          68.0   68.00  84.87    4.04    -16.87    -1.89 X
31          54.0   90.00  69.23    1.59     20.77     2.15R
```

This indicates that the point (54,90) is still an outlier, and there is now a new potential influential observation at (68,68).

4.67 (a) Using Minitab, with the data in columns named LOT SIZE and VALUE EXPECTANCY, choose **Graph ▶ Scatterplot...**, select the **Simple** version and click **OK**. Enter VALUE in Row 1 of the **Y variables** column and 'LOT SIZE' in the **X variables** column, and click **OK**. The result is

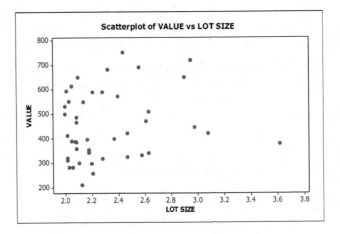

(b) The data points do not have a linear pattern, there is at least one influential observation, and there are likely some outliers. Therefore, find a regression line for these data is not reasonable.

4.69 (a) Using Minitab, with the data in columns named HIGH and LOW, choose **Graph ▶ Scatterplot...**, select the **Simple** version and click **OK**. Enter LOW in Row 1 of the **Y variables** column and HIGH in the **X variables** column, and click **OK**. The result is

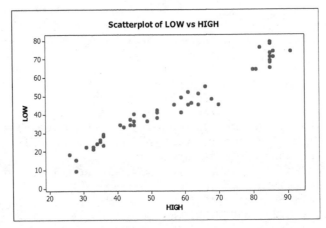

(b) It is reasonable to find a regression line for these data. There is a linear pattern without influential observations.

Copyright © 2012 Pearson Education, Inc. Publishing as Addison-Wesley.

(c) Choose **Stat ▶ Regression ▶ Regression**, enter <u>LOW</u> in the **Response** text box, and <u>HIGH</u> in the **Predictors** text box. Click **OK**. The results are

```
The regression equation is
LOW = - 7.57 + 0.917 HIGH

Predictor      Coef   SE Coef      T      P
Constant     -7.572     1.786   -4.24  0.000
HIGH        0.91685   0.02967   30.91  0.000

S = 4.15111   R-Sq = 95.2%   R-Sq(adj) = 95.1%

Analysis of Variance

Source           DF      SS      MS       F      P
Regression        1   16459   16459  955.17  0.000
Residual Error   48     827      17
Total            49   17286

Unusual Observations

Obs   HIGH     LOW     Fit   SE Fit  Residual  St Resid
 15   70.0  45.000  56.607    0.705   -11.607     -2.84R
 17   82.0  76.000  67.610    0.949     8.390      2.08R
 30   85.0  79.000  70.360    1.021     8.640      2.15R
 45   28.0   9.000  18.100    1.038    -9.100     -2.26R

R denotes an observation with a large standardized residual.
```

As one would expect, there is positive linear relationship between the high and low average temperatures.

(d) Observations 15, 17, 30 and 45 are all potential outliers. There are no influential observations.

(e) Delete the four data points mentioned in part (d), starting with number 45 and ending with number 15. Repeat the regression procedure. The result is

```
The regression equation is
LOW = - 5.66 + 0.884 HIGH

Predictor      Coef   SE Coef      T      P
Constant     -5.655     1.442   -3.92  0.000
HIGH        0.88407   0.02433   36.34  0.000

S = 3.18323   R-Sq = 96.8%   R-Sq(adj) = 96.7%

Analysis of Variance

Source           DF      SS      MS       F      P
Regression        1   13381   13381  1320.50  0.000
Residual Error   44     446      10
Total            45   13826

Unusual Observations

Obs   HIGH     LOW     Fit   SE Fit  Residual  St Resid
 32   68.0  48.000  54.462    0.552    -6.462     -2.06R
 37   85.0  78.000  69.491    0.847     8.509      2.77R

R denotes an observation with a large standardized residual.
```

Copyright © 2012 Pearson Education, Inc. Publishing as Addison-Wesley.

Deleting the outliers resulted in an increase in the intercept and a slight decrease in the slope of the regression line. It also resulted in two new potential outliers.

(f) Not applicable

4.71 (a) Using Minitab, with the data in columns named INCOME and BEER, choose

Graph ▶ Scatterplot..., select the **Simple** version and click **OK**.

Enter <u>BEER</u> in Row 1 of the **Y variables** column and <u>INCOME</u> in the **X variables** column, and click **OK**. The result is

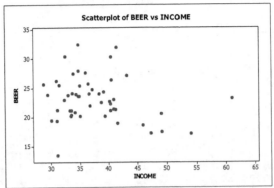

(b) There is no linear pattern in these data, so it is inappropriate to carry out a linear regression. Parts (c)-(f) are omitted.

4.73 (a) Using Minitab, with the data in columns named MANAGERS and ASSETS,

choose **Graph ▶ Scatterplot...**, select the **Simple** version and click

OK. Enter <u>MANAGERS</u> in Row 1 of the **Y variables** column and <u>ASSETS</u> in the **X variables** column, and click **OK**. When you see the scatterplot, you might wonder just where the regression line will go. To see, put

the mouse cursor on the graph and right click. Select Add ▶

Regression fit…, select the button for Linear and check the box for Fit intercept. Click OK twice. The result is (You could reverse the roles of ASSETS and MANAGERS, but the basic conclusion will remain the same.)

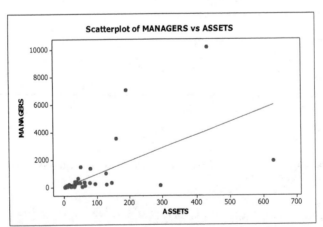

(b) Clearly, there will be outliers and influential data points if we fit a line to these data. A linear regression fit will be of little use, so we omit parts (c-f). [You can do them if you wish just to confirm this opinion.]

Copyright © 2012 Pearson Education, Inc. Publishing as Addison-Wesley.

4.75 (a) From Exercise 4.47,

$$n = 4; \quad \sum x_i = 10; \quad \sum y_i = 8; \quad \sum x_i y_i = 30; \quad \sum x_i^2 = 30$$

Therefore, $s_{xy} = \dfrac{\sum x_i y_i - \left(\sum x_i\right)\left(\sum y_i\right)/n}{n-1} = \dfrac{30 - (10)(8)/4}{3} = 10/3 = $ Covariance

$$s_x^2 = \dfrac{\sum x_i^2 - \left(\sum x_i\right)^2/n}{n-1} = \dfrac{30 - (10)^2/4}{3} = 5/3$$

(b)

$$b_1 = s_{xy}/s_x^2 = (10/3)/(5/3) = 2; \quad b_0 = \bar{y} - b_1\bar{x} = 2 - 2(10/4) = -3$$

$$\hat{y} = -3 + 2x$$

This is the same result as was obtained in Exercise 4.47.

4.77 (a) Using Minitab, with the data in columns named YEAR and POPULATION,
choose **Graph ▶ Scatterplot...**, select the **Simple** version and click **OK**.
Enter POPULATION in Row 1 of the **Y variables** column and YEAR in the **X
variables** column, and click **OK**.   The result is

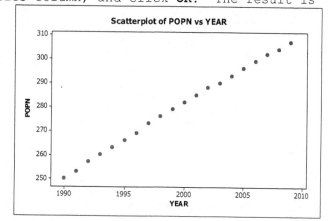

(b) Choose **Stat ▶ Regression ▶ Regression**, enter POPULATION in the
Response text box, and YEAR in the **Predictors** text box. Click **OK**. The
results are

```
The regression equation is
  POPN = - 5719 + 3.00 YEAR

Predictor        Coef   SE Coef        T       P
Constant     -5718.90     57.21   -99.96   0.000
YEAR          3.00000   0.02861   104.85   0.000

S = 0.737865    R-Sq = 99.8%    R-Sq(adj) = 99.8%

Analysis of Variance

Source            DF       SS      MS        F       P
Regression         1   5985.0  5985.0  10992.86   0.000
Residual Error    18      9.8     0.5
Total             19   5994.8
```

The regression equation above indicates that the population is
increasing 3.00 million per year.

(c) The population forecast for 2010 is -5719 + 3.00(2010) 311.00 million
The population forecast for 2011 is -5719 + 3.00(2011) 314.00 million

Copyright © 2012 Pearson Education, Inc. Publishing as Addison-Wesley.

Exercises 4.3

4.79 (a) The coefficient of determination is a descriptive measure of the utility of the regression equation for making predictions. The symbol for the coefficient of determination is r^2.

(b) Two interpretations of the coefficient of determination are
(i) It represents the percentage reduction obtained in the total squared error by using the regression equation to predict the observed values, instead of simply using the mean of the y-values.
(ii) It represents the percentage of variation in the observed y-values that is explained by the regression.

4.81 (a) The coefficient of determination is r^2 = SSR/SST = 7626.6/8291.0 = 0.9199. 91.99% of the variation in the observed values of the response variable is explained by the regression. The fact that r^2 is near 1 indicates that the regression equation is extremely useful for making predictions.

(b) SSE = SST - SSR = 8291.0 - 7626.6 = 664.4

4.83 To use the defining formulas, we begin with the following table.

x	y	$(y-\bar{y})$	$(y-\bar{y})^2$	$\hat{y}$	$(\hat{y}-\bar{y})$	$(\hat{y}-\bar{y})^2$	$(y-\hat{y})$	$(y-\hat{y})^2$
3	-4	-1.0	1	-5	-2.0	4.00	1	1
1	0	3.0	9	-1	2.0	4.00	1	1
2	-5	-2.0	4	-3	0.0	0.00	-2	4
6	-9	0.0	14		0.0	8.00	0	6

Each $\hat{y}$ value is obtained by substituting the respective value of x into the regression equation $\hat{y}=1-2x$. (This equation was derived in Exercise 4.45).

(a) $SST = \sum(y-\bar{y})^2 = 14$ $SSR = \sum(\hat{y}-\bar{y})^2 = 8$ $SSE = \sum(y-\hat{y})^2 = 6$

(b) 14 = 8 + 6

(c) $r^2 = 1 - \dfrac{SSE}{SST} = 1 - \dfrac{6}{14} = \dfrac{4}{7} = 0.571$ or $r^2 = \dfrac{SSR}{SST} = \dfrac{8}{14} = 0.571$

(d) The percentage of variation in the observed y-values that is explained by the regression is 57.1%.

(e) Based on the answer in part (d), the regression equation is somewhat useful for making predictions.

4.85 To use the defining formulas, we begin with the following table.

x	y	$(y-\bar{y})$	$(y-\bar{y})^2$	$\hat{y}$	$(\hat{y}-\bar{y})$	$(\hat{y}-\bar{y})^2$	$(y-\hat{y})$	$(y-\hat{y})^2$
3	4	2.0	4	3	1.0	1.00	1	1
4	5	3.0	9	5	3.0	9.00	0	0
1	0	-2.0	4	-1	-3.0	9.00	1	1
2	-1	-3.0	9	1	-1.0	1.00	-2	4
10	8	0.0	26		0.0	20.00	0	6

Copyright © 2012 Pearson Education, Inc. Publishing as Addison-Wesley.

Each $\hat{y}$ value is obtained by substituting the respective value of x into the regression equation $\hat{y} = -3 + 2x$. (This equation was derived in Exercise 4.47).

(a) $SST = \sum(y - \bar{y})^2 = 26 \qquad SSR = \sum(\hat{y} - \bar{y})^2 = 20 \qquad SSE = \sum(y - \hat{y})^2 = 6$

(b) $46 = 40 + 6$

(c) $r^2 = 1 - \dfrac{SSE}{SST} = 1 - \dfrac{6}{26} = \dfrac{10}{13} = 0.769 \quad or \quad r^2 = \dfrac{SSR}{SST} = \dfrac{20}{26} = 0.769$

(d) The percentage of variation in the observed y-values that is explained by the regression is 76.9%.

(e) Based on the answer in part (d), the regression equation is useful for making predictions.

4.87 To use the defining formulas, we begin with the following table.

x	Y	$(y - \bar{y})$	$(y - \bar{y})^2$	$\hat{y}$	$(\hat{y} - \bar{y})$	$(\hat{y} - \bar{y})^2$	$(y - \hat{y})$	$(y - \hat{y})^2$
0	4	3	9	2.875	1.875	3.51563	1.125	1.26563
2	2	1	1	1.625	0.625	0.39063	0.375	0.14063
2	0	-1	1	1.625	0.625	0.39063	-1.625	2.64062
5	-2	-3	9	-0.250	-1.250	1.56250	-1.750	3.06250
6	1	0	0	-0.875	-1.875	3.51563	1.875	3.51563
15	5	0	20		0	9.375	0	10.625

Each y value is obtained by substituting the respective value of x into the regression equation $\hat{y} = 2.875 - 0.625x$. (This equation was derived in Exercise 4.49).

(a) $SST = \sum(y - \bar{y})^2 = 20 \qquad SSR = \sum(\hat{y} - \bar{y})^2 = 9.375 \qquad SSE = \sum(y - \hat{y})^2 = 10.625$

(b) $20 = 9.365 + 10.625$

(c) $r^2 = 1 - \dfrac{SSE}{SST} = 1 - \dfrac{10.625}{20} = 0.46875 \quad or \quad r^2 = \dfrac{SSR}{SST} = \dfrac{6.375}{20} = 0.46875$

(d) The percentage of variation in the observed y-values that is explained by the regression is 46.875%.

(e) Based on the answer in part (c), the regression equation appears to be moderately useful for making predictions.

4.89 To use the computing formulas, we begin with the following table. Notice that columns 1-4 are presented in the solution to part (a) of Exercise 4.51, so that only the column for y^2 needs to be developed.

x	y	xy	x^2	y^2
6	290	1740	36	84100
6	280	1680	36	78400
6	295	1770	36	87025
2	425	850	4	180625
2	384	768	4	147456
5	315	1575	25	99225
4	355	1420	16	126025
5	328	1640	25	107584
1	425	425	1	180625
4	325	1300	16	105625
41	3422	13168	199	1196690

Copyright © 2012 Pearson Education, Inc. Publishing as Addison-Wesley.

(a) Using the last row of the table and Formula 4.2 of the text, we obtain the three sums of squares as follows.

$$SST = S_{yy} = \sum y^2 - (\sum y)^2 / n = 1196690 - 3422^2 / 10 = 25681.6$$

$$SSR = \frac{S_{xy}^2}{S_{xx}} = \frac{[\sum xy - (\sum x)(\sum y)/n]^2}{\sum x^2 - (\sum x)^2 / n} = \frac{[13168 - (41)(3422)/10]^2}{199 - 41^2 / 10} = 24057.9$$

$$SSE = S_{yy} - \frac{S_{xy}^2}{S_{xx}} = 25681.6 - 24057.9 = 1623.7$$

(b) $r^2 = \dfrac{SSR}{SST} = \dfrac{24057.9}{25681.6} = 0.9368$

(c) The percentage of variation in the observed y-values that is explained by the regression is 93.68% In words, 93.68% of the variation in the tax efficiency data is explained by the percentage of the mutual funds investments that are in energy securities.

(d) Based on the answers to parts (b) and (c), the regression equation appears to be very useful for making predictions.

4.91 To use the computing formulas, we begin with the following table. Notice that columns 1-4 are presented in the solution to part (a) of Exercise 4.53, so that only the column for y^2 needs to be developed.

x	y	xy	x^2	y^2
57	8.0	456.0	3249	64.00
85	22.0	1870.0	7225	484.00
57	10.5	598.5	3249	110.25
65	22.5	1462.5	4225	506.25
52	12.0	624.0	2704	144.00
67	11.5	770.5	4489	132.25
62	7.5	465.0	3844	56.25
80	13.0	1040.0	6400	169.00
77	16.5	1270.5	5929	272.25
53	21.0	1113.0	2809	441.00
68	12.0	816.0	4624	144.00
723	156.5	10486.0	48747	2523.25

(a) Using the last row of the table and Formula 4.2 of the text, we obtain the three sums of squares as follows.

$$SST = S_{yy} = \sum y^2 - (\sum y)^2 / n = 2523.25 - 156.5^2 / 11 = 296.68$$

$$SSR = \frac{S_{xy}^2}{S_{xx}} = \frac{[\sum xy - (\sum x)(\sum y)/n]^2}{\sum x^2 - (\sum x)^2 / n} = \frac{[10486 - (723)(1565)/11]^2}{48747 - 723^2 / 11} = 32.52$$

$$SSE = S_{yy} - \frac{S_{xy}^2}{S_{xx}} = 296.68 - 32.52 = 264.16$$

(b) $r^2 = \dfrac{SSR}{SST} = \dfrac{32.52}{7296.68} = 0.1096$

(c) The percentage of variation in the observed y-values that is explained by the regression is 10.96%. In words, only 10.96% of the variation in the price data is explained by size.

(d) Based on the answers to parts (b) and (c), the regression equation appears to be useless for making predictions.

4.93 To use the computing formulas, we begin with the following table. Notice that columns 1-4 are presented in the solution to part (a) of Exercise 4.55, so that only the column for y^2 needs to be developed.

Copyright © 2012 Pearson Education, Inc. Publishing as Addison-Wesley.

x	y	xy	x^2	y^2
10	92	920	100	8464
15	81	1215	225	6561
12	84	1008	144	7056
20	74	1480	400	5476
8	85	680	64	7225
16	80	1280	256	6400
14	84	1176	196	7056
22	80	1760	484	6400
117	660	9519	1869	54638

(a) Using the last row of the table and Formula 4.2 of the text, we obtain the three sums of squares as follows.

$$SST = S_{yy} = \sum y^2 - \left(\sum y\right)^2 / n = 54638 - 660^2/8 = 188$$

$$SSR = \frac{S_{xy}^2}{S_{xx}} = \frac{\left[\sum xy - \left(\sum x\right)\left(\sum y\right)/n\right]^2}{\sum x^2 - \left(\sum x\right)^2/n} = \frac{[9519 - (117)(660)/8]^2}{1869 - 117^2/8} = 112.89$$

$$SSE = S_{yy} - \frac{S_{xy}^2}{S_{xx}} = 188 - 112.89 = 75.11$$

(b) $r^2 = \dfrac{SSR}{SST} = \dfrac{112.89}{188} = 0.6005$

(c) The percentage of variation in the observed y-values that is explained by the regression is 60.05%. In words, 60.05% of the variation in the score data is explained by study-time.

(d) Based on the answers to parts (b) and (c), the regression equation appears to be somewhat useful for making predictions.

4.95 (a) In Exercise 4.65, we saw from the scatterplot that finding a regression line for these data was reasonable.

(b) From the regression output in that exercise, the coefficient of determination (shown as R-Sq) was found to be 0.371.

(c) The percentage of variation in the observed scores explained by the regression line is 37.1%; i.e., 37.1% of the variation in the death ages was explained by inauguration ages.

(d) Although the pattern is linear, the variation around the fitted line is so great that the regression equation is only marginally useful for making predictions.

4.97 (a) In Exercise 4.67, we saw from the scatterplot that finding a regression line for these data was not reasonable. Parts (b-d) are omitted.

4.99 (a) In Exercise 4.69, we saw from the scatterplot that finding a regression line for these data was very reasonable.

(b) From the regression output in that exercise, the coefficient of determination (shown as R-Sq) was found to be 0.952.

(c) The percentage of variation in the observed values explained by the regression line is 95.2%; i.e., 95.2% of the variation in the average low temperatures was explained by the average high temperatures.

(d) The regression equation is very useful for making predictions.

4.101 (a) Using Minitab, with the data in columns named INCOME and BEER, we

choose **Graph ▶ Scatterplot...**, select the **Simple** version and click **OK**.

Enter BEER in the first row of the **Y variables** column and INCOME in the first row of the **X variables** column. Click **OK**. The result is

Copyright © 2012 Pearson Education, Inc. Publishing as Addison-Wesley.

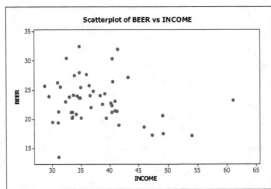

There does not appear to be a linear relationship between beer consumption and income. It is not reasonable to find a regression line. Parts (b-d) are omitted.

4.103(a) In Exercise 4.74, we saw from the scatterplot that finding a regression line for these data was not reasonable because the data exhibited an increasing concave upward curved patternn. Parts (b-d) are omitted.

4.105(a) Using Minitab, with the data in columns named ESTRIOL and WEIGHT, we choose **Graph ▶ Scatterplot...**, select the **Simple** version and click **OK**. Enter WEIGHT in the first row of the **Y variables** column and ESTRIOL in the first row of the **X variables** column. Click **OK**. The result is

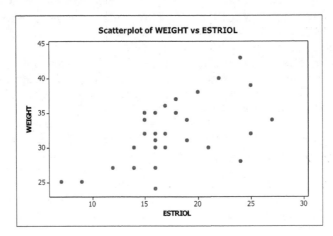

It appears that there is a linear relationship between Estriol and Weight. It is reasonable to find a linear regression equation.

(b) Using Minitab, with the data in columns named ESTRIOL and WEIGHT, we choose **Stat ▶ Regression ▶ Regression...**, select WEIGHT in the **Response** text box, select ESTRIOL in the **Predictors** text box, and click **OK**. The results are

The regression equation is
WEIGHT = 21.5 + 0.608 ESTRIOL

Predictor	Coef	StDev	T	P
Constant	21.523	2.620	8.21	0.000
ESTRIOL	0.6082	0.1468	4.14	0.000

S = 3.821   **R-Sq = 37.2%**   R-Sq(adj) = 35.0%

Copyright © 2012 Pearson Education, Inc. Publishing as Addison-Wesley.

Analysis of Variance

Source	DF	SS	MS	F	P
Regression	1	250.57	250.57	17.16	0.000
Residual Error	29	423.43	14.60		
Total	30	674.00			

Thus the coefficient of determination is $r^2 = 0.372$.

(c) Only 37.2% of the variation in the observed values of birth weight is explained by a linear relationship with the predictor variable, estriol level.

(d) Based solely on the coefficient of determination, the regression equation is not very useful for making predictions.

4.107 (a) $r^2 = 1 - SSE/SST = (SST - SSE)/SST$. If the mean were used to predict the observed values of the response variable, the total squared error would be SST. By using the regression line to predict the observed values of the response variable, the sum of squares of the differences between the predicted values and the mean is SSR = SST - SSE. This is a reduction in the error sum of squares. Dividing this quantity by SST (and converting to a percentage) gives us the percentage reduction in the squared error when we use the regression equation instead of the mean to predict the observed values of the response variable.

(b) From Exercise 4.65, $r^2 = 0.9368$, so the percentage reduction obtained in the total squared error by using the regression equation instead of the mean of observed prices to predict the observed prices is 93.68%.

Exercises 4.4

4.109 Pearson product moment correlation coefficient

4.111 (a) $+1$ (b) not very useful

4.113 False. It is possible that both variables are effects from a third variable. Correlation does not imply causation.

4.115 The sign of the slope and of r are always the same. This can be seen

from $r = \dfrac{S_{xy}}{\sqrt{S_{xx}S_{yy}}} = \dfrac{S_{xy}}{S_{xx}}\dfrac{\sqrt{S_{xx}}}{\sqrt{S_{yy}}} = b_1\dfrac{\sqrt{S_{xx}}}{\sqrt{S_{yy}}}$.

Since the ratio of square roots in the last term is always positive, r will have the same sign as b_1. Since the slope of the regression line

is -3.58, r will also be negative. Therefore, $r = -\sqrt{r^2} = -\sqrt{0.709} = -0.842$

4.117 (a) To compute the linear correlation coefficient using the defining formula, begin with the following table.

x	y	$x - \bar{x}$	$y - \bar{y}$	$(x - \bar{x})^2$	$(y - \bar{y})^2$	$(x - \bar{x})(y - \bar{y})$
3	-4	1	-1	1	1	-1
1	0	-1	3	1	9	-3
2	-5	0	-2	0	4	0
6	-9			2	14	-4

The columns after the first two have used $\bar{x} = 2$ and $\bar{y}=-3$.
We will need s_x and s_y. These are given by

$$s_x = \sqrt{\frac{\sum(x_i - \bar{x})^2}{n-1}} = \sqrt{\frac{2}{2}} = 1 \;\; and \;\; s_y = \sqrt{\frac{\sum(y_i - \bar{y})^2}{n-1}} = \sqrt{\frac{14}{2}} = 2.646$$

$$r = \frac{\frac{1}{n-1}\sum(x_i - \bar{x})(y_i - \bar{y})}{s_x s_y} = \frac{\frac{1}{2}(-4)}{(1)(2)} = \frac{-2}{2.646} = -0.76$$

Copyright © 2012 Pearson Education, Inc. Publishing as Addison-Wesley.

(b) To compute the linear correlation coefficient using the computing formula, begin with the following table.

x	y	xy	x^2	y^2
3	-4	-12	9	16
1	0	0	1	0
2	-5	-10	4	25
6	-9	-22	14	41

$$r = \frac{\sum x_i y_i - \left(\sum x_i\right)\left(\sum y_i\right)/n}{\sqrt{\left[\sum x_i^2 - \left(\sum x_i\right)^2/n\right]\left[\sum y_i^2 - \left(\sum y_i\right)^2/n\right]}}$$

$$= \frac{-22 - (6)(-9)/3}{\sqrt{\left[14 - 6^2/3\right]\left[41 - (-9)^2/3\right]}} = \frac{-4}{\sqrt{28}} = -0.76$$

The answers to both the defining and computing formula are equivalent.

4.119 (a) To compute the linear correlation coefficient using the defining formula, begin with the following table.

x	y	$x - \bar{x}$	$y - \bar{y}$	$(x - \bar{x})^2$	$(y - \bar{y})^2$	$(x - \bar{x})(y - \bar{y})$
3	4	0.5	2	0.25	4	1.0
4	5	1.5	3	2.25	9	4.5
1	0	-1.5	-2	2.25	4	3.0
2	-1	-0.5	-3	0.25	9	1.5
10	8			5.00	26	10.0

The columns after the first two have used $\bar{x} = 2.5$ and $\bar{y} = 2$.
We will need s_x and s_y. These are given by

$$s_x = \sqrt{\frac{\sum (x_i - \bar{x})^2}{n-1}} = \sqrt{\frac{5}{3}} = 1.291 \ \text{and} \ s_y = \sqrt{\frac{\sum (y_i - \bar{y})^2}{n-1}} = \sqrt{\frac{26}{3}} = 2.944$$

$$r = \frac{\frac{1}{n-1}\sum (x_i - \bar{x})(y_i - \bar{y})}{s_x s_y} = \frac{\frac{1}{3}(10)}{(1.291)(2.944)} = \frac{10/3}{3.801} = 0.877$$

(b) To compute the linear correlation coefficient using the computing formula, begin with the following table.

x	y	xy	x^2	y^2
4	5	20	16	25
3	4	12	9	16
1	0	0	1	0
2	-1	-2	4	1
10	8	30	30	42

$$r = \frac{\sum x_i y_i - \left(\sum x_i\right)\left(\sum y_i\right)/n}{\sqrt{\left[\sum x_i^2 - \left(\sum x_i\right)^2/n\right]\left[\sum y_i^2 - \left(\sum y_i\right)^2/n\right]}}$$

$$= \frac{30 - (10)(8)/4}{\sqrt{\left[30 - 10^2/4\right]\left[42 - (8)^2/4\right]}} = \frac{10}{\sqrt{130}} = 0.877$$

The answers to both the defining and computing formula are equivalent.

Copyright © 2012 Pearson Education, Inc. Publishing as Addison-Wesley.

4.121 (a) To compute the linear correlation coefficient using the defining formula, begin with the following table.

x	y	$x-\bar{x}$	$y-\bar{y}$	$(x-\bar{x})^2$	$(y-\bar{y})^2$	$(x-\bar{x})(y-\bar{y})$
0	4	-3	3	9	9	-9
2	2	-1	1	1	1	-1
2	0	-1	-1	1	1	1
5	-2	2	-3	4	9	-6
6	1	3	0	9	0	0
15	5			24	20	-15

The columns after the first two have used $\bar{x}=3$ and $\bar{y}=1$.
We will need s_x and s_y. These are given by

$$s_x = \sqrt{\frac{\sum (x_i - \bar{x})^2}{n-1}} = \sqrt{\frac{24}{4}} = 2.449 \quad and \quad s_y = \sqrt{\frac{\sum (y_i - \bar{y})^2}{n-1}} = \sqrt{\frac{20}{4}} = 2.236$$

$$r = \frac{\frac{1}{n-1}\sum (x_i - \bar{x})(y_i - \bar{y})}{s_x s_y} = \frac{\frac{1}{4}(-15)}{(2.449)(2.236)} = \frac{-3.75}{5.476} = -0.685$$

(b) To compute the linear correlation coefficient using the computing formula, begin with the following table.

x	y	xy	x^2	y^2
0	4	0	0	16
2	2	4	4	4
2	0	0	4	0
5	-2	-10	25	4
6	1	6	36	1
15	5	0	69	25

$$r = \frac{\sum x_i y_i - \left(\sum x_i\right)\left(\sum y_i\right)/n}{\sqrt{\left[\sum x_i^2 - \left(\sum x_i\right)^2/n\right]\left[\sum y_i^2 - \left(\sum y_i\right)^2/n\right]}}$$

$$= \frac{0-(15)(5)/5}{\sqrt{\left[69-15^2/5\right]\left[25-(5)^2/5\right]}} = \frac{-15}{\sqrt{480}} = -0.685$$

The answers to both the defining and computing formula are equivalent.

4.123 To compute the linear correlation coefficient, return to the table presented at the beginning of the solution to Exercise 4.89. Use the last row of this table to perform the calculations in part (a).

(a) The linear correlation coefficient r is computed using the formula in Definition 4.6 of the text:

$$r = \frac{S_{xy}}{\sqrt{S_{xx}S_{xy}}} = \frac{\sum xy - \left(\sum x\right)\left(\sum y\right)/n}{\sqrt{[\sum x^2 - (\sum x)^2/n][\sum y^2 - (\sum y)^2/n]}}$$

$$= \frac{13168 - (41)(3422)/10}{\sqrt{[199-41^2/10][1196690-3422^2/10]}} = -0.967872$$

(b) The value of r in part (a) suggests a strong negative linear relationship between age and price of Corvettes.

(c) Data points are clustered closely about the regression line.

(d) $r^2 = (-0.967872)^2 = 0.9368$. This matches the coefficient of determination that was calculated in part (b) of Exercise 4.89.

Copyright © 2012 Pearson Education, Inc. Publishing as Addison-Wesley.

4.125   To compute the linear correlation coefficient, return to the table presented at the beginning of the solution to Exercise 4.91. Use the last row of this table to perform the calculations in part (a).

(a)   The linear correlation coefficient r is computed using the formula in Definition 4.6 of the text:

$$r = \frac{S_{xy}}{\sqrt{S_{xx}S_{xy}}} = \frac{\sum xy - (\sum x)(\sum y)/n}{\sqrt{[\sum x^2 - (\sum x)^2/n][\sum y^2 - (\sum y)^2/n]}}$$

$$= \frac{10486 - (723)(156.5)/11}{\sqrt{[48747 - 723^2/11][2523.252 - 156.5^2/11]}} = 0.331067$$

(b)   The value of r in part (a) suggests a weak positive linear relationship between potato plant weight and quantity of volatile emissions.

(c)   Data points are clustered very loosely about the regression line.

(d) $r^2 = (0.331037)^2 = 0.1096$. This matches the coefficient of determination that was calculated in part (b) of Exercise 4.91.

4.127 To compute the linear correlation coefficient, return to the table presented at the beginning of the solution to Exercise 4.93. Use the last row of this table to perform the calculations in part (a).

(a)   The linear correlation coefficient r is computed using the formula in Definition 4.6 of the text:

$$r = \frac{S_{xy}}{\sqrt{S_{xx}S_{xy}}} = \frac{\sum xy - (\sum x)(\sum y)/n}{\sqrt{[\sum x^2 - (\sum x)^2/n][\sum y^2 - (\sum y)^2/n]}}$$

$$= \frac{9519 - (117)(660)/8}{\sqrt{[1869 - 117^2/8][54638 - 660^2/8]}} = -0.7749$$

(b)   The value of r in part (a) suggests a fairly strong negative linear relationship between study-time and score for calculus students.

(c)   Data points are clustered fairly closely about the regression line.

(d) $r^2 = (-0.7749)^2 = 0.6005$. This matches the coefficient of determination that was calculated in part (b) of Exercise 4.93.

4.129 To compute the linear correlation coefficient, begin with the following table.

x	y	xy	x^2	y^2
-3	9	-27	9	81
-2	4	-8	4	16
-1	1	-1	1	1
0	0	0	0	0
1	1	1	1	1
2	4	8	4	16
3	9	27	9	81
0	28	0	28	196

(a)   The linear correlation coefficient r is computed using the formula in Definition 4.6 of the text:

Copyright © 2012 Pearson Education, Inc. Publishing as Addison-Wesley.

$$r = \frac{S_{xy}}{\sqrt{S_{xx}S_{xy}}} = \frac{\sum xy - (\sum x)(\sum y)/n}{\sqrt{[\sum x^2 - (\sum x)^2/n][\sum y^2 - (\sum y)^2/n]}}$$

$$= \frac{0 - (0)(282)/7}{\sqrt{[28 - 0^2/7][196 - 28^2/7]}} = 0.0$$

(b)   We cannot conclude from the result in part (a) that x and y are unrelated. We can conclude only that there is no *linear* relationship between x and y.

(c)   Graph for part (e)

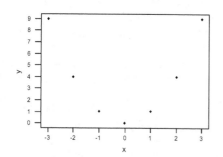 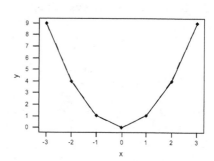

(d)   It is not appropriate to use the linear correlation coefficient as a descriptive measure for the data because the data points are not scattered about a line.

(e)   For each data point (x, y), we have y = x². (See columns 1 and 2 of the previous table.) See the graph above at the right.

4.131(a)   r is zero.   (b)   r is negative.   (c)   r is positive.

4.133(a)   In Exercise 4.65, we saw that a linear relationship between the age at inauguration and the age at death was reasonable. The scatterplot showed that the line relating the two variables had a positive slope.

(b)   From the output in Exercise 4.65, the coefficient of determination was 0.371. Since this is r² and r is positive because of the positive slope of the line, $r = \sqrt{0.371} = 0.609$. You can also obtain the correlation coefficient in Minitab by choosing **Stat ▶ Basic Statistics ▶ Correlation...**, select INAUGURATION and DEATH in the **Variables** text box, and click **OK**. The result is
Pearson correlation of INAUGURATION and DEATH = 0.609

(c)   The correlation coefficient is positive, indicating that the linear relationship has a positive slope. The coefficient is not very close to 1, indicating that the linear relationship is not very strong.

4.135(a)   In Exercise 4.67, we saw that a linear relationship between LOT SIZE and VALUE was not reasonable. Parts (b-c) are omitted.

4.137(a)   In Exercise 4.69, we saw that a linear relationship between HIGH and LOW was reasonable.

(b)   In Minitab, choose **Stat ▶ Basic Statistics ▶ Correlation...**, select HIGH and LOW in the **Variables** text box, and click **OK**. The result is
Pearson correlation of HIGH and LOW = 0.976

(c)   The correlation coefficient is positive, indicating that the linear relationship has a positive slope. The coefficient is very close to 1, indicating that the linear relationship is very strong.

Copyright © 2012 Pearson Education, Inc. Publishing as Addison-Wesley.

4.139 (a) In Exercise 4.71, we saw that a linear relationship between INCOME and BEER was not reasonable as there was no linear pattern in the data. Parts (b-c) are omitted.

4.141 (a) In Exercise 4.74, we saw that there was an increasing relationship between diameter and volume, but the relationship was concave upward, not linear. Parts (b-c) are omitted.

4.143 (a) In Exercise 4.105, we saw that there was an increasing relationship between estriol levels of pregnant women and birth weights of their children. Use of the correlation coefficient is appropriate as a descriptive measure of the strength of the linearity of that relationship.

(b) Using Minitab, with the data in columns named ESTRIOL and WEIGHT, we choose **Stat ▶ Basic Statistics ▶ Correlation...**, select ESTRIOL and WEIGHT in the **Variables** text box, and click **OK**. The result is

Pearson correlation of ESTRIOL and WEIGHT = 0.610, P-Value = 0.000

(c) There is a moderately positive linear relationship between estriol concentration and birth weight. The data points will be moderately clustered about the regression line.

4.145 (a) No. We only know that the linear correlation coefficient is the square root of the coefficient of determination or it is the negative of that square root.

(b) No. The slope is positive if r is positive and negative if r is negative, but we can not determine the sign of r.

(c) Yes. $r = -\sqrt{0.716} = -0.846$

(d) Yes. $r = \sqrt{0.716} = 0.846$

4.147 (a) Using Minitab, choose **Graph ▶ Scatterplot**, select the **Simple** version, enter SCORE in the first row of the **Y variables** column and TIME in the first row of the **X variables** column, and click **OK**. The result is

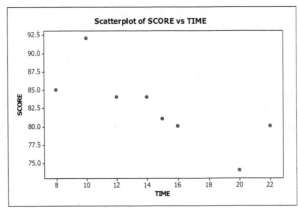

(b) Score decreases as time increases, so the using the rank correlation coefficient is reasonable.

(c) Score decreases linearly as time increases, so the using the linear correlation coefficient is reasonable.

(d) To find the rank correlation coefficient using Minitab, we must first find the ranks for each of the data values in the SCORE and TIME columns. This is done for TIME by choosing **Data ▶ Rank**, entering TIME in the **Rank data in** text box and RANKTIME in the **Store ranks in** text box and clicking **OK**. Then choose **Data ▶ Rank**, enter SCORE in the **Rank**

Copyright © 2012 Pearson Education, Inc. Publishing as Addison-Wesley.

data in text box and RANKSCORE in the **Store ranks in** text box and click

OK. Now choose **Stat ▶ Basic Statistics ▶ Correlation...**, select

RANKTIME and RANKSCORE in the **Variables** text box, and click **OK**. The
result is
    Pearson correlation of RANKTIME and RANKSCORE = -0.928
    P-Value = 0.001
The rank correlation coefficient is -0.928, indicating that the ranks
are strongly negatively correlated and that SCORE decreases as TIME
increases.

Review Problems for Chapter 4

1.   (a) x (b) y (c) b_1 (d) b_0
2.   (a) It intersects the y-axis when $x = 0$. Therefore $y = 4$.
     (b) It intersects the x-axis when $y = 0$. Therefore $x = 4/3$.
     (c) slope = -3
     (d) The y-value decreases by 3 units when x increases by 1 unit.
     (e) The y-value increases by 6 units when x decreases by 2 units.
3.   (a) True. The y-intercept is determined by b_0 and that value is independent
         of b_1, the slope.
     (b) False. A horizontal line has a slope of zero.
     (c) True. If the slope is positive, the x-values and y-values increase and
         decrease together.
4.   Scatterplot or scatter diagram
5.   A regression equation can be used to make predictions of the response
     variable for specific values of the predictor variable within the range of
     the observed values of the predictor variable.
6.   (a) predictor variable or explanatory variable
     (b) response variable
7.   (a) Based on the least-squares criterion, the line that best fits a set of
         data points is the one have the smallest possible sum of squared
         errors.
     (b) The line that best fits a set of data points according to the least-
         squares criterion is called the regression line.
     (c) Using a regression equation to make predictions for values of the
         predictor variable outside the range of the observed values of the
         predictor variable is called extrapolation.
8.   (a) An outlier is a data point that lies far from the regression line
         relative to the other data points.
     (b) An influential observation is a data point whose removal causes the
         regression equation to change considerably. Often this is a data point
         for which the x-value is considerably to the left or right of the rest
         of the data points.
9.   The coefficient of determination is the percentage of the total variation in
     the y-values that is explained by the regression equation.
10. (a) SST is the total sum of squares and measures the variation in the
         observed values of the response variable.
     (b) SSR is the regression sum of squares and measures the variation in the
         observed values of the response variable that is explained by the
         regression. It can also be thought of as the variation in the
         predicted values of the response variable corresponding to the x-values
         in the data points.
     (c) SSE is the error sum of squares and measures the variation in the
         observed values of the response variable that is not explained by the
         regression.
11. (a) One use of the linear correlation coefficient is as a descriptive
         measure of the strength of the linear relationship between two
         variables.

Copyright © 2012 Pearson Education, Inc. Publishing as Addison-Wesley.

    (b) A positive linear relationship between two variables means that one variable tends to increase linearly as the other <u>increases</u>.

    (c) A value of r close to -1 suggests a strong <u>negative</u> linear relationship between the variables.

    (d) A value of r close to <u>zero</u> suggests at most a weak linear relationship between the variables.

12. True. It is quite possible that both variables are strongly affected by one (or more) other variables (called lurking variables).

13. (a) $y = 72 - 12x$ (b) $b_0 = 72$, $b_1 = -12$

    (c) The line slopes downward since $b_1 < 0$.

    (d) After two years: $y = 72 - 12(2) = \$48$ hundred $= \$4800$
        After five years: $y = 72 - 12(5) = \$12$ hundred $= \$1200$.

    (e)

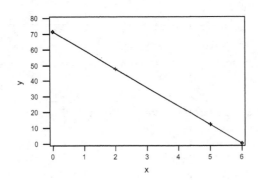

    (f) From the graph, we estimate the value to be about $2500 after 4 years. The actual value is $y = 7200 - 1200(4) = \$2400$.

14. (a)

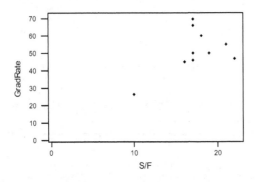

    (b) It is moderately reasonable to find a regression line for the data because the data points appear to be scattered about a line. This perception is, however, heavily influenced by the single data point at $x = 10$.

    (c) The regression equation can be determined by calculating its slope and intercept. Formulas for the slope (b_1) and intercept (b_0) of the regression equation are, respectively:

$$b_1 = \frac{\sum xy - \sum x \sum y / n}{\sum x^2 - (\sum x)^2 / n} \qquad b_0 = \frac{1}{n}\left(\sum y - b_1 \sum x\right)$$

Copyright © 2012 Pearson Education, Inc. Publishing as Addison-Wesley.

To compute b_0 and b_1, construct a table of values for x, y, xy, x^2, and their sums. (Note: A column for y^2 is also presented. This is used in Problems 15 and 16.)

x	y	xy	x^2	y^2
16	45	720	256	2,025
20	55	1,100	400	3,025
17	70	1,190	289	4,900
19	50	950	361	2,500
22	47	1,034	484	2,209
17	46	782	289	2,116
17	50	850	289	2,500
17	66	1,122	289	4,356
10	26	260	100	676
18	60	1,080	324	3,600
173	515	9,088	3,081	27,907

Thus,

$$b_1 = \frac{9088 - 173(515)/10}{3081 - 173^2/10} = 2.02611 \qquad b_0 = \frac{1}{11}(515 - 2.02611(173)) = 16.448$$

and the regression equation is: $\hat{y} = 16.448 + 2.02611x$.

To graph the regression equation, begin by selecting two x-values within the range of the x-data. For the x-values 10 and 22, the calculated values for y are, respectively:

$$\hat{y} = 16.448 + 2.02611(10) = 36.71$$

$$\hat{y} = 16.448 + 2.02611(22) = 61.02$$

The regression equation can be graphed by plotting the pairs (10, 36.71) and (22, 61.02) and connecting these points with a line. This equation and the original set of 10 data points are presented as follows:

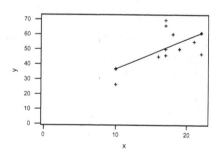

(d) Graduation rate tends to increase as the student-to-faculty ratio increases.

(e) Graduation rate increases an estimated 2.026% for each additional 1 unit increase in the student-to-faculty ratio.

(f) The predicted graduation rate for a university with a student-to-faculty ratio of 17 is 16.448 + 2.0261(17) = 50.89%.

(g) There is one potential influential observation, (10, 26) and no outliers. Looking at the scatterplot in part (c), we suspect that there may be little relationship between the student-to-faculty ratio

Copyright © 2012 Pearson Education, Inc. Publishing as Addison-Wesley.

and the graduation rate if the influential observation is removed from the data.

15. (a) To compute SST, SSR, and SSE using the computing formulas, begin with the table presented in Problem 14(c). Using the last row of this table and the formula in Definition 4.6 of the text, we obtain the three sums of squares as follows.

$$SST = S_{yy} = \sum y^2 - \left(\sum y\right)^2 / n = 27907 - 515/10 = 1384.5$$

$$SSR = \frac{S_{xy}^2}{S_{xx}} = \frac{\left[\sum xy - \left(\sum x\right)\left(\sum y\right)/n\right]^2}{\sum x^2 - \left(\sum x\right)^2 / n} = \frac{[9088 - (173)(515)/10]^2}{3081 - 173^2/10} = 361.66$$

$$SSE = S_{yy} - \frac{S_{xy}^2}{S_{xx}} = 1384.50 - 361.66 = 1022.846$$

$$r^2 = \frac{SSR}{SST} = \frac{361.66}{1384.50} = 0.261$$

(b) The percentage reduction obtained in the total squared error by using the regression equation, instead of the sample mean y, to predict the observed costs is 26.1%.

(c) The percentage of the variation in the graduation rate that is explained by the student-to-faculty ratio is 26.1%.

(d) The regression equation appears to be not very useful for making predictions.

16. (a) To compute the linear correlation coefficient, begin with the table presented in Problem 14(c). Using the last row of this table and the formula in Definition 4.6 of the text, we get

$$r = \frac{S_{xy}}{\sqrt{S_{xx} S_{xy}}} = \frac{\sum xy - \left(\sum x\right)\left(\sum y\right)/n}{\sqrt{[\sum x^2 - \left(\sum x\right)^2/n][\sum y^2 - \left(\sum y\right)^2/n]}}$$

$$= \frac{9088 - (173)(515)/10}{\sqrt{[3081 - 173^2/10][27907 - 515^2/10]}} = 0.511$$

(b) The value of r suggests a weak to moderate positive linear correlation.

(c) Data points are clustered about the regression line, but not very closely.

(d) $r^2 = (0.511)^2 = 0.261$

17. Using Minitab, with the data in three columns named POPULATION, AREA, and PLANTS, choose **Stat ▶ Basic statistics ▶ Correlation**, enter POPULATION, AREA, and PLANTS in the **Variables** text **box**. Make sure that the **Display P-values** box is checked. **Click OK**. The output is

	POPULATION	AREA
AREA	0.109	
	0.453	
PLANTS	0.721	-0.309
	0.000	0.029

Cell Contents: Pearson correlation
                P-Value

(a) The correlation coefficient between population and area is 0.109.

(b) The correlation coefficient between population and number of exotic plants is 0.721.

(c) The correlation coefficient between area and number of exotic plants is -0.309.

(d) Any linear association between population and area is very weak since the correlation coefficient is near zero. There is a moderately strong

Copyright © 2012 Pearson Education, Inc. Publishing as Addison-Wesley.

positive linear relationship between population and number of exotic plants. There is a fairly weak negative linear association between area and number of exotic plants. One possible explanation of the stronger positive relationship between population and number of exotic plants might be that the more people there are, the greater the possibility that exotic plants are introduced into the state by those people, intentionally or unintentionally.

18.   (a) Using Minitab, with the data in columns named IMR and LE, choose **Graph ▶ Scatterplot...**, select the **Simple** version, enter 'LE' in row 1 of the **Y variables** column and IMR in the **X variables** column and click **OK**.

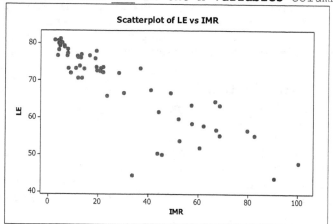

Life expectancy appears to decrease linearly as IMR increases.

(b) It appears to be reasonable to find a regression line for the data, although we expect that there will be both outliers and influential observations.

(c) Choose **Stat ▶ Regression ▶ Regression**, enter 'LE' in the **Response box** and IMR in the **Predictor box**. Click on the Options button and enter 30 in the Prediction intervals for new observations text box. Click **OK** twice. The results are

```
The regression equation is
LE = 79.4 - 0.354 IMR

Predictor        Coef   SE Coef        T      P
Constant       79.396     1.009    78.72  0.000
IMR           -0.35374   0.02601   -13.60  0.000

S = 5.17598    R-Sq = 76.1%    R-Sq(adj) = 75.7%

Analysis of Variance

Source           DF       SS       MS       F      P
Regression        1   4956.2   4956.2  185.00  0.000
Residual Error   58   1553.9     26.8
Total            59   6510.1

Unusual Observations

Obs   IMR       LE      Fit   SE Fit   Residual   St Resid
 11    46   49.890   63.251    0.795    -13.361     -2.61R
 25   100   47.430   43.894    1.971      3.536      0.74 X
 34    34   44.280   67.418    0.680    -23.138     -4.51R
 48    44   50.160   63.828    0.773    -13.668     -2.67R
 52    91   43.450   47.365    1.734     -3.915     -0.80 X
```

Copyright © 2012 Pearson Education, Inc. Publishing as Addison-Wesley.

```
            R denotes an observation with a large standardized residual.
            X denotes an observation whose X value gives it large leverage.

            Predicted Values for New Observations

            New Obs    Fit  SE Fit        95% CI            95% PI
                 1  68.784   0.669  (67.445, 70.122)  (58.337, 79.231)

            Values of Predictors for New Observations

            New Obs   IMR
                 1  30.0
```

The regression equation, **LE = 79.4 - 0.354 IMR**, indicates that life expectancy decreases as the infant mortality rate increases.

(d) For a country with an IMR = 30, the predicted life expectancy is 79.4 – 0.354(30) = 68.78 years. The output above also has the predicted value of 68.784 years

(e) Choose **Stat ▶ Basic statistics ▶ Correlation**, enter 'LE' and IMR in the **Variables** text **box**. Click **OK**. The results are
```
            Pearson correlation of IMR and LE = -0.873
            P-Value = 0.000
```
Life expectancy and IMR are highly negatively correlated. Thus, there is strong negative linear association between the two variables.

(f) The output shown in part (c) indicates that observations 11, 34, and 48 are potential outliers and observations 25 and 52 are influential observations.

19. (a) Using Minitab, with the data in columns named HIGH and PRECIPITAION, choose **Graph ▶ Scatterplot...**, select the **Simple** version, enter 'PRECIPITATIONN' in row 1 of the **Y variables** column and HIGH in the **X variables** column and click **OK**.

The graph at the right does not show any relationship between average July precipitation and average July high temperature.

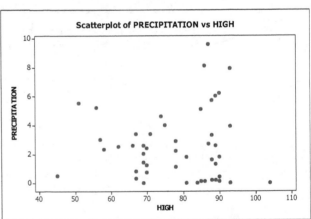

(b) The graph in part (a) does not indicate that there is any linear relationship between average precipitation and average high temperature in July for 48 cities. Parts (c-f) are omitted.

20. (a) Using Minitab, with the data in columns named FAT and DEATHRATE, choose **Graph ▶ Scatterplot...**, select the **Simple** version, enter DEATHRATE in row 1 of the **Y variables** column and FAT in the **X variables** column and click **OK**.

Copyright © 2012 Pearson Education, Inc. Publishing as Addison-Wesley.

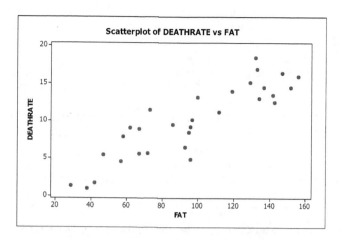

The scatter diagram shows that the death rate from prostate cancer tends to increase linearly with an increase in fat consumption.

(b) Obtaining a linear regression equation for the data appears reasonable. There is no noticeable curvature in the data, and there is an increasing trend along a line.

(c) Choose **Stat ▸ Regression ▸ Regression**, enter DEATHRATE in the **Response box** and FAT in the **Predictor box**. Click on the Options button and enter 92 in the **Prediction intervals for new observations** text box. Click **OK** twice. The results are

```
The regression equation is
DEATHRATE = - 1.06 + 0.113 FAT

Predictor      Coef   SE Coef      T       P
Constant     -1.063     1.170   -0.91   0.371
FAT         0.11336   0.01126   10.06   0.000

S = 2.29453    R-Sq = 78.3%   R-Sq(adj) = 77.6%

Analysis of Variance

Source           DF       SS       MS       F       P
Regression        1   533.31   533.31  101.30   0.000
Residual Error   28   147.42     5.26
Total            29   680.73

Unusual Observations

Obs   FAT   DEATHRATE      Fit   SE Fit   Residual   St Resid
  5    96       4.800    9.820    0.419     -5.020     -2.23R
 30   132      18.400   13.901    0.575      4.499      2.03R

R denotes an observation with a large standardized residual.

Predicted Values for New Observations

New
Obs     Fit   SE Fit        95% CI            95% PI
  1   9.366    0.423   (8.500, 10.232)   (4.587, 14.145)

Values of Predictors for New Observations

New
Obs    FAT
  1   92.0
```

Copyright © 2012 Pearson Education, Inc. Publishing as Addison-Wesley.

The slope of the regression line is 0.11336. This means that for each 1 unit increase in fat consumption, the death rate goes up by 0.11336. Note that this is an observational study, and therefore this relationship cannot be construed as a cause and effect relationship. There may be other lurking variables. For example, the countries with the highest fat consumption and death rates tend to be those in which the life expectancy is greatest. Since prostate cancer is typically a disease that strikes older men, some of the countries with lower death rates from prostate cancer may be experiencing higher rates of other diseases that cause death before a man can experience prostate cancer.

(d) For a country with a per capita fat consumption of 92 grams per day, the predicted prostate cancer death rate, from the output above, is 9.366 per 100.000 males. This can also be obtained by substituting 92 for FAT in the regression equation.

(e) Choose **Stat ▶ Basic statistics ▶ Correlation, enter** DEATHRATE and FAT in the **Variables** text **box.** Click **OK.** The results are
Pearson correlation of FAT and DEATHRATE = 0.885 P-Value = 0.000

The correlation coefficient is quite high, indicating a fairly strong linear association between fat consumption and prostate cancer death rate.

(f) From the regression output in part (c), we see that there are two potential outliers, observations 5 and 30, which arise from the countries of Greece and Sweden. The data point for Greece lies below the fitted line and the one for Sweden lies above the line. No observations are identified as potential influential observations.

21. (a) Using Minitab, with the data in columns named FIRST ROUND and SECOND ROUND, choose **Graph ▶ Scatterplot...**, select the **Simple** version, enter SECOND ROUND in row 1 of the **Y variables** column and FIRST ROUND in the **X variables** column and click **OK.**

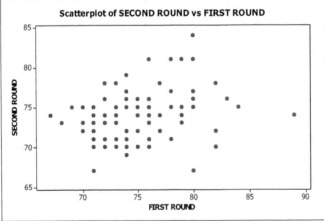

(b) Obtaining a linear regression does not seem reasonable. There is no noticeable linear relationship in the data. Therefore, we will skip steps (c) – (f).

Copyright © 2012 Pearson Education, Inc. Publishing as Addison-Wesley.

CHAPTER 5 ANSWERS

Exercises 5.1

5.1 An experiment is an action the result of which cannot be predicted with certainty. An event is a specified result that may or may not occur when the experiment is performed.

5.3 There is no difference.

5.5 The frequentist interpretation of probability is that the probability of an event is the proportion of times the event occurs in a large number of repetitions of the experiment.

5.7 The following could not possibly be probabilities:

(b) -0.201: A probability cannot be negative.

(e) 3.5: A probability cannot exceed 1.

5.9 (a) G,L,S,A   G,L,S,T   G,L,A,T   G,S,A,T   L,S,A,T

(b) Two samples include the governor, attorney general, and treasurer. Therefore, the probability is $f/N = 2/5$.

(c) Three samples include the governor and treasurer. Therefore, the probability is $f/N = 3/5$.

(d) Four samples include the governor. Thus, the probability is $f/N = 4/5$.

5.11 (a) There are 3 chips that are red. Therefore, the probability is $f/N = 3/12 = 1/4$.

(b) There are $3 + 4 = 7$ chips that are red or white. Therefore, the probability is $f/N = 7/12$.

(c) There are $3 + 5 = 8$ chips that are not white. Therefore, the probability is $f/N = 8/12 = 2/3$.

5.13 (a) There was a total of 66,529,566 votes cast. Therefore, the probability that a randomly selected voter voted for Vladimir Putin is $f/n = 49,565,238/66,529,566 = 0.745$.

(b) There were $1,405,315 + 524,324 = 1,929,639$ votes cast for Malyshkin or Mironov. . Therefore, the probability that a randomly selected voter voted for Malyshkin or Mironov is $f/n = 1,929,639/66,529,566 = 0.029$.

(c) The number of votes not for Putin is $66,529,566 - 49,565,238 = 16,964,328$. Thus, the probability that a randomly selected voter did not vote for Putin is $16,964,328/66,529,566 = 0.255$.

5.15 The total number of units (in thousands) is $N = 124,378$.

(a) The probability that the unit has 4 rooms is $f/N = 22774/124378 = 0.183$ (to three decimal places).

(b) The probability that the unit has more than 4 rooms is $f/N = (28619 + 25325 + 15284 + 19399)/124378 = 88627/124378 = 0.713$.

(c) The probability that the unit has 1 or 2 rooms is $f/N = (637 + 1399)/124378 = 2036/124378 = 0.016$.

(d) The probability that the unit has fewer than one room is $f/N = 0/124378 = 0.000$.

(e) The probability that the unit has one or more rooms is $f/N = 124378/124378 = 1.000$.

Copyright © 2012 Pearson Education, Inc. Publishing as Addison-Wesley.

5.17 The total number of graduate students is 49,088.

(a) The probability that his occupation is service is $f/N = 9274/49088 = 0.189$.

(b) The probability that his occupation is administrative is $f/N = (2197 + 6450)/49088 = 8647/49088 = 0.176$.

(b) The probability that his occupation is manufacturing is $f/N = (2197 + 2166 + 1640 + 5721)/49088 = 11724/49088 = 0.239$

(c) The number whose occupation is not manufacturing is $49088 - 11724 = 37364$. The probability that his occupation is not manufacturing is $f/N = 37364/49088 = 0.761$.

5.19 (a) The total number of graduate science students in doctorate-granting institutions is 320.0 thousand. Of these, 46.7 thousand are in psychology. Thus, the probability that a randomly selected graduated student in science is in the field of psychology is $f/N = 46.7/320.0 = 0.146$.

(b) There are $35.4 + 87.8 = 123.2$ thousand $= 123200$ graduate students in physical or social science. Thus, the probability that a randomly selected graduated student in science is in physical or social science is $f/N = 123.2/320.0 = 0.385$.

(c) There are 44.3 thousand graduate students in computer science and $320.0 - 44.3 = 275.7$ thousand graduate students not in computer science. Thus, the probability that a randomly selected graduated student in science is not in computer science is $f/N = 275.7/320.0 = 0.862$.

5.21 (a) There are five ways $[(1,5), (2,4), (3,3), (4,2), (5,1)]$ among the 36 possibilities that sum to 6. Thus, $f/N = 5/36 = 0.139$.

(b) There are 18 ways among the 36 possibilities that provide a sum that is even. Thus, $f/N = 18/36 = 0.500$.

(c) There are six ways among the 36 possibilities that sum to 7 and two ways among the 36 possibilities that sum to 11. Thus, $f/N = (6 + 2)/36 = 8/36 = 0.222$.

(d) There is one way among the 36 possibilities that sums to 2, two ways in which the sum is 3, and one way in which the sum is 12. Thus, $f/N = (1 + 2 + 1)/36 = 4/36 = 0.111$.

5.23 (a) The event in part (e), that the housing unit has one or more rooms, is a certainty. The event in part (d), that the housing unit has fewer than one room, is impossible.

(b) The event that is a certainty has a probability of 1. The event that is impossible has a probability of 0.

5.25 Answers will vary

5.27 (a) In a large sample of women, approximately 31.4% will have a gestation period that is longer than 9 months. Therefore, 31.4% of 4000 is 1256. Approximately 1,256 will exceed 9 months.

(b) In a large sample of horse races, approximately 66.7% of the favorites will finish in the money. Therefore, 66.7% of 500 is 333.5. Approximately 334 of the favorites will finish in the money.

(c) In a large sample of traffic fatalities, approximately 40% involve an intoxicated or alcohol-impaired driver or nonoccupant. Therefore, 40% of 389 is 155.6. Approximately 156 fatalities involved an intoxicated or alcohol-impaired driver or nonoccupant.

Copyright © 2012 Pearson Education, Inc. Publishing as Addison-Wesley.

5.29 When rolling two balanced dice and observing the sum, there are 36 possible equally likely outcomes, not 11. The probability that the sum is 12 equals 1/36, not 1/11.

5.31 (a) Answers will vary. For my experiment, here are the outcomes from tossing a coin five times and keeping track of the running totals and running proportions.

Toss	Outcome	Number of Heads	Proportion of heads
1	H	1	1.00
2	H	2	1.00
3	T	2	0.67
4	T	2	0.50
5	T	2	0.40

(b) Based on my five tosses, I would estimate that the probability of a head when this coin is tossed is 0.40 because that was my running total for the proportion of heads after tossing the coin five times.

(c)

Toss	Outcome	Number of Heads	Proportion of heads
1	H	1	1.00
2	H	2	1.00
3	T	2	0.67
4	T	2	0.50
5	T	2	0.40
6	H	3	0.50
7	T	3	0.43
8	H	4	0.50
9	T	4	0.44
10	H	5	0.50

Based on my ten tosses, I would estimate that the probability of a head when this coin is tossed is 0.50 because that was my running total for the proportion of heads after tossing the coin ten times.

(d)

Toss	Outcome	Number of Heads	Proportion of heads
1	H	1	1.00
2	H	2	1.00
3	T	2	0.67
4	T	2	0.50
5	T	2	0.40
6	H	3	0.50
7	T	3	0.43
8	H	4	0.50
9	T	4	0.44
10	H	5	0.50
11	T	5	0.45
12	T	5	0.42
13	T	5	0.38
14	T	5	0.36
15	H	6	0.40
16	T	6	0.38
17	H	7	0.41
18	T	7	0.39
19	H	8	0.42
20	H	9	0.45

Copyright © 2012 Pearson Education, Inc. Publishing as Addison-Wesley.

> Based on my twenty tosses, I would estimate that the probability of a head when this coin is tossed is 0.45 because that was my running total for the proportion of heads after tossing the coin twenty times.

(d) For my answers in part (b)-(d), the frequency interpretation cannot be used as the definition of probability because the number of trials was small. The frequency interpretation requires a large number of trials before it can be used as the definition of probability.

5.33 If p is the probability that a randomly selected adult woman believes that a "cyber affair" is cheating, then p = 0.75. The odds <u>against</u> selecting such a woman are in the ratio of 1-p to p, or 0.25 to 0.75. This is normally expressed in integers as 1 to 3.

5.35 If p is the probability that a randomly selected person age 18-34 years has cursed at their computer, then p = 0.46. The odds <u>against</u> selecting such a person are in the ratio of 1-p to p, or 0.54 to 0.46. This is normally expressed in integers and reduced as 27 to 23.

Exercises 5.2

5.37 Venn diagrams are useful for portraying events and relationships between events.

5.39 Two events are mutually exclusive if they cannot occur at the same time, i.e., they have no outcomes in common. Three events are mutually exclusive if no two of them can occur at the same time, i.e., no pair of the events has any outcomes in common.

5.41 A = {2, 4, 6}; B = {4, 5, 6}; C = {1, 2}; D = {3}

5.43 A contains the outcomes JM, JS, JH, JB, WM, WS, WH, WJ

B contains the outcomes HM, HS, HJ, HW

C contains the outcomes MW, SW, HW, JW

D contains the outcomes MS, MH, SM, SH, HM, HS

5.45 (a) (not A) = {1, 3, 5} = the event the die comes up odd.

(b) (A & B) = {4, 6} = the event the die comes up four or six.

(c) (B or C) = {1, 2, 4, 5, 6} = the event the die does *not* come up three.

5.47 (a) (not A) = (MS, MH, MJ, MW, SM, SH, SJ, SW, HM, HS, HJ, HW) = the event that a female is appointed chairperson

(b) (B & D) = (HM, HS) = the event that Holly is appointed chairperson and a woman is appointed secretary

(c) (B or C) = (HM, HS, HJ, HW, MW, SW, JW) = the event that Holly is appointed chairperson or Will is appointed secretary.

5.49 (a) (not C) = the event that a state has a diabetes prevalence percentage less than 5% or is 10% or more. There are 8 + 1 = 9 states in (not C).

(b) (A&B) = the event that a state has a diabetes prevalence percentage at least 8% and is less than 7%. There are no states in this event.

(c) (C or D) = the event that a state has a diabetes prevalence percentage that is less than 10%. There are 49 states in this event.

(d) (C&B) = the event that a state has a diabetes prevalence percentage of at least 5% and less than 7%. There are 10 + 15 = 25 states in this event.

Copyright © 2012 Pearson Education, Inc. Publishing as Addison-Wesley.

5.51 (a) (A or D) = the event that the bill was paid by Medicare or by the patient or a charity. There are 9983 + 5512 = 15,495 bills in this event.

(b) (not C) = the event that private insurance **did** pay the bill. There are 26,825 bills in this event.

(c) (B & (not A)) = the event that some government agency other than Medicare paid the bill. There are 8142 + 1777 = 9919 bills in this event.

(d) Since D is included in C, (C or D) is just C = the event that private insurance did not pay the bill. Therefore (not(C or D)) = (not C) = the event that private insurance did pay the bill. There are 26,825 bills in this event.

5.53 (a) (not A) is the event that the unit has more than 4 rooms. There are 28,619 + 25,325 + 15,284 + 19,399 = 88,627 (thousand) such units.

(b) (A&B) is the event that the unit has at most 4 rooms and at least 2 rooms, i.e., has 2 or 3 or 4 rooms. There are 1,399 + 10,941 + 22,774 = 35,114 (thousand) such units.

(c) The event (C or D) is the event that the unit has between 5 and 7 rooms inclusive or more than seven rooms, i.e., has 5 or more rooms. This is the same as the event (not A) and therefore there are 88,627 (thousand) such units.

5.55 (a) Events A and B are not mutually exclusive. They have a four and a six in common.

(b) Events B and C are mutually exclusive. They have no outcomes in common.

(c) Events A, C, and D are not mutually exclusive. The outcome two is common to A and C.

(d) Among A, B, C, and D, there are three mutually exclusive events. These are B, C, and D. Among B, C, and D, there are no outcomes in common. There are not, however, four mutually exclusive events. The outcome two is common to A and C, four is common to A and B, and six is common to A and B.

5.57 The groups that are mutually exclusive are A and C; A and D; C and D; and A, C, and D.

5.59 Below is a Venn diagram portraying four mutually exclusive events.

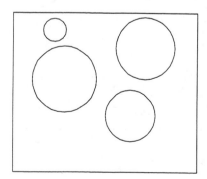

5.61 (a) The experiment is to choose the number of men on the jury. You cannot have the outcomes zero through three because there are only eight

Copyright © 2012 Pearson Education, Inc. Publishing as Addison-Wesley.

possible women to choose and you need to choose a total of twelve people. Therefore, the sample space for number of men chosen would be 4, 5, 6, 7, 8, 9, 10.

(b) The event (A or B) is the event that at least half of the jurors are men or at least half of the 8 women are one the jury. This consists of all possible outcomes. Thus, the outcomes in (A or B) are 4, 5, 6, 7, 8, 9, 10.

The event (A & B) is the event that at least half of the jurors are men and at least half of 8 women are on the jury. This means that there are at least 6 men on the jury and there are at least 4 women on the jury. At least 4 women on the jury means that there are at most 8 men on the jury. At least 6 and at most 8 intersect at the outcomes 6, 7, and 8. Thus, the outcomes in (A & B) are 6, 7, 8.

The event (A & (not B)) is the event that at least half of the jurors are mean and less than half of the 8 women are on the jury. This means that there are at least 6 men on the jury and there are at most 3 women on the jury. At most 3 women on the jury means that there are at least 9 men on the jury. At least 6 and at least 9 intersect at the outcomes 9 and 10. Thus, the outcomes in (A & (not B)) are 9 and 10.

(c) The events A and B are not mutually exclusive. You can have at least 6 men and at least 4 women on the jury. For example, the jury can consist of 6 men and 6 women.

The events A and (not B) are not mutually exclusive. You can have at least 6 men and less than 4 women on the jury. For example, the jury can consist of 9 men and 3 women.

The events (not A) and (not B) are mutually exclusive. You cannot have less than 6 men on the jury and less than 4 women on the jury. That would not allow the total number of jurors to be twelve.

5.63 In the following Venn diagram, events A, B, and C are mutually exclusive; events A, B, and D are mutually exclusive; but no other three of the four events are mutually exclusive.

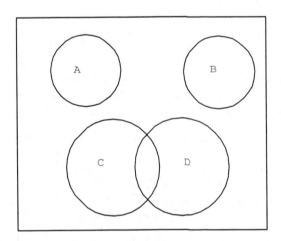

Copyright © 2012 Pearson Education, Inc. Publishing as Addison-Wesley.

5.65

Event	Description
(A & B)	Both A and B occur
(A or B)	At least one of A and B occur
(A & (not B))	A occurs but B does not occur
((not A) and (not B))	Neither A nor B occur
(A or B or C)	At least one of A, B, and C occur
(A & B & C)	All three of A, B, and C occur
(A & (not B) & (not C)) or (B & (not A) & (not C)) or (C & (not A) & (not B))	Exactly one of A, B, and C occur
(A & B & (not C)) or (A & C & (not B)) or (B & C & (not A))	Exactly two of A, B, and C occur
not((A & B) or (A & C) or (B & C))	At most one of A, B, and C occur

Exercises 5.3

5.67 There are 5 chips that are blue. Therefore, the probability is f/N = 5/12. P(B) = 5/12.

5.69 (a) P(S) = f/N = (12 + 33 + 32)/100 = 0.77.

(b) S = (A or B or C)

(c) P(A) = 12/100 = 0.12;   P(B) = 33/100 = 0.33;

P(C) = 32/100 = 0.32.

(d) P(S) = P(A or B or C) = P(A) + P(B) + P(C) = 0.12 + 0.33 + 0.32 = 0.77. This result is identical to the one found in part (a).

5.71 (a) The percentage who smoked within the last 30 days is 9.8 + 7.8 + 5.3 + 2.8 + 0.7 + 0.3 = 26.7. Therefore, the probability that a randomly selected twelfth grader smoked is 0.267.

(b) The percentage who smoked at least one cigarette per day within the last 30 days is 7.8 + 5.3 + 2.8 + 0.7 + 0.3 = 16.9. Therefore, the probability that a randomly selected twelfth grader smoked at least one cigarette per day is 0.169.

(d) The percentage who smoked between 6 and 34 cigarettes per day, inclusive, within the last 30 days is 5.3 + 2.8 + 0.7 = 8.8. Therefore, the probability that a randomly selected twelfth grader smoked between 6 and 34 cigarettes per day is 0.088.

5.73 (a) P(Spill occurs in an ocean) = P(Atlantic Ocean or Pacific Ocean) = P(Atlantic Ocean) + P(Pacific Ocean) = 0.008 + 0.037 = 0.045. That is, 4.5% of the spills occur in an ocean.

(b) P(Spill occurs in a lake or harbor) = P(Great Lakes or Other lakes or Harbors) = P(Great Lakes) + P(Other lakes) + P(Harbors)

= 0.020 + 0.002 + 0.161 = 0.183. That is, 18.3% of the spills occur in a lake or harbor.

(c) P(Spill does not occur in a lake, ocean, river, or canal) = P(Gulf of

Copyright © 2012 Pearson Education, Inc. Publishing as Addison-Wesley.

Mexico or Bays and Sounds or Harbor or Other) = P(Gulf of Mexico) + P(Bays and Sounds) + P(Harbor) + P(Other) = 0.233 + 0.146 + 0.161 + 0.027 = 0.567. That is, 56.7% of the spills do not occur in a lake, ocean, river, or canal.

5.75 (a) P(senator is at least 50) = $\dfrac{33}{100} + \dfrac{32}{100} + \dfrac{18}{100} + \dfrac{5}{100} = 0.88$.

This is accomplished more easily using the complementation rule:

P(senator is at least 50) = 1 - P(senator is under 50) = $1 - \dfrac{12}{100} = 0.88$.

(b) P(senator is under 70) = $\dfrac{12}{100} + \dfrac{33}{100} + \dfrac{32}{100} = \dfrac{77}{100} = 0.77$.

This is accomplished more easily using the complementation rule:

P(senator is under 70) = 1 - P(senator is at least 70) =

$1 - \left(\dfrac{18}{100} + \dfrac{5}{100}\right) = 1 - \dfrac{23}{100} = \dfrac{77}{100} = 0.77$.

5.77 (a) The percentage of day laborers who have lived in the United States either between 1 and 20 years, inclusive, or less than 11 years is just the sum of the percentages in the first five categories, or 93%. Thus the probability is 0.93.

(b) P(1 to 20) = (30 + 21 + 12 + 13)/100 = 0.76

P(Less than 11) = (17 + 30+ 21 + 12)/100 = 0.80

P(1 to 20 & Less than 11) = (30 + 21 + 12)/100 = 0.63

P(1 to 20 or Less than 11) = P(1 to 20) + P(Less than 11) - P(1 to 20 & Less than 11) = 0.76 + 0.80 - 0.63 = 0.93

(c) In this case, it was easier in part (a). If the three probabilities needed in part (b) had been provided, the general addition rule would have been the easier method.

5.79 (a) $P(A) = \dfrac{6}{36} = 0.167$ $P(B) = \dfrac{2}{36} = 0.056$ $P(C) = \dfrac{1}{36} = 0.028$ $P(D) = \dfrac{2}{36} = 0.056$

$P(E) = \dfrac{1}{36} = 0.028$ $P(F) = \dfrac{5}{36} = 0.139$ $P(G) = \dfrac{6}{36} = 0.167$

(b) P(7 or 11)= P(A) + P(B) = 0.167 + 0.056 = 0.223.

(c) P(2 or 3 or 12) = P(C) or P(D) or P(E).

= 0.028 + 0.056 + 0.028 = 0.112

(d) P(8 or doubles):   Using Figure 4.1: $\dfrac{10}{36} = 0.278$

(e) P(8) + P(doubles) - P(8 & doubles) = 0.139 + 0.167 - 0.028 = 0.278.

5.81 P(Public school or college) = P(Public school) + P(College) – P(Public school & college) = 0.848 + 0.230 - 0.177 = 0.901. 90.1% of students attend either public school or college.

5.83 (a) For A and B to be mutually exclusive, P(A & B) must equal zero. Recalling the general addition rule and making the appropriate

Copyright © 2012 Pearson Education, Inc. Publishing as Addison-Wesley.

substitutions, we have:

P(A or B) = P(A) + P(B) - P(A & B)

or

1/2 = 1/4 + 1/3 - P(A & B).

Rearranging terms and solving for P(A & B), we find:

P(A & B) = 1/4 + 1/3 - 1/2 = 3/12 + 4/12 - 6/12 = 1/12.

Thus, A and B are not mutually exclusive because

P(A & B) = 1/12 ≠ 0.

(b) From part (a), P(A & B) = 1/12 = 0.083.

5.85 Let H = the event that the household gets the *Herald,* E = the event that the household gets the *Examiner,* and T = the event that the household gets the *Times.* The event that a household gets at least one of the three major newspapers can be written (H or E or T).

The general addition rule for three events is

P(H or E or T) = P(H) + P(E) + P(T) - P(H&E) - P(H&T) - P(E&T) + P(H&E&T).

Substituting in the values given in the problem we have

P(H or E or T) = 0.334 + 0.346 + 0.470 - 0.104 - 0.119 - 0.151 + 0.048 = 0.824.

Therefore, the probability that a household gets at least one of the three major newspapers is 0.824.

Exercises 5.4

5.87 (a) probability

(b) probability

5.89 The notation {X=3} denotes an event that occurs when X=3, whereas P(X=3) denotes the probability that the event {X=3} will occur. Another way of thinking about the difference is that the first is an event and the second is a number between zero and one.

5.91 This table will resemble the probability distribution of the random variable.

5.93 (a) X = 2, 3, 4, 5, 6, 7, 8

(b) {X = 7}

(c) The total number of shuttle missions is 96. P(X = 4) = 2/96 = 0.021, so 2.1% of the shuttle crews consist of exactly four people.

(d)

Size of Crew	Probability
2	0.042
3	0.010
4	0.021
5	0.375
6	0.188
7	0.344
8	0.021

(e)

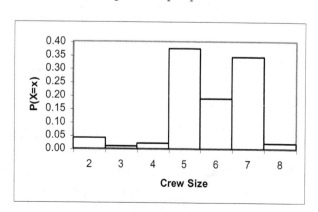

Copyright © 2012 Pearson Education, Inc. Publishing as Addison-Wesley.

5.95 (a) $\{Y \geq 1\}$ (b) $\{Y = 2\}$ (c) $\{1 \leq Y \leq 3\}$

(d) $\{Y = 1\}$ or $\{Y = 3\}$ or $\{Y = 5\}$

(e) $P(Y \geq 1) = P(Y = 1) + P(Y = 2) + P(Y = 3) + P(Y = 4) = P(Y = 5)$
$= 0.376 + 0.371 + 0.167 + 0.061 + 0.016 = 0.991$

(f) $P(Y = 2) = 0.371$

(g) $P(1 \leq y \leq 3) = P(Y = 1) + P(Y = 2) + P(Y = 3) = 0.376 + 0.371 + 0.167$
$= 0.914$

(h) $P(Y = 1 \text{ or } 2 \text{ or } 3) = P(Y = 1) + P(Y = 3) + P(Y = 5) = 0.376 + 0.167 + 0.016 = 0.559$

5.97 (a) $Y = 2, 3, 4, 5, 6, 7, 8, 9, 10, 11, 12$

(b) $\{Y = 7\}$

(c) $P(Y = 7) = 6/36 = 1/6 = 0.167$

(d)

Y	P(Y)
2	1/36
3	2/36
4	3/36
5	4/36
6	5/36
7	6/36
8	5/36
9	4/36
10	3/36
11	2/36
12	1/36

(e)

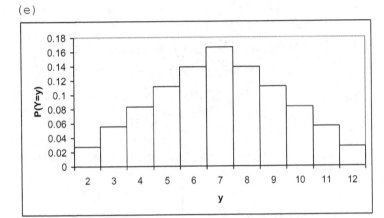

5.99 (a) The area of the square is $6^2 = 36$ square feet. The area of the bull's eye is $\pi(1)^2 = 3.1416$ square feet. The area of the ring is the area of a circle with radius 2 less the area of the bull's eye, or $\pi(2)^2 - \pi(1)^2 = 12.5664 - 3.1416 = 9.4248$. The remaining area of the square outside the ring has area $36 - 12.5664 = 23.4336$. Thus, $P(S = 10) = 3.1416/36 = 0.087$; $P(S = 5) = 9.4248/36 = 0.262$; and $P(S = 0) = 23.3226/36 = 0.651$. These are shown in the table below.

S	P(S = s)
0	0.651
5	0.262
10	0.087

Assuming that all points in the six foot square are equally likely, the archer has almost a 2/3 chance of scoring 0 points and about a 1/12 chance of scoring 10 points.

(b) $P(S = 5) = 0.262$

$P(S > 0) = P(S = 5 \text{ or } 10) = 0.262 + 0.087 = 0.349$

$P(S \leq 7) = P(S = 0 \text{ or } 5) = 0.651 + 0.262 = 0.913$

$P(5 < S \leq 15) = P(S = 10) = 0.087$

$P(S < 15) = P(S = 0 \text{ or } 5 \text{ or } 10) = 1$; This event always occurs.

$P(S < 0) = 0.000$; This event cannot occur.

5.101 $P(Z \leq 1.96) + P(Z > 1.96) = 1$.

Since $P(Z > 1.96) = 0.025$,

Copyright © 2012 Pearson Education, Inc. Publishing as Addison-Wesley.

```
P(Z ≤ 1.96) + 0.025 = 1, or

P(Z ≤ 1.96) = 1 - 0.025 = 0.975.
```

5.103 (a) $P(X ≤ c) + P(X > c) = 1.$

Since $P(X > c) = α,$

$P(X ≤ c) + α = 1,$ or

$P(X ≤ c) = 1 - α.$

(b) $P(Y < -c) + P(-c ≤ Y ≤ c) + P(Y > c) = 1.$

Since $P(Y < -c) = P(Y > c) = α/2,$

$α/2 + P(-c ≤ Y ≤ c) + α/2 = 1,$ or

$P(-c ≤ Y ≤ c) = 1 - α/2 - α/2 = 1 - α.$

(c) $P(T < -c) + P(-c ≤ T ≤ c) + P(T > c) = 1.$

Since $P(-c ≤ T ≤ c) = 1 - α,$

$P(T < -c) + (1 - α) + P(T > c) = 1,$ or

$P(T < -c) + P(T > c) = 1 - 1 + α = α.$

Since $P(T < -c) = P(T > c),$

$2· [P(T > c)] = α$ or $P(T > c) = α/2.$

Exercises 5.5

5.105 The mean of a discrete random variable generalizes the concept of a population mean.

5.107 The required calculations are

x	P(X=x)	xP(X=x)	x^2	x^2 P(X=x)
2	0.042	0.084	4.000	0.168
3	0.010	0.030	9.000	0.090
4	0.021	0.084	16.000	0.336
5	0.375	1.875	25.000	9.375
6	0.188	1.128	36.000	6.768
7	0.344	2.408	49.000	16.856
8	0.021	0.168	64.000	1.344
		5.777		34.937

(a) $μ_x = \sum xP(X=x) = 5.777.$ The average number of persons in a shuttle crew is about 5.8.

(b) $σ = \sqrt{\sum x^2 P(X=x) - μ_x^2} = \sqrt{34.937 - 5.777^2} = 1.250$

Copyright © 2012 Pearson Education, Inc. Publishing as Addison-Wesley.

(c)

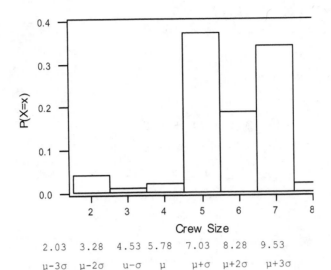

2.03   3.28   4.53 5.78   7.03   8.28   9.53

$\mu-3\sigma$ $\mu-2\sigma$ $u-\sigma$ μ $\mu+\sigma$ $\mu+2\sigma$ $\mu+3\sigma$

5.109 The required calculations are

y	P(Y=y)	yP(Y=y)	y^2	y^2P(Y=y)
0	0.009	0.000	0	0.000
1	0.376	0.376	1	0.376
2	0.371	0.742	4	1.484
3	0.167	0.501	9	1.503
4	0.061	0.244	16	0.976
5	0.016	0.080	25	0.400
	1	1.943		4.739

$\mu_y = \sum yP(Y=y) = 1.943$. The mean number of color TVs owned by households with annual incomes between \$15,000 and \$29,999 is 1.943.

(b) $\sigma = \sqrt{\sum y^2 P(Y=y) - \mu_y^2} = \sqrt{4.739 - 1.943^2} = 0.982$. Roughly speaking, the number of color TVs per one of these households averages about 1 away from the mean of 1.943.

(c)

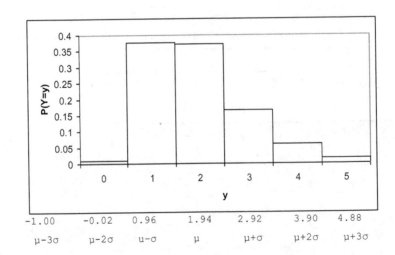

-1.00   -0.02   0.96   1.94   2.92   3.90   4.88

$\mu-3\sigma$ $\mu-2\sigma$ $u-\sigma$ μ $\mu+\sigma$ $\mu+2\sigma$ $\mu+3\sigma$

Copyright © 2012 Pearson Education, Inc. Publishing as Addison-Wesley.

5.111 The required calculations are

y	$P(Y=y)$	$yP(Y=y)$	$y-\mu_y$	$(y-\mu_y)^2$	$(y-\mu_y)^2 P(Y=y)$	y^2	$y^2 P(Y=y)$
2	1/36	2/36	-5	25	25/36	4	4/36
3	2/36	6/36	-4	16	32/36	9	18/36
4	3/36	12/36	-3	9	27/36	16	48/36
5	4/36	20/36	-2	4	16/36	25	100/36
6	5/36	30/36	-1	1	5/36	36	180/36
7	6/36	42/36	0	0	0/36	49	294/36
8	5/36	40/36	1	1	5/36	64	320/36
9	4/36	36/36	2	4	16/36	81	324/36
10	3/36	30/36	3	9	27/36	100	300/36
11	2/36	22/36	4	16	32/36	121	242/36
12	1/36	12/36	5	25	25/36	144	144/36
		252/36=7			210/36		1,974/36

(a) $\mu_y = \sum yP(Y=y) = 252/36 = 7$. When rolling two balanced dice, the mean of their sum is 7.

(b) $\sigma = \sqrt{\sum (y-\mu_y)^2 P(Y=y)} = \sqrt{210/36} = 2.415$

$\sigma = \sqrt{\sum y^2 P(Y=y) - \mu_y^2} = \sqrt{\dfrac{1974}{36} - 7^2} = 2.415$

(c)

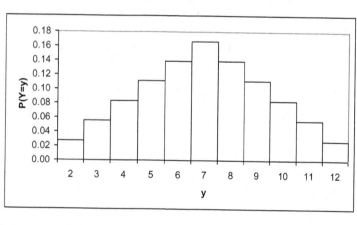

-0.2	2.2	4.6	7	9.4	11.8	14.2
$\mu-3\sigma$	$\mu-2\sigma$	$u-\sigma$	μ	$\mu+\sigma$	$\mu+2\sigma$	$\mu+3\sigma$

5.113 (a) The necessary calculations are

s	P(S=s)	sP(S=s)	s^2	$s^2P(S=s)$
0	0.651	0.000	0	0.000
5	0.262	1.310	25	6.550
10	0.087	0.870	100	8.700
		2.180		15.250

On average, the archer will score 2.180 points per arrow shot.

Copyright © 2012 Pearson Education, Inc. Publishing as Addison-Wesley.

(b) $\sigma = \sqrt{\sum s^2 P(S=s) - \mu_s^2} = \sqrt{15.250 - 2.180^2} = 3.24$. Roughly speaking, on average, the number of points on a shot will differ from the mean by 3.24 points.

5.115 (a)

$$P(X=1) = \frac{18}{38} = 0.474$$

$$P(X=-1) = \frac{20}{38} = 0.526; \quad 0.474 + 0.526 = 1.000$$

(b) $\mu = \sum x \cdot P(X=x) = (1)(0.474) + (-1)(0.526) = -0.052$

(c) On the average, you will lose 5.2¢ per play.

(d) If you bet \$1 on red 100 times, you can expect to lose $100 \cdot 0.052 = \$5.20$. If you bet \$1 on red 1000 times, you can expect to lose \$52.00.

(e) Roulette is not a profitable game for a person to play. Parts (c) and (d) demonstrate that, no matter how much you play, you can expect to lose. Also, part (a) shows that a higher probability is associated with losing rather than winning.

5.117 (a) $\mu = \sum x \cdot P(X=x) = (0)(0.95) + (10)(0.045) + (50)(0.004) + 100(0.0009) + (200)(0.0001) = 0.76$. The expected annual claim amount per homeowner is 0.76 thousand dollars, or \$760.

(b) They want an average net profit of \$50 per policy, so they need to charge \$50 more than the expected claim amount of \$760. Therefore, they should charge \$810.

5.119

w	P(W=w)	w·P(w)	w^2	w^2·P(W=w)
0	0.80	0.00	0	0.00
1	0.15	0.15	1	0.15
2	0.05	0.10	4	0.20
		0.25		0.35

(a)

$$\mu_w = \sum wP(W=w) = 0.25$$

$$\sigma_w = \sqrt{w^2 P(W=w) - \mu_w^2} = \sqrt{0.350 - 0.25^2} = \sqrt{0.2875} = 0.536$$

(b) On the average, there are 0.25 breakdowns per day.

(c) Assuming 250 workdays per year, the number of breakdowns expected per year is (0.25)(250) = 62.5.

5.121 (a) The necessary calculations are

y	P(Y=y)	yP(Y=y)	y^2	y^2P(Y=y)
0	0.424	0.000	0	0.000
1	0.161	0.161	1	0.161
2	0.134	0.268	4	0.536
3	0.111	0.333	9	0.999
4	0.093	0.372	16	1.488
5	0.077	0.385	25	1.925
		1.519		5.109

Copyright © 2012 Pearson Education, Inc. Publishing as Addison-Wesley.

The mean of Y, the mean number of customers waiting is 1.519. On the average, there are about 1.5 customers waiting at any given time.

(b) In a large number of independent observations, 1.519 customers will be waiting, on average.

(c) We will use Minitab to simulate these observations. Enter the values of Y in column C1 and their probabilities in column C2. Then select

Calc ▶ Random data ▶ Discrete, enter 500 in the **Generate Rows of data**

text box, enter Customers in the **Store in Column(s)** text box, then enter C1 in the **Values in** text box and C2 in the **Probabilities in** text box and click **OK**. This will produce 500 random values in a column

named Customeers. Now select **Stat ▶ Basic Statistics ▶ Display**

Descriptive Statistics, enter Customers in the **Variables** text box and click **OK**. The result appears in the Sessions Window. Our result is

Variable	N	N*	Mean	SE Mean	StDev	Minimum	Q1	Median
Customers	500	0	1.5800	0.0767	1.7151	0.000000000	0.000000000	1.0000

Variable	Q3	Maximum
Customers	3.0000	5.0000

(d) The mean for our observations was 1.58, a little bit higher than the exact mean of 1.519. Your data and mean will likely be different.

(e) Part (d) illustrates that simulations yield sample means that are close to the population mean.

Exercises 5.6

5.123 (1) Randomly selected pieces of identical rope are tested by subjecting each to a 1000 pound force. Each rope either breaks or it doesn't. (2) People are selected at random by ticket numbers at a large convention and their gender is noted. Each is either male or female.

5.125 $3! = 3 \cdot 2 \cdot 1 = 6;$ $7! = 7 \cdot 6 \cdot 5 \cdot 4 \cdot 3 \cdot 2 \cdot 1 = 5,040;$

$8! = 8 \cdot 7 \cdot 6 \cdot 5 \cdot 4 \cdot 3 \cdot 2 \cdot 1 = 40,320;$ $9! = 9 \cdot 8 \cdot 7 \cdot 6 \cdot 5 \cdot 4 \cdot 3 \cdot 2 \cdot 1 = 362,880$

5.127 (a) $\begin{pmatrix} 5 \\ 2 \end{pmatrix} = \dfrac{5!}{2!(5-2)!} = \dfrac{5 \cdot 4 \cdot 3 \cdot 2 \cdot 1}{2 \cdot 1 \cdot 3 \cdot 2 \cdot 1} = 10$

(b) $\begin{pmatrix} 7 \\ 4 \end{pmatrix} = \dfrac{7!}{4!(7-4)!} = \dfrac{7 \cdot 6 \cdot 5 \cdot 4 \cdot 3 \cdot 2 \cdot 1}{4 \cdot 3 \cdot 2 \cdot 1 \cdot 3 \cdot 2 \cdot 1} = 35$

(c) $\begin{pmatrix} 10 \\ 3 \end{pmatrix} = \dfrac{10!}{3!(10-3)!} = \dfrac{10 \cdot 9 \cdot 8 \cdot 7 \cdot 6 \cdot 5 \cdot 4 \cdot 3 \cdot 2 \cdot 1}{3 \cdot 2 \cdot 1 \cdot 7 \cdot 6 \cdot 5 \cdot 4 \cdot 3 \cdot 2 \cdot 1} = 120$

(d) $\begin{pmatrix} 12 \\ 5 \end{pmatrix} = \dfrac{12!}{5!(12-5)!} = \dfrac{12 \cdot 11 \cdot 10 \cdot 9 \cdot 8 \cdot 7 \cdot 6 \cdot 5 \cdot 4 \cdot 3 \cdot 2 \cdot 1}{5 \cdot 4 \cdot 3 \cdot 2 \cdot 1 \cdot 7 \cdot 6 \cdot 5 \cdot 4 \cdot 3 \cdot 2 \cdot 1} = 792$

5.129 (a) $\begin{pmatrix} 4 \\ 1 \end{pmatrix} = \dfrac{4!}{1!(4-1)!} = \dfrac{4 \cdot 3 \cdot 2 \cdot 1}{1 \cdot 3 \cdot 2 \cdot 1} = 4$

(b) $\begin{pmatrix} 6 \\ 2 \end{pmatrix} = \dfrac{6!}{2!(6-2)!} = \dfrac{6 \cdot 5 \cdot 4 \cdot 3 \cdot 2 \cdot 1}{2 \cdot 1 \cdot 4 \cdot 3 \cdot 2 \cdot 1} = 15$

Copyright © 2012 Pearson Education, Inc. Publishing as Addison-Wesley.

(c) $\dbinom{8}{3} = \dfrac{8!}{3!(8-3)!} = \dfrac{8\cdot7\cdot6\cdot5\cdot4\cdot3\cdot2\cdot1}{3\cdot2\cdot1\cdot5\cdot4\cdot3\cdot2\cdot1} = 56$

(d) $\dbinom{9}{6} = \dfrac{9!}{6!(9-6)!} = \dfrac{9\cdot8\cdot7\cdot6\cdot5\cdot4\cdot3\cdot2\cdot1}{6\cdot5\cdot4\cdot3\cdot2\cdot1\cdot3\cdot2\cdot1} = 84$

5.131 (a) Each trial consists of observing whether the child is cured of the pinworm infestation by pyrantel pamoate. Each is either cured or not cured. The results are independent assuming that the children are not in the same families. Since a success s is that the child is cured, the success probability p is 0.90.

(b)

Outcome	Probability
sss	(0.90)(0.90)(0.90)=0.729
ssf	(0.90)(0.90)(0.10)=0.081
sfs	(0.90)(0.10)(0.90)=0.081
sff	(0.90)(0.10)(0.10)=0.009
fss	(0.10)(0.90)(0.90)=0.081
fsf	(0.10)(0.90)(0.10)=0.009
ffs	(0.10)(0.10)(0.90)=0.009
fff	(0.10)(0.10)(0.10)=0.001

(c)

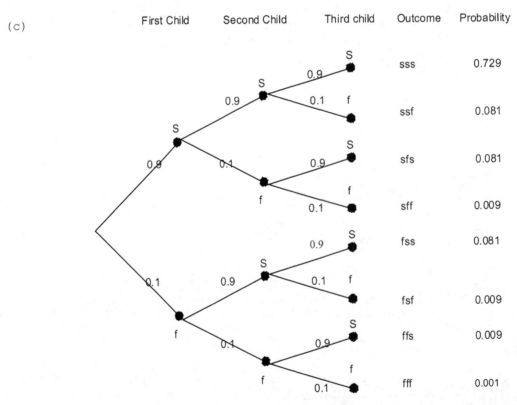

(d) The outcomes in which exactly two of the three children are cured are ssf, sfs, and fss.

Copyright © 2012 Pearson Education, Inc. Publishing as Addison-Wesley.

(e) Each outcome in part (d) has the probability 0.081. This probability is the same for each outcome because each probability is obtained by multiplying two success probabilities of 0.90 and one failure probability of 0.10.

(f) P(exactly two children are cured) = P(ssf) + P(sfs) + P(fss)

$$= 0.081 + 0.081 + 0.081 = 0.243$$

(g) P(exactly one child is cured) = P(sff) + P(fsf) + P(ffs)

$$= 0.009 + 0.009 + 0.009 = 0.027$$

P(exactly three children are cured) = P(sss) = 0.729, and

P(exactly zero children are cured) = P(fff) = 0.001.

Thus the complete probability distribution of X, the number of children out of three that are cured is

X	0	1	2	3
P (X=x)	0.001	0.027	0.243	0.729

5.133 (a) Plugging $n = 4$, $p = 0.3$, and $x = 2$ into formula 5.1, we have

$$P(X = 2) = \binom{4}{2}(0.3)^2(1-0.3)^{4-2} = 0.265 .$$

(b) Using Table XI1 in Appendix A, look for $n = 4$, $x = 2$, and then navigate over to the column with $p = 0.3$. The answer is 0.265. This is the same answer as that in part (a).

5.135 (a) Plugging $n = 6$, $p = 0.5$, and $x = 4$ into formula 5.1, we have

$$P(X = 4) = \binom{6}{4}(0.5)^4(1-0.5)^{6-4} = 0.234 .$$

(b) Using Table XI1 in Appendix A, look for $n = 6$, $x = 4$, and then navigate over to the column with $p = 0.5$. The answer is 0.234. This is the same answer as that in part (a).

5.137 (a) Plugging $n = 5$, $p = 3/4 = 0.75$, and $x = 4$ into formula 5.1, we have

$$P(X = 4) = \binom{5}{4}(0.75)^4(1-0.75)^{5-4} = 0.396 .$$

(b) Using Table XI1 in Appendix A, look for $n = 5$, $x = 4$, and then navigate over to the column with $p = 0.75$. The answer is 0.396. This is the same answer as that in part (a).

5.139 Step 1: A success is that a treated child is cured.

Step 2: The success probability is p = 0.90.

Step 3: The number of trials is n = 3.

Step 4: The formula for x successes is

$$P(X = x) = \binom{3}{x}(.90)^x(.10)^{3-x} .$$ For x = 0, 1, 2, and 3, the probabilities

are

Copyright © 2012 Pearson Education, Inc. Publishing as Addison-Wesley.

$$P(X=0)=\binom{3}{0}(.90)^{0}(.10)^{3}=\frac{3!}{0!\,3!}(.90)^{0}(.10)^{3}=0.001$$

$$P(X=1)=\binom{3}{1}(.90)^{1}(.10)^{2}=\frac{3!}{1!\,2!}(.90)^{1}(.10)^{2}=0.027$$

$$P(X=2)=\binom{3}{2}(.90)^{2}(.10)^{1}=\frac{3!}{2!\,1!}(.90)^{2}(.10)^{1}=0.243$$

$$P(X=3)=\binom{3}{3}(.90)^{3}(.10)^{0}=\frac{3!}{0!\,3!}(.90)^{3}(.10)^{0}=0.729$$

5.141 (a) Since the graph is symmetric, p = 0.5.

(b) Since the graph is right skewed, p is less than 0.5.

5.143 Step 1: A success is that the coin will come up heads.

Step 2: The success probability is p = 0.5.

Step 3: The number of trials is n = 10.

Step 4: The formula for y successes is

$$P(Y=y)=\binom{10}{y}(.50)^{y}(.50)^{10-y}\ .$$

We want the probability of exactly half of 10, or 5 successes

$$P(Y=5)=\binom{10}{5}(.50)^{5}(.50)^{10-5}=0.246$$

5.145 The calculations required to answer all parts of this exercise are

$$P(0)=\binom{5}{0}\cdot(.67)^{0}(.33)^{5}=0.004 \quad P(3)=\binom{5}{3}\cdot(.67)^{3}(.33)^{2}=0.328$$

$$P(1)=\binom{5}{1}\cdot(.67)^{1}(.33)^{4}=0.040 \quad P(4)=\binom{5}{4}\cdot(.67)^{4}(.33)^{1}=0.332$$

$$P(2)=\binom{5}{2}\cdot(.67)^{2}(.33)^{3}=0.161 \quad P(5)=\binom{5}{5}\cdot(.67)^{5}(.33)^{0}=0.135$$

(a)   P(2) = 0.161   (b)   P(4) = 0.332

(c)   P(X ≥ 4) = P(4) + P(5) = 0.332 + 0.135 = 0.467 (actually 0.468 if rounding is done after the addition).

(d)   P(2 ≤ X ≤ 4) = P(2) + P(3) + P(4) = 0.161 + 0.328 + 0.332 = 0.821

Copyright © 2012 Pearson Education, Inc. Publishing as Addison-Wesley.

(e)

x	P(X=x)
0	0.004
1	0.040
2	0.161
3	0.328
4	0.332
5	0.135

(f)   Left skewed

(g)

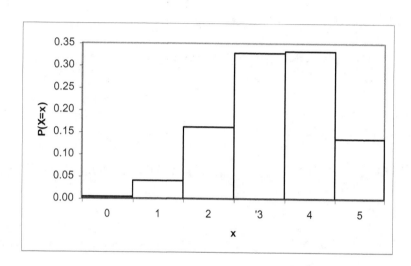

(h)

X	P(X=x)	xP(X=x)	x^2	x^2P(X=x)
0	0.004	0.000	0	0.000
1	0.040	0.040	1	0.040
2	0.161	0.322	4	0.644
3	0.328	0.984	9	2.952
4	0.332	1.328	16	5.312
5	0.135	0.675	25	3.375
		3.349		12.323

μ = 3.349

σ^2 = 12.323 - 3.349^2 = 1.107; $\sigma = \sqrt{1.107} = 1.052$

(i) μ = np = 5(0.67) = 3.35

σ^2 = np(1-p) = 5(0.67)(0.33) = 1.1055

$\sigma = \sqrt{1.1055} = 1.051$

(j)   Out of any five races, the average number of favorites that finish in the money is 3.35.

Copyright © 2012 Pearson Education, Inc. Publishing as Addison-Wesley.

5.147 (a) n = 8; p = 0.40

$$P(\text{Exactly } 3) = \binom{8}{3}(0.4)^3(0.6)^5 = 0.0241$$

$$P(3 \text{ or more}) = 1 - P(0 \text{ or } 1 \text{ or } 2) = 1 - \binom{8}{0}(0.4)^0(0.6)^8 - \binom{8}{1}(0.4)^1(0.6)^7 - \binom{8}{2}(0.4)^2(0.6)^6$$

$$= 1 - 0.0168 - 0.0896 - 0.2090 = 0.6846$$

$$P(\text{At most } 3) = P(0) + P(1) + P(2) + P(3)$$

$$= \binom{8}{0}(0.4)^0(0.6)^8 + \binom{8}{1}(0.4)^1(0.6)^7 + \binom{8}{2}(0.4)^2(0.6)^6 + \binom{8}{3}(0.4)^3(0.6)^5$$

$$= 0.0168 + 0.0896 + 0.2090 + 0.2787 = 0.5941$$

(b)

$$P(2 \leq Y \leq 4) = P(2) + P(3) + P(4)$$

$$= \binom{8}{2}(0.4)^2(0.6)^6 + \binom{8}{3}(0.4)^3(0.6)^5 + \binom{8}{4}(0.4)^4(0.6)^4 = 0.2092 + 0.2787 + 0.2322 = 0.7201$$

(If rounding is done after the addition, you will get 0.7200.)

(c) The mean of Y is μ = np = 8(0.4) = 3.2. On average, of eight traffic fatalities, 3.2 will involve an intoxicated or alcohol-impaired driver or non-occupant.

(d) σ^2 = np(1 - p) = 8(0.4)(0.6) = 19.2, so σ = 4.3418.

5.149 (a) n = 9, p = 0.35, so

$$P(\text{Exactly } 5) = \binom{9}{5}(0.35)^5(0.65)^4 = 0.1181$$

$$P(\text{At least } 5) = \binom{9}{5}(0.35)^5(0.65)^4 + \binom{9}{6}(0.35)^6(0.65)^3 + \binom{9}{7}(0.35)^7(0.65)^2$$

$$+ \binom{9}{8}(0.35)^8(0.65)^1 + \binom{9}{9}(0.35)^9(0.65)^1$$

$$= 0.1181 + 0.0424 + 0.0098 + 0.0013 + 0.0001 = 0.1717$$

$$P(\text{At most } 5) = \binom{9}{0}(0.35)^0(0.65)^9 + \binom{9}{1}(0.35)^1(0.65)^8 + \binom{9}{2}(0.35)^2(0.65)^7$$

$$+ \binom{9}{3}(0.35)^3(0.65)^6 + \binom{9}{4}(0.35)^4(0.65)^5 + \binom{9}{5}(0.35)^5(0.65)^4$$

$$= 0.0207 + 0.1004 + 0.2162 + 0.2716 + 0.2194 + 0.1181 = 0.9464$$

Copyright © 2012 Pearson Education, Inc. Publishing as Addison-Wesley.

$$P(\text{At least } 1 \text{ correct}) = 1 - P(0 \text{ Correct}) = 1 - \binom{9}{0}(0.35)^0(0.65)^9 = 1 - 0.0207 = 0.9793$$

(b) $P(\text{At most } 1 \text{ correct}) = P(0) + P(1) = \binom{9}{0}(0.35)^0(0.65)^9 + \binom{9}{1}(0.35)^1(0.65)^8$

$$= 0.0207 + 0.1004 = 0.1211$$

(c) $P(6 \text{ or } 7 \text{ or } 8) = \binom{9}{6}(0.35)^6(0.65)^3 + \binom{9}{7}(0.35)^7(0.65)^2 + \binom{9}{8}(0.35)^8(0.65)^1$

$$= 0.0424 + 0.0098 + 0.0013 = 0.0535$$

(d)   The formula is $P(X=x) = \binom{9}{x}(0.35)^x(0.65)^{9-x}$ for x = 0, 1, 2, ..., 9

Applying this formula for each value of x results in the table below.

x	P(X=x)
0	0.0207
1	0.1004
2	0.2162
3	0.2716
4	0.2194
5	0.1181
6	0.0424
7	0.0098
8	0.0013
9	0.0001

(e)   The sampling was actually done without replacement, so the trials are not independent and the success probability changes very slightly from trial to trial. The exact probability distribution is called a hypergeometric distribution.

5.151 (a)   n = 4 and p = 0.165, so the formula for the probability distribution is

$$P(X=x) = \binom{4}{x}(0.165)^x(0.835)^{4-x} \text{ for } x = 0, 1, 2, 3, 4$$

Applying this formula for each value of x results in the table below

x	P(X=x)
0	0.4059
1	0.4011
2	0.1585
3	0.0313
4	0.0031

(b)   The mean μ = np = 4(0.165) = 0.66. Thus, on the average, we would expect 0.66 people out of four under the age of 65 to have no health insurance.

(c)   Yes. If, in fact, 16.5% of people under 65 have no health insurance, then finding that 3 or more out of 4 had no health insurance would mean that an event with probability 0.0313 + 0.0031 = 0.0344 had just occurred. This is a fairly rare event if p is really 0.165, but would be less rare if p were greater than 0.165. Rather than believe that we

Copyright © 2012 Pearson Education, Inc. Publishing as Addison-Wesley.

had just observed a rare event, we would be inclined to conclude that our assumption (p = 0.165) was incorrect and that p has increased from the 16.5% rate of 2002.

(d) No. If, in fact, 16.5% of people under 65 have no health insurance, then finding that two or more out of four had no health insurance would mean that an event with probability 0.1585 + 0.0313 + 0.0031 = 0.1929 had just occurred. This is a fairly common event if p is really 0.165, so we would not be inclined to take 2 out of 4 as evidence that the uninsured rate has increased from the 16.5% rate in 2002.

5.153 (a) n = 4, p = 18/38 = 0.4737

$$P(2) = \binom{4}{2}(0.4737)^2(0.5263)^2 = 0.3729$$

(b) $$P(\text{At least } 1) = 1 - P(0) = 1 - \binom{4}{0}(0.4737)^0(0.5263)^4 = 1 - 0.0767 = 0.9233$$

5.155 (a) P(Both parents pass hemoglobin S to an offspring) = (0.5)(0.5) = 0.25

(b) n = 5, p = 0.25; X = number of children with sickle cell anemia.

$$P(X \geq 1) = 1 - P(X = 0) = 1 - \binom{5}{0}(0.25)^0(0.75)^5 = 1 - 0.2373 = 0.7627$$

(c)

x	P (X=x)
0	0.2373
1	0.3955
2	0.2637
3	0.0879
4	0.0146
5	0.0010

(d)

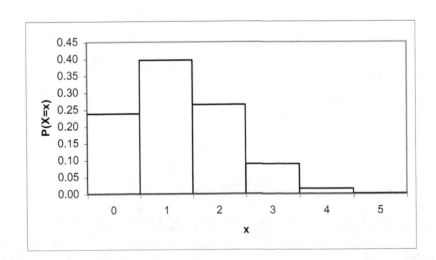

(e) On average, they can expect $\mu = np = 5(.25) = 1.25$ children to have sickle cell anemia.

5.157 The probability that a given party will show is 0.96. If n parties make a reservation, the probability that all n parties will show is

Copyright © 2012 Pearson Education, Inc. Publishing as Addison-Wesley.

$P(n) = \binom{n}{n}(0.96)^n (0.04)^0 = (0.96)^n$. We want to find the largest n for which this quantity is still 0.80 or larger.

n	P(n)
1	0.9600
2	0.9216
3	0.8847
4	0.8493
5	0.8154
6	0.7827

Thus, we see from the table that 5 is the largest value of n for which the owner can be at least 80% certain that all parties that make a reservation will show.

5.159 (a) P(both are male) = (17/40)(17/40) = 289/1600 = 0.180625 if sampling is done with replacement.

(b) P(both are male) = (17/40)(16/39) = 272/1560 = 0.174359 if sampling is done without replacement.

(c) The probability without replacement is slightly smaller than the probability with replacement. There is about a 3% difference in the probabilities.

(d) P(both are male) = (170/400)(170/400) = 28900/160000 = 0.180625 if sampling is done with replacement. P(both are male) = (170/400)(169/399) = 28730/159600 = 0.180013 if sampling is done without replacement. The probability without replacement is just barely smaller than the probability with replacement. There is only about a 0.3% difference in the probabilities.

(e) In the second case, there is less difference between sampling without and with replacement. This is because removing the first male from the population makes less difference in the probability of the second person being a male when there are 400 people in the class than when there are only 40 people in the class. In other words, 169/399 is closer to 170/400 than 16/39 is to 17/40.

5.161 (a) p = 0.0290647 or 0.029 to three decimals; $P(X = x) = 0.029(0.971)^{x-1}$

This formula gives the probability that a person will first win a prize in week x.

(b) $P(3) = 0.029(0.971)^2 = 0.0273$

P(X ≤ 3) = P(1) + P(2) + P(3) = 0.0290 + 0.0282 + 0.0273 = 0.0845

P(X ≥ 3) = 1 - P(X ≤ 2) = 1 - [0.0290 + 0.0282] = 1 - 0.0572

= 0.9428

(c) On the average, it will take μ = 1/p = 1/0.029 = 34.5 weeks until you win a prize.

Review Problems for Chapter 5

1. Probability theory enables us to control and evaluate the likelihood that a statistical inference is correct, and it provides the mathematical basis for statistical inference.

Copyright © 2012 Pearson Education, Inc. Publishing as Addison-Wesley.

2.   (a) The equal-likelihood model is used for computing probabilities when an experiment has N possible outcomes, all equally likely.

      (b) If an experiment has N equally likely outcomes and an event can occur in f ways, then the probability of the event is f/N.

3.   In the frequentist interpretation of probability, the probability of an event is the relative frequency of the event in a large number of experiments.

4.   The numbers (b) -0.047 and (c) 3.5 cannot be probabilities because probabilities must always be between 0 and 1.

5.   The Venn diagram is a common graphical technique for portraying events and relationships among events.

6.   Two or more events are mutually exclusive if no two of the events have any outcomes in common, i.e., no two of the events can occur at the same time.

7.   (a) P(E)

      (b) P(E) = 0.436

8.   (a) False. For any two events, the probability that one or the other occurs equals the sum of the two individual probabilities minus the probability that both events occur.

      (b) True. Either the event occurs or it does not. Therefore, the probability that the event occurs plus the probability that it does not occur equals 1. Thus the probability that it occurs is 1 minus the probability that it does not occur.

9.   Frequently, it is quicker to compute the probability of the complement of an event than it is to compute the probability of the event itself. The complement rule allows one to use the probability of the complement to obtain the probability of the event itself.

10.   (a) $P(A) = \dfrac{25352}{134372} = 0.0.189$

      (b) $P(D \text{ or } E \text{ or } F) = P(D) + P(E) + P(F) = \dfrac{13940 + 10619 + 28801}{134372} = \dfrac{53360}{134372} = 0.397$

      (b)

Event	Probability
A	0.189
B	0.169
C	0.138
D	0.104
E	0.079
F	0.214
G	0.107

11.   (a) (not J) is the event that the return selected shows an adjusted gross income of at least $100,000. There are 14,376 thousand such returns.

      (b) (H & I) is the event that the return selected shows an adjusted gross income of between $20,000 and $50,000. There are 43,081 thousand such returns.

      (c) (H or K) is the event that the return selected shows an adjusted gross income of at least $20,000. There are 86,258 thousand such returns.

Copyright © 2012 Pearson Education, Inc. Publishing as Addison-Wesley.

(d) (H & K) is the event that the return selected shows an adjusted gross income of between \$50,000 and \$100,000. There are 28,801 thousand such returns.

12. (a) Events H and I are not mutually exclusive. Both have events C, D, and E in common.

(b) Events I and K are mutually exclusive. They have no events in common.

(c) Events H and (not J) are mutually exclusive. H = (C or D or E or F) while (not J) = G. H and (not J) have no events in common.

(d) Events H, (not J), and K are not mutually exclusive. While H and (not J) have no events in common, a part of K (i.e., event F) is common to event H, and the other part of K (i.e., event G) is common to event (not J).

13. (a)

$$P(H) = \frac{18522 + 13940 + 10619 + 28801}{134372} = \frac{71882}{134372} = 0.535$$

$$P(I) = \frac{25352 + 22762 + 18522 + 13940 + 10619}{134372} = \frac{91195}{134372} = 0.679$$

$$P(J) = \frac{134372 - 14376}{134372} = \frac{119996}{134372} = 0.893$$

$$P(K) = \frac{28801 + 14376}{134372} = \frac{43177}{134372} = 0.321$$

(b) H = (C or D or E or F)

I = (A or B or C or D or E)

J = (A or B or C or D or E or F)

K = (F or G)

(c) P(H) = P(C) + P(D) + P(E) + P(F) = 0.138 + 0.104 + 0.079 + 0.214

= 0.535

P(I) = P(A) + P(B) + P(C) + P(D) + P(E)

= 0.189 + 0.169 + 0.138 + 0.104 + 0.079 = 0.679

P(J) = P(A) + P(B) + P(C) + P(D) + P(E) + P(F)

= 0.189 + 0.169 + 0.138 + 0.104 + 0.079 + 0.214 = 0.893

P(K) = P(F) + P(G) = 0.214 + 0.107 = 0.321

14. (a)

$$P(not\ J) = \frac{14376}{134372} = 0.107$$

$$P(H\ \&\ I) = \frac{43081}{134372} = 0.321$$

$$P(H\ or\ K) = \frac{86258}{134372} = 0.642$$

$$P(H\ \&\ K) = \frac{28801}{134372} = 0.214$$

(b) P(J)· = 1 - P(not J) = 1 - 0.107 = 0.893

(c) P(H or K) = P(H) + P(K) - P(H & K) = 0.535 + 0.321 - 0.214

= 0.642.

(d) The answer in (c) agrees with that in (a).

Copyright © 2012 Pearson Education, Inc. Publishing as Addison-Wesley.

15. (a) random variable

    (b) are finite (or countably infinite)

16. A probability distribution of a discrete distribution of a discrete random variable gives us a listing of the possible values of the random variable and their probabilities; or a formula for the probabilities.

17. Probability histogram

18. 1

19. (a) $P(X = 2) = 0.386$

    (b) 38.6%

    (c) $50(0.386) = 19.3$ or about 19; $500(0.386) = 193$

20. 3.6

21. X is more likely to take a value close to its mean because it has a smaller standard deviation and therefore less variation.

22. Each trial must have the same two possible outcomes (success and failure), the trials must be independent, and they must have a probability of success p that remains constant for all trials.

23. The binomial distribution is a probability distribution for the number of successes in a sequence of n Bernoulli trials.

24. $\dbinom{10}{3} = \dfrac{10\,!}{3\,!\,7\,!} = \dfrac{10(9)\,(8)}{3(2)\,(1)} = 120$

25. Definition 5.4 for the mean and definition 5.5 for the standard deviation are applied to the binomial and Poisson distribution. For example, in Definition 5.4, the formula for the binomial probability function is substituted for $P(X = x)$ to get the mean.

26. (a) Binomial distribution

    (b) Hypergeometric distribution

    (c) The hypergeometric distribution may be approximated by the binomial distribution if the population size N is much greater than the sample size n. When n is no more than 5% of N, the probability of a success does not change much from trial to trial.

27. (a) X = 1, 2, 3, 4   (b)   {X = 3}

    (c) $P(X = 3) = 13,957/52,883 = 0.2639.$

       26.4% of the undergraduates at this university are juniors.

    (d)

Class level x	Probability P(X=x)
1	0.2080
2	0.2121
3	0.2639
4	0.3160

Copyright © 2012 Pearson Education, Inc. Publishing as Addison-Wesley.

(e)

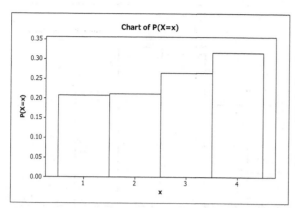

28.   (a)   {Y = 4}

(b)   {Y ≥ 4}

(c)   {2 ≤ Y ≤ 4}

(d)   {Y ≥ 1}

(e)   P(Y = 4) = 0.174

(f)   P(Y ≥ 4) = P(4) + P(5) + P(6) = 0.174 + 0.105 + 0.043 = 0.322

(g)   P(2 ≤ Y ≤ 4) = P(2) + P(3) + P(4) = 0.232 + 0.240 + 0.174 = 0.646

(h)   P(Y ≥ 1) = 1 - P(Y ≤ 0) = 1 - 0.052 = 0.948

29.   The required calculations are

y	P(Y=y)	yP(Y=y)	y^2	y^2P(Y=y)
0	0.052	0.000	0	0.000
1	0.154	0.154	1	0.154
2	0.232	0.464	4	0.928
3	0.240	0.720	9	2.160
4	0.174	0.696	16	2.784
5	0.105	0.525	25	2.625
6	0.043	0.258	36	1.548
		2.817		10.199

(a)   (a) $\mu_Y = \sum yP(Y=y) = 2.817$.

(b)   On the average, the number of busy lines is about 2.817.

(c)

$$\sigma_y = \sqrt{\sum (Y - \mu_y)^2 P(Y = y)} = \sqrt{2.2635} = 1.504, \text{ or}$$

$$\sigma_y = \sqrt{\sum Y^2 P(Y = y) - \mu_y^2} = \sqrt{10.199 - 2.817^2} = 1.504$$

Copyright © 2012 Pearson Education, Inc. Publishing as Addison-Wesley.

(d)

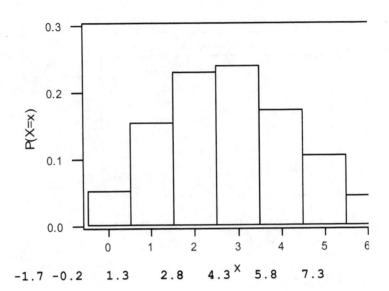

$\mu - 3\sigma$ $\quad \mu - 2\sigma$ $\quad \mu - \sigma$ $\quad \mu$ $\quad \mu + \sigma$ $\quad \mu + 2\sigma$ $\quad \mu + 3\sigma$

30. 0! = 1   3! = 3·2·1 = 6   4! = 4·3·2·1 = 24

7! = 7·6·5·4·3·2·1 = 5040

31.

$(a)\ \dbinom{8}{3} = \dfrac{8!}{3!5!} = 56$ $(b)\ \dbinom{8}{5} = \dfrac{8!}{5!3!} = 56$ $(c)\ \dbinom{6}{6} = \dfrac{6!}{6!6!} = 1$

$(d)\ \dbinom{10}{2} = \dfrac{10!}{2!8!} = 45$ $(e)\ \dbinom{40}{4} = \dfrac{40!}{4!36!} = 91390$ $(f)\ \dbinom{100}{0} = \dfrac{100!}{0!100!} = 1$

32.   (a) The success probability is p = 0.493.

(b) The outcomes and their probabilities are shown in the table.

Outcome	Probability
WWW	0.120
WWL	0.123
WLW	0.123
WLL	0.127
LWW	0.123
LWL	0.127
LLW	0.127
LLL	0.130

See the following tree diagram for the details of each probability.

Copyright © 2012 Pearson Education, Inc. Publishing as Addison-Wesley.

(c)

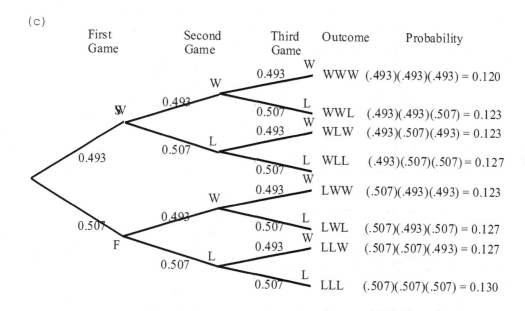

(d) The outcomes in which the player wins two out of the three times are WWL, WLW, and LWW.

(e) Each outcome in part (d) has probability 0.123. The probabilities are equal for each outcome because each probability is obtained by multiplying two WIN probabilities of 0.493 and one LOSS probability of 0.507.

(f) P(2 wins) = P(WWL) + P(WLW) + P(LWW) = 0.123 + 0.123 + 0.123

        = 0.369

(g) P(0 wins) = P(LLL) = 0.130

P(1 win) = P(WLL) + P(LWL) + P(LLW) = 0.123 + 0.123 + 0.123 = 0.369

P(3 wins) = P(WWW) = 0.120

Y	P(Y=y)
0	0.120
1	0.381
2	0.369
3	0.130

(h)   Binomial distribution with parameters n = 3 and p = 0.493.

33. The calculations required to answer all parts of this exercise are

$$P(0) = \binom{4}{0} \cdot (0.6)^0 (0.4)^4 = 0.0256 \qquad P(1) = \binom{4}{1} \cdot (0.6)^1 (0.4)^3 = 0.1536$$

$$P(2) = \binom{4}{2} \cdot (0.6)^2 (0.4)^2 = 0.3456 \qquad P(3) = \binom{4}{3} \cdot (0.6)^3 (0.4)^1 = 0.3456$$

$$P(4) = \binom{4}{4} \cdot (0.6)^4 (0.4)^0 = 0.1296$$

(a) P(3) = 0.3456

Copyright © 2012 Pearson Education, Inc. Publishing as Addison-Wesley.

(b) P(X ≥ 3) = P(3) + P(4) = 0.3456 + 0.1296 = 0.4752

(c) P(X ≤ 3) = P(0) + P(1) + P(2) + P(3)

           = 0.0256 + 0.1536 + 0.3456 + 0.3456 = 0.8704

(d)

Number with Pets x	Probability P(X=x)
0	0.0256
1	0.1536
2	0.3456
3	0.3456
4	0.1296

(e) Left skewed, since p > 0.5.

(f)

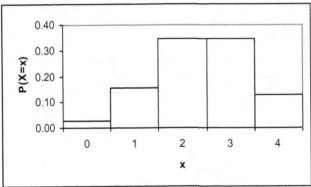

(g) The distribution is only approximate for several reasons: the actual distribution is hypergeometric, based on sampling without replacement; and the success probability p = 0.60 is probably based on a sample.

(h) μ = np = 4(.6) = 2.4; On the average, 2.4 out of 4 households have one or more pets.

(i) σ^2 = np(1-p) = 4(.6)(.4) = 0.96; $\sigma = \sqrt{0.96}$ = 0.98

34. (a) p < 0.5 since the distribution is left-skewed.

     (b) p = 0.5 since the distribution is symmetric.

Copyright © 2012 Pearson Education, Inc. Publishing as Addison-Wesley.

CHAPTER 6 ANSWERS

Exercises 6.1

6.1 A density curve is a smooth curve that identifies the shape of a distribution for a variable.

6.3 The percentage of all possible observations of a variable that lie within a specified range is approximately equal to the corresponding area under its density curve.

6.5 The percentage of all possible observations of the variable that lie to the right of 4 equals the area under its density curve to the right of 4, expressed as a percentage.

6.7 (a) Since the area to the right of 15 is 0.324, 32.4% of all possible observations of the variable exceed 15.

(b) Since the total area under the density curve is one and the area to the right of 15 is 0.324, the area to the left of 15 must be 1 - 0.324 = 0.676. Therefore, 67.6% of all possible observations of the variable are at most 15.

6.9 Since the area under the curve that lies between 15 and 20 is 0.414 and the total area under the density curve is one, the area <u>not</u> between 15 and 20 is 1 - 0.414 = 0.586. The area not between 15 and 20 would be the area to the left of 15 or the area to the right of 20. Therefore, 58.6% of all possible observations of the variable are either less than 15 or greater than 20.

6.11 (a) Since 28.4% of all possible observations of the variable are less than 11, the area under the density curve that lies to the left of 11 is 0.284.

(b) Since the total area under the density curve is one and the area to the left of 11 is 0.284, the area to the right of 11 must be 1 - 0.284 = 0.716.

6.13 No, because the total area to the right and left of 65 should be 1, not 0.9.

6.15 The histogram will be roughly bell-shaped.

6.17 Their distributions are identical. The mean and standard deviation completely determine the shape of a normal distribution. Thus if two normally distributed variables have the same mean and standard deviation, they also have the same distribution.

6.19 (a) True. Both normal curves have the same shape. Spread (or shape) is represented by σ. For each normal curve, $\sigma = 3$.

(b) False. The parameter μ affects where the normal curve is centered. Since this parameter is different for each normal curve, each normal curve is centered at a different place.

6.21 True. The value of the parameter μ has no effect on the shape of a normal curve. The parameter μ affects only where the normal curve is centered. The shape of the normal curve is determined by the parameter σ.

Copyright © 2012 Pearson Education, Inc. Publishing as Addison-Wesley.

6.23

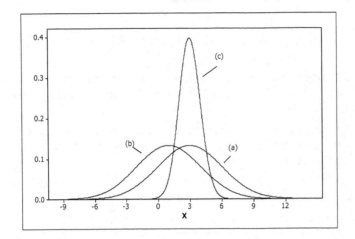

6.25 The percentage of all possible observations of a normally distributed variable that lie between 2 and 3 is the same as the area under the associated normal curve between 2 and 3. If the variable is only approximately normally distributed, the percentage of all possible observations between 2 and 3 is only approximately the area under the associated normal curve between 2 and 3.

6.27 If the area under a particular normal curve to the left of 105 is 0.6227, then 62.27% of all possible observations of the variable lie to the left of 105.

6.29 (a)   The percentage of female students who are between 60 and 65 inches tall is $100(0.0450 + 0.0757 + 0.1170 + 0.1480 + 0.1713) = 55.70\%$.

(b)   The area under the normal curve with parameters $\mu = 64.4$ and $\sigma = 2.4$ between 60 and 65 is approximately 0.5570. This is only an estimate because the distribution of heights is only approximately normally distributed.

6.31 (a)

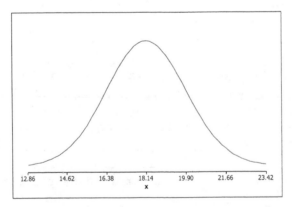

(b) $z = (x - 18.14)/1.76$

Copyright © 2012 Pearson Education, Inc. Publishing as Addison-Wesley.

(c) z has a standard normal distribution (μ = 0 and σ = 1).

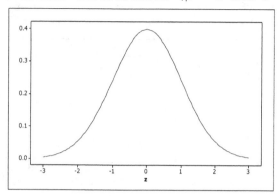

(d) The percentage of adult males *G. mollicoma* that have carapace lengths between 16 mm and 17 mm is equal to the area under the standard normal curve between <u>-1.22</u> and <u>-0.65</u>.

(e) The percentage of adult males *G. mollicoma* that have carapace lengths exceeding 19 mm is equal to the area under the standard normal curve that lies to the <u>right</u> of <u>0.49</u>.

6.33 (a)

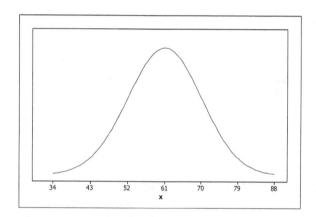

(b) z = (x - 61)/9

(c) z has a standard normal distribution (μ = 0 and σ = 1).

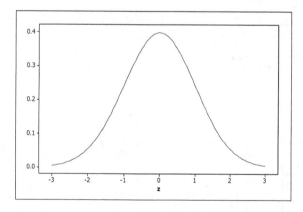

(d) The percentage of finishers with times between 50 and 70 minutes is equal to the area under the standard normal curve between <u>-1.22</u> and <u>1.00</u>.

Copyright © 2012 Pearson Education, Inc. Publishing as Addison-Wesley.

(e) The percentage of finishers with times less than 75 minutes is equal to the area under the standard normal curve that lies to the <u>left</u> of <u>1.56</u>.

6.35 (a)

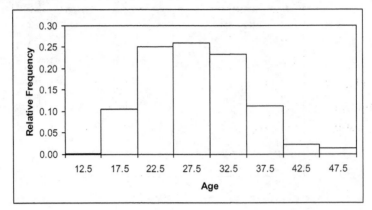

(b) Yes. The distribution appears to be slightly right skewed, but could be approximated quite well by a normal distribution.

6.37 (a)

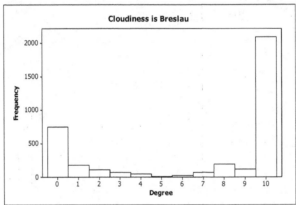

(b) The data is definitely not normal. The distribution instead looks bimodal with the most frequent values of 0 and 10 occurring at the far left and far right of the distribution. The rest of the distribution looks evenly distributed or uniform.

6.39 (a) Using Minitab, we obtain the data from the WeissStats CD and choose

Graph ▶ Histogram, select **Simple** from the first row, and click **OK**.

Then enter <u>VERBAL</u> and <u>MATH</u> in the Graphs variables text box and click OK. The resulting graphs follow.

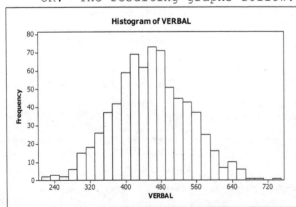
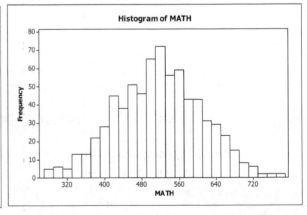

(a) Yes. The SAT verbal scores appearing in the left hand histogram have roughly a bell shaped distribution.

Copyright © 2012 Pearson Education, Inc. Publishing as Addison-Wesley.

(b) Yes. The SAT verbal scores appearing in the right hand histogram have roughly a bell shaped distribution.

6.41 The number of chips per bag could <u>not</u> be exactly normally distributed because that number is a discrete random variable, whereas any variable having a normal distribution is a continuous random variable.

6.43 (a) Using Minitab we choose **Calc ▶ Make patterned data ▶ Simple set of numbers...**, enter x in the **Store patterned data** in text box, type <u>218</u> in the **From first value** text box, type <u>314</u> in the **To last value** text box, type <u>.5</u> in the **In steps of** text box, and click **OK**. This will provide x values within 3 standard deviations on both sides of the mean. Now choose **Calc ▶ Probability distribution ▶ Normal...**, click on **Probability density**, type <u>266</u> in the **Mean** text box and <u>16</u> in the **Standard deviation** text box, click in the **Input column** text box and select x, click in the **Optional storage** text box and type <u>P(x)</u>, and click **OK**. Now choose **Graph ▶ Scatterplot...**, select the **With connect line** version, select P(X) in the **Y** column in row **1** and X for the **X** column. Click on the **Data view** button and check only the **Connect line** box. Click **OK** twice. The result is

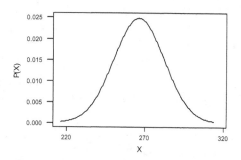

(b) Following the procedure in Example 6.2, we choose **Calc ▶ Random Data ▶ Normal...**, type <u>1000</u> in the **Generate rows of data** text box, click in the **Store in column(s)** text box and type <u>DAYS</u>, click in the **Mean** text box and type <u>266</u>, click in the **Standard deviation** text box and type <u>16</u>, and click **OK**.

(c) We would expect the sample mean and sample deviation to be approximately equal to the mean and standard deviation of the population, 266 and 16 respectively. This is because we expect the sample to reflect approximately the characteristics of the population.

(d) We choose **Calc ▶ Column Statistics...**, click on the **Mean** button, click in the **Input variable** text box and select DAYS, and click **OK**. Then repeat the process, selecting the **Standard deviation** button. The results are shown in the Session Window. The results we obtained were Mean of DAYS = 266.88 and Standard deviation of DAYS = 16.048. Your results will vary from ours.

(e) We would expect a histogram of the 1000 observations to look roughly like a normal curve with mean 266 and standard deviation 16.

(f) Choose **Graph ▶ Histogram...**, select the **Simple** version, select DAYS in the **Graph variables** text box and click **OK**. The result shown is as

Copyright © 2012 Pearson Education, Inc. Publishing as Addison-Wesley.

expected.

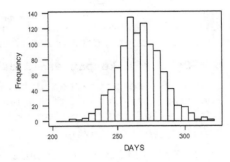

Exercises 6.2

6.45 Finding areas under the standard normal curve is important because for <u>any</u> normally distributed variable, we can obtain the percentage of all possible observations that lie within any specified range by first converting x values to z-scores and then finding the corresponding area under the standard normal curve.

6.47 The total area under the curve is 1, and the standard normal curve is symmetric about 0. Therefore, the area to the left of 0 is 0.5, and the area to the right of 0 is 0.5.

6.49 The area under the standard normal curve to the right of 0.43 is 1 - the area to the left of 0.43. The area to the left of 0.43 is 0.6664. Therefore, the area to the right of 0.43 is 1 - 0.6664 = 0.3336.

6.51 The area to the left of z = 3.00 is 0.9987 and the area to the left of -3.00 is 0.0013. Therefore the area between -3.00 and 3.00 is 0.9987 - 0.0013 = 0.9974 (99.74%).

6.53 (a) Locate the row (tenths digit) and column (hundredths digit) of the specified z-score. The corresponding table entry is the area under the standard normal curve that lies to the left of the z-score.

(b) The area that lies under the standard normal curve to the right of a specified z-score is 1 - (area to the left of the z-score).

(c) The area that lies under the standard normal score between two specified z-scores, say a and b, where a < b, is found by subtracting the area to the left of a from the area to the left of b.

6.55 (a)   (b)

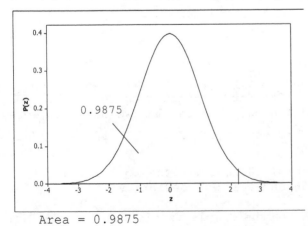

Area = 0.9875

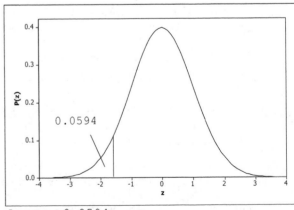

Area = 0.0594

Copyright © 2012 Pearson Education, Inc. Publishing as Addison-Wesley.

(c)

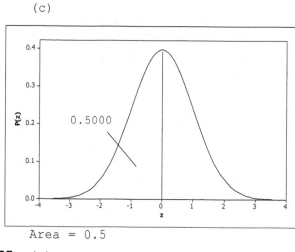

Area = 0.5

(d)

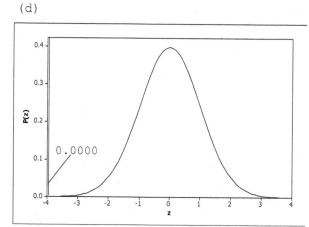

Area = 0.0000 (to 4 dp)

6.57 (a)

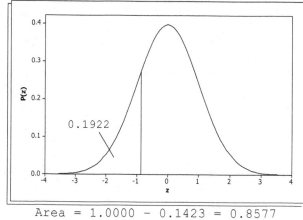

Area = 1.0000 − 0.1423 = 0.8577

(b)

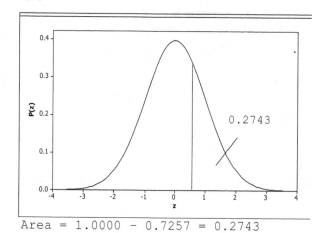

Area = 1.0000 − 0.7257 = 0.2743

(c)

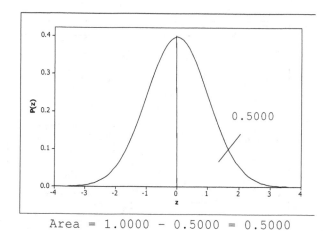

Area = 1.0000 − 0.5000 = 0.5000

(d)

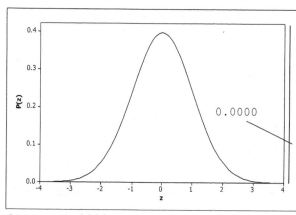

Area = 1.0000 − 1.0000 = 0.0000

Copyright © 2012 Pearson Education, Inc. Publishing as Addison-Wesley.

6.59 (a)

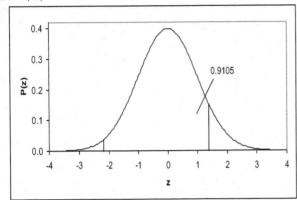

Area = 0.9251 − 0.0146 = 0.9105

(b)

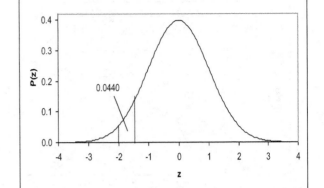

Area = 0.0668 − 0.0228 = 0.0440

(c)

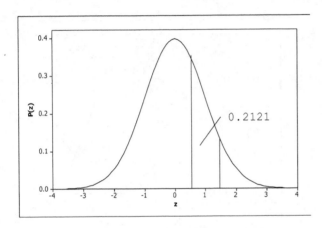

Area = 0.9345 − 0.7224 = 0.2121

(d)

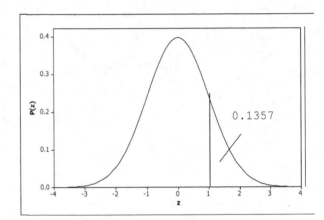

Area = 1.0000 − 0.8643 = 0.1357

6.61 (a)

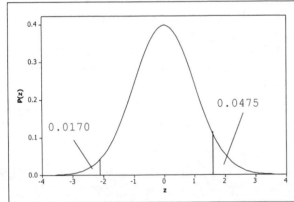

Area = 0.0170 + (1 − 0.9525) = 0.0645

(b)

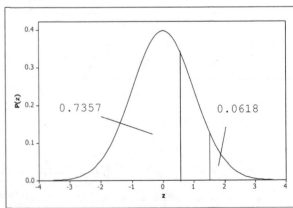

Area = 0.7357 + (1 − 0.9382) = 0.7975

6.63 (a) The area to the left of z = 1.28 is 0.8997. The area to the left of z = −1.28 is 0.1003. The area between z = −1.28 and z = 1.28 is 0.8997 − 0.1003 = 0.7994.

(b) The area to the left of z = 1.64 is 0.9495. The area to the left of z = −1.64 is 0.0505. The area between z = −1.64 and z = 1.64 is 0.9495 − 0.0505 = 0.8990.

Copyright © 2012 Pearson Education, Inc. Publishing as Addison-Wesley.

(c) The area to the left of z = -1.96 is 0.0250. The area to the right of z = 1.96 is 1.0000 - 0.9750 = 0.0250. The area either to the left of z = -1.96 or to the right of z = 1.96 is 0.0250 + 0.0250 = 0.0500.

(d) The area to the left of z = -2.33 is 0.0099. The area to the right of z = 2.33 is 1.0000 - 0.9901 = 0.0099. The area either to the left of z = -2.33 or to the right of z = 2.33 is 0.0099 + 0.0099 = 0.0198.

6.65 (a)

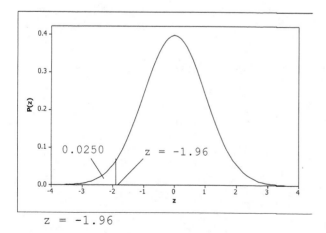

Area = 0.8413 - 0.1587 = 0.6826

(b)

Area = 0.9772 - 0.0228 = 0.9544

(c)

Area = 0.9987 - 0.0013 = 0.9974

6.67

z = -1.96

Copyright © 2012 Pearson Education, Inc. Publishing as Addison-Wesley.

6.69

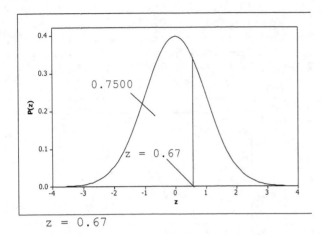

z = 0.67

6.71

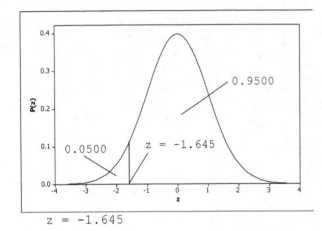

z = -1.645

Note: The area 0.0500 was exactly between
the area of two z-scores on the table.
Therefore, the two z-scores of -1.64 and
-1.65 were averaged.

6.73

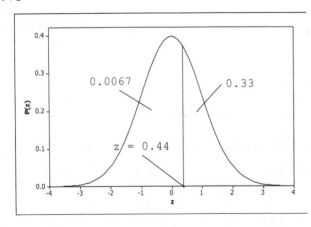

z = 0.44

Copyright © 2012 Pearson Education, Inc. Publishing as Addison-Wesley.

6.75 (a)

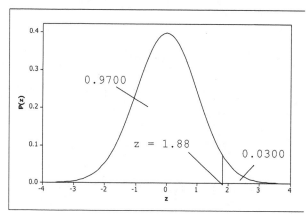

z = 1.88

(b)

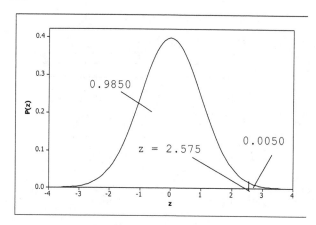

z = 2.575

6.77

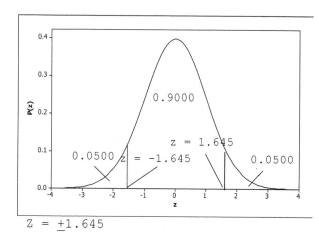

Z = ±1.645

6.79

$z_{0.10}$	$z_{0.05}$	$z_{0.025}$	$z_{0.01}$	$z_{0.005}$
1.28	1.645	1.96	2.33	2.575

Copyright © 2012 Pearson Education, Inc. Publishing as Addison-Wesley.

6.81 (a) z_α (b) $-z_\alpha$ (c) $\pm z_{\alpha/2}$

(d)

For part (a):   For part (b):

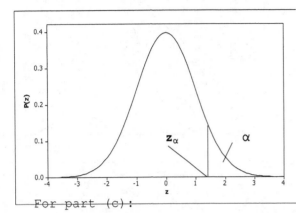 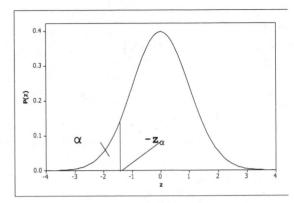

For part (c):

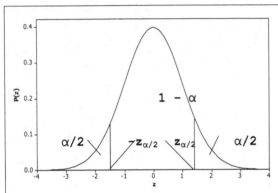

Exercises 6.3

6.83 The x values delimiting the interval within two standard deviations either side of the mean are $\mu - 2\sigma$ and $\mu + 2\sigma$. When these are standardized using $z = (x - \mu)/\sigma$, the former becomes $z = [(\mu - 2\sigma) - \mu]/\sigma = -2$ and the latter becomes $z = [(\mu + 2\sigma) - \mu]/\sigma = +2$.

6.85 (a) For the variable values 1 and 7, the z-values are

$$z = \frac{1-6}{2} = -2.50 \text{ and } z = \frac{7-6}{2} = 0.50$$

The area to the left of $z = -2.50$ is 0.0062 and the area to the left of $z = 0.50$ is 0.6915. Therefore the area between $z = -2.50$ and $z = 0.50$ is $0.6915 - 0.0062 = 0.6853$. Thus the percentage of all possible values of the variable that lie between 1 and 7 is 68.53%.

(b) For a variable value of 5, the z value is

$$z = \frac{5-6}{2} = -0.50$$

The area to the left of $z = -0.50$ is 0.3085. The area to right of $z = -0.50$ is $1 - 0.3085 = 0.6915$. Thus the percentage of values of the variable exceeding 5 is 69.15%.

(c) For a variable value of 4, the z value is

Copyright © 2012 Pearson Education, Inc. Publishing as Addison-Wesley.

$$z = \frac{4-6}{2} = -1.00$$

The area to the left of z = -1.00 is 0.1587. Thus the percentage of values of the variable that are less than 4 is 15.87%.

6.87 (a) For the variable values 6 and 7, the z-values are

$$z = \frac{6-10}{3} = -1.33 \text{ and } z = \frac{7-10}{3} = -1.00$$

The area to the left of z = -1.33 is 0.0918 and the area to the left of z = -1.00 is 0.1587. Therefore the area between z = -1.33 and z = -1.00 is 0.1587 - 0.0918 = 0.0669. Thus the percentage of all possible values of the variable that lie between 6 and 7 is 6.69%.

(b) For a variable value of 10, the z value is

$$z = \frac{10-10}{3} = 0.00$$

The area to the left of z = 0.00 is 0.5000. The area to right of z = 0.00 is   1 - 0.5000 = 0.5000. Thus the percentage of values of the variable that are at least 10 is 50.00%.

(c) For a variable value of 17.5, the z value is

$$z = \frac{17.5-10}{3} = 2.50$$

The area to the left of z = 2.50 is 0.9938. Thus the percentage of values of the variable that are at most 17.5 is 99.38%.

6.89 (a) Using Table II, we find that an area of 0.2500 lies to left of z = -0.67, an area of .5000 lies to the left of z = 0.00 and an area of 0.7500 lies to the left of 0.67. We convert each of these z-values to x-values using x = μ + zσ. Thus 25% of the values of the variable lie to the left of the first quartile, x = 6 + (-0.67)(2) = 4.66. Half (50%) of the values of the variable lie to the left of the second quartile (median), x = 6 + (0.00)(2) = 6. Finally, 75% of the values of the variable lie to the left of the third quartile, x = 6 + (0.67)(2) = 7.34.

(b) Using Table II, we find that an area of 0.8500 lies to the left of z = 1.04. We convert this z-value to an x-value using x = μ + zσ. Thus 85% of the values of the variable lie to the left of the 85th percentile, x = 6 + (1.04)(2) = 8.08.

(c) If 65% of the possible values exceed the value, then 35% of the possible values are less than the value. Using Table II, we find that an area of 0.3500 lies to the left of z = -0.39. We convert this z-value to an x-value using x = μ + zσ. Thus 65% of the values of the variable exceed x = 6 + (-0.39)(2) = 5.22.

(d) The two values that divide the area into a middle area of 0.95 and two outside areas of 0.025 have 2.5% of the curve and 97.5% of the curve to their left. Using Table II, we find that an area of 0.0250 lies to the left of z = -1.96 and an area of 0.9750 lies to the left of z = 1.96. We convert these z-values to x-values using x = μ + zσ. Thus 95% of the variable lie between x = 6 + (-1.96)(2) = 2.08 and x = 6 + (1.96)(2) = 9.92.

6.91 (a) Using Table II, we find that an area of 0.2500 lies to left of z = -0.67, an area of .5000 lies to the left of z = 0.00 and an area of

Copyright © 2012 Pearson Education, Inc. Publishing as Addison-Wesley.

0.7500 lies to the left of 0.67. We convert each of these z-values to x-values using $x = \mu + z\sigma$. Thus 25% of the values of the variable lie to the left of the first quartile, $x = 10 + (-0.67)(3) = 7.99$. Half (50%) of the values of the variable lie to the left of the second quartile (median), $x = 10 + (0.00)(3) = 10$. Finally, 75% of the values of the variable lie to the left of the third quartile, $x = 10 + (0.67)(3) = 12.01$.

(b) The seventh decile is the same as the 70th percentile. Using Table II, we find that an area of 0.7000 lies to the left of $z = 0.52$. We convert this z-value to an x-value using $x = \mu + z\sigma$. Thus 70% of the values of the variable lie to the left of the seventh decile, $x = 10 + (0.52)(3) = 11.56$.

(c) If 35% of the possible values exceed the value, then 65% of the possible values are less than the value. Using Table II, we find that an area of 0.6500 lies to the left of $z = 0.39$. We convert this z-value to an x-value using $x = \mu + z\sigma$. Thus 35% of the values of the variable exceed $x = 10 + (0.39)(3) = 11.17$.

(d) The two values that divide the area into a middle area of 0.99 and two outside areas of 0.005 have 0.5% of the curve and 99.5% of the curve to their left. Using Table II, we find that an area of 0.0050 lies to the left of $z = -2.575$ and an area of 0.9950 lies to the left of $z = 2.575$. We convert these z-values to x-values using $x = \mu + z\sigma$. Thus 99% of the variable lie between $x = 10 + (-2.575)(3) = 2.275$ and $x = 10 + (2.575)(3) = 17.725$.

6.93 (a) For carapace lengths 16 mm and 17 mm, the z-values are

$$z = \frac{16 - 18.14}{1.76} = -1.22 \quad \text{and} \quad z = \frac{17 - 18.14}{1.76} = -0.65$$

The area to the left of $z = -1.22$ is 0.1112 and the area to the left of $z = -0.65$ is 0.2578. Therefore the area between $z = -1.22$ and $z = -0.65$ is 0.2578 - 0.1112 = 0.1466. Thus the percentage of carapace lengths that are between 16 mm and 17 mm is 14.66%.

(b) For a carapace length of 19 mm, the z value is

$$z = \frac{19 - 18.14}{1.76} = 0.49$$

The area to the left of $z = 0.49$ is 0.6879. The area to right of 0.49 is 1 - 0.6879 = 0.3121. Thus the percentage of adult *G. mollicoma* with carapace lengths exceeding 19 mm is 31.21%.

(c) Using Table II, we find that an area of 0.2500 lies to left of $z = -0.67$, an area of .5000 lies to the left of $z = 0.00$ and an area of 0.7500 lies to the left of 0.67. We convert each of these z-values to x-values using $x = \mu + z\sigma$. Thus 25% of the carapace lengths of adult *G. mollicoma* lie to the left of the first quartile, $x = 18.14 + (-0.67)(1.76) = 16.96$ mm. Half (50%) of the carapace lengths of adult *G. mollicoma* lie to the left of the second quartile (median), $x = 18.14 + (0.00)(1.76) = 18.14$ mm. Finally. 75% of the carapace lengths of adult *G. mollicoma* lie to the left of the third quartile, $x = 18.14 + (0.67)(1.76) = 19.32$ mm.

(d) Using Table II, we find that an area of 0.9500 lies to the left of $z = 1.645$. We convert this z-value to an x-value using $x = \mu + z\sigma$. Thus, the 95th percentile of the carapace lengths of adult *G. mollicoma* is $x = 18.14 + (1.645)(1.76) = 21.04$ mm.

Copyright © 2012 Pearson Education, Inc. Publishing as Addison-Wesley.

6.95 (a) For finishers with times of 50 and 70 minutes, the z-values are

$$z = \frac{50-61}{9} = -1.22 \text{ and } z = \frac{70-61}{9} = 1.00$$

The area to the left of z = -1.225 is 0.1112 and the area to the left of z = 1.00 is 0.8413. Therefore the area between z = -1.22 and z = 1.00 is 0.8413 - 0.1112 = 0.7301. Thus the percentage of finishers with times between 50 minutes and 70 minutes in the New York City 10 km run is 73.01%.

(b) For a finishing time of 75 minutes, the z value is

$$z = \frac{75-61}{9} = 1.56$$

The area to the left of z = 1.56 is 0.9406. Thus the percentage of finishers with times less than 765 minutes is 94.06%.

(c) Using Table II, we find that an area of 0.4000 lies to left of z = -0.25. We convert this z-value to an x-value using x = μ + zσ. Thus 40% of the finishers had times less than the 40th percentile, x = 61 + (-0.25)(9) = 58.75 minutes.

(d) The eighth decile is the same as the 80th percentile. Using Table II, we find that an area of 0.8000 lies to the left of z = 0.84. We convert this z-value to an x-value using x = μ + zσ. Thus, 80% of the finishing times were less than x = 61 + (0.84)(9) = 68.56 minutes.

6.97 (a) For x = 260 and 280, the z-values are

$$z = \frac{260 - 272.2}{8.12} = -1.50 \text{ and } z = \frac{280 - 272.2}{8.12} = 0.96$$

The area to the left of z = -1.50 is 0.0668 and the area to the left of z = 0.96 is 0.8315. Therefore the area between z = -1.50 and z = 0.96 is 0.8315 - 0.0668 = 0.7647. Thus the percentage of tee shots that went between 260 and 280 yards is 76.47%.

(b) For x = 300, the z-value is

$$z = \frac{300 - 272.2}{8.12} = 3.42$$. The area to the left of z = 3.42 is 0.9997.

Thus the area to the right of z = 3.42 is 1 - 0.9997 = 0.0003, implying that only 0.03% of tee shots went more than 300 yards.

6.99

Part	Standard deviations to either side of the mean	Area under normal curve	Percent
(a)	1	0.3413 x 2 = 0.6826	68.26
(b)	2	0.4772 x 2 = 0.9544	95.44
(c)	3	0.4987 x 2 = 0.9974	99.74

6.101 (a) 68.26% of Swedish men have brain weights between <u>1.29</u> and <u>1.51</u> kg.

(b) 95.44% of Swedish men have brain weights between <u>1.18</u> and <u>1.62</u> kg.

(c) 99.74% of Swedish men have brain weights between <u>1.07</u> and <u>1.73</u> kg.

Copyright © 2012 Pearson Education, Inc. Publishing as Addison-Wesley.

(d)

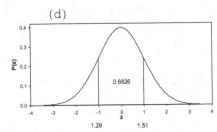 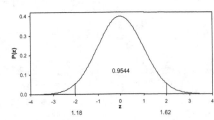 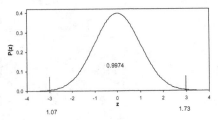

6.103 (a) From Table 6.1, exactly 11.70% of the heights of female students are between 62 and 63. For heights of 62 and 63 inches, the z values are

$$z = \frac{62 - 64.4}{2.4} = -1.00 \text{ and } z = \frac{63 - 64.4}{2.4} = -0.58$$

The area to the left of z = -1.00 is 0.1587 and the area to the left of z = -0.58 is 0.2810. Thus, the area between z = -1.00 and z = -0.58 is 0.2810 - 0.1587 = 0.1223 (12.23%). The two percentages are quite close.

(b) From Table 6.1, exactly = 0.1575 + 0.1100 + 0.0735 + 0.0374 + 0.0199 = 0.3983 (39.83%) of the heights of female students are between 65 and 70 inches. For heights of 65 and 70 inches, the z values are

$$z = \frac{65 - 64.4}{2.4} = 0.25 \text{ and } z = \frac{70 - 64.4}{2.4} = 2.33$$

The area to the left of z = 0.25 is 0.5987 and the area to the left of z = 2.33 is 0.9901. Thus, the area between z = 0.25 and a = 2.33 is 0.9901 - 0.5987 = 0.3914 (39.14%). The two percentages are quite close.

6.105 (a) $P(X \le 3) = P(Z \le \frac{3 - 6.1}{1.3}) = P(Z \le -2.38) = 0.0087$. We interpret this to mean that only about 0.87% of female polychaete worms have a length less than or equal to 3 mm.

(b)

$$P(5 < X < 7) = P(\frac{5 - 6.1}{1.3} < Z < \frac{7 - 6.1}{1.3}) = P(-0.85 < Z < 0.69)$$
$$= 0.7549 - 0.1977 = 0.5572$$

We interpret this to mean that about 55.72% of female polychaete worms have a length that is between 5 and 7 mm.

6.107 (a) 95% of the population values lie within 1.96 standard deviations to either side of the mean.

(b) 89.9% of the population values lie within 1.64 standard deviations to either side of the mean.

6.109 If the times between patients were approximately normally distributed, then approximately 15.87% of the inter-arrival times would be negative since P(Time < 0) = P(z<(0-8.7)/8.7) = P(z < -1) = 0.1587. Thus about one out of every six inter-arrival times would be negative. Since this is not possible, the distribution of times must not be normal.

6.111 (a) $Q_1 = \mu - 0.67\sigma$

$Q_2 = \mu$

Copyright © 2012 Pearson Education, Inc. Publishing as Addison-Wesley.

$$Q_3 = \mu + 0.67\sigma$$

$$P_k = \mu + z_{(1-k/100)}\sigma$$

Exercises 6.4

6.113 Decisions about whether a variable is normally distributed often are important in subsequent analyses such as percentile calculations or in determining the type of statistical inference procedure to be used.

6.115 If the normal probability plot is roughly linear except for a small number of points that lie well outside the overall pattern of the plot, it is possible that some or all of those points are outliers. If there is sufficient reason to remove the potential outliers from the sample and doing so results in a plot that is linear without outliers, the analysis can often be carried out with the remaining observations.

6.117 The variable is approximately normally distributed because the normal probability plot is approximately linear.

6.119 The variable is not normally distributed because the normal probability plot is not linear and is instead curved.

6.121 The variable is not normally distributed because the normal probability plot is not linear and is instead curved.

6.123 (a)

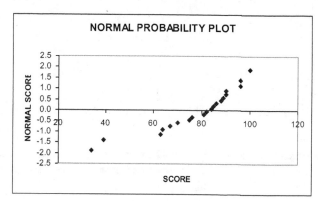

Exam	Normal Score
34	-1.87
39	-1.40
63	-1.13
64	-0.92
67	-0.74
70	-0.59
75	-0.45
76	-0.31
81	-0.19
82	-0.06
84	0.06
85	0.19
86	0.31
88	0.45
89	0.59
90	0.74
90.01	0.92
96	1.13
96.01	1.40
100	1.87

Copyright © 2012 Pearson Education, Inc. Publishing as Addison-Wesley.

   (b) Based on the probability plot, there appear to be two outliers in the sample: 34 and 39.

   (c) Based on the probability plot, the sample does not appear to come from a normally distributed population.

6.125 (a)

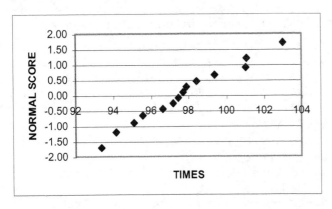

Times X	Normal Score y
93.37	-1.71
94.15	-1.20
95.10	-0.90
95.57	-0.66
96.63	-0.45
97.19	-0.27
97.47	-0.09
97.73	0.09
97.91	0.27
98.44	0.45
99.38	0.66
101.05	0.90
101.09	1.20
103.02	1.71

   (b) Based on the probability plot, there do not appear to be any outliers in the sample.

   (c) Based on the probability plot, the sample appears to be from an approximately normally distributed population.

6.127 (a) We enter the data in Excel in a column which we named "x" at the top.

Highlight the name and the data with your mouse and then choose **DDXL** ▶ **Charts and Plots**. Select **Normal Probability Plot** from the drop down **Function type** box and enter x in both the **Quantitative variable** and **Label variable** boxes. Click **OK**. The resulting plot is

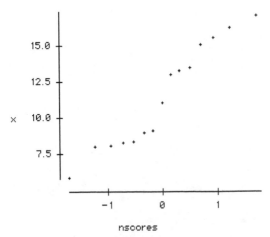

   (b) From the plot, there do not appear to be any outliers.

   (c) From the plot, normality seems to be a reasonable assumption.

6.129 (a) We enter the data in Excel in a column which we named "DOU" at the top.

Copyright © 2012 Pearson Education, Inc. Publishing as Addison-Wesley.

Highlight the name and the data with your mouse and then choose **DDXL ▶**
Charts and Plots. Select **Normal Probability Plot** from the drop down
Function type box and enter <u>DOU</u> in both the **Quantitative variable** and
Label variable boxes. Click **OK.** The resulting plot is

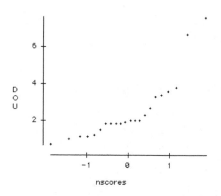

(b) From the plot, it appears that there may be two outliers, 7.6 and 6.7.
(c) From the plot, the data do not appear to be normally distributed.
     However, if the two outliers were excluded from the data, the remaining
     data may be normally distributed.

6.131 (a) Using Minitab and assuming that the data are in a column named 'TEMP',

     choose **Graph ▶ Histogram...**, select the **Simple** plot and click **OK.**
     Specify TEMP in the **Graph Variables** text box and click **OK.** The
     resulting plot is

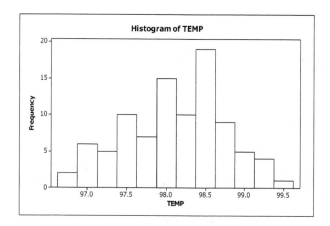

Aside from the unusual 'saw tooth' pattern of the left side of the
graph, an assumption of normality for this data appears to be
reasonable.

(b) Choose **Graph ▶ Probability Plot...**, select the **Single** plot and click
     OK. Specify TEMP in the Graph Variables text box. Click on the
     Distribution button, click the **Data Display** tab, select the **Symbols**
     only option button from the **Data Display** list and click **OK.** Now click
     the **Scale** button, click on the **Y-Scale Type** tab, select the **Score**
     option button from the **Y-Scale Type** list, and click **OK** twice. The
     resulting plot is

Copyright © 2012 Pearson Education, Inc. Publishing as Addison-Wesley.

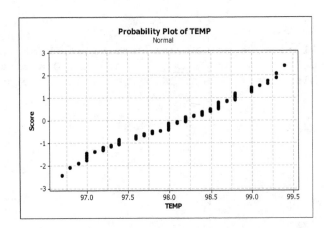

(b) There are no apparent outliers and the graph is roughly linear, so it seems reasonable to assume normality for these data.

(c) We arrived at the same conclusion from both graphs.

6.133 Choose **Graph ▶ Probability Plot...**, select the **Single** plot and click **OK**. Specify CHIPS in the **Graph Variables** text box. Click on the **Distribution** button, click the **Data Display** tab, select the **Symbols only** option button from the **Data Display** list and click **OK**. Now click the **Scale** button, click on the **Y-Scale Type** tab, select the **Score** option button from the **Y-Scale Type** list, and click **OK** twice. The resulting plot is

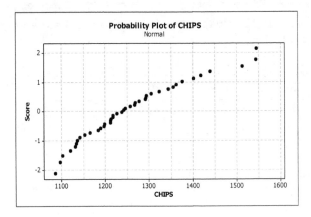

Yes. There do not appear to be any outliers in these data. Since the plot follows a fairly straight line, normality of the chips data seems to be a reasonable assumption.

6.135 (a) Using Minitab, we will generate four columns of fifty observations each by choosing **Calc ▶ Random data ▶ Normal...**, entering 50 in the **Generate rows of data** text box, clicking in the **Store in column(s)** text box and typing C1-C4, clicking in the **Mean** text box and entering 266, clicking in the **Standard deviation** text box and entering 16, and clicking **OK**. Now click in the worksheet column title row and name the four columns GEST1, GEST2, GEST3, and GEST4.

Next we select **Graph ▶ Probability Plot**, select the **Single** version and click **OK**, then enter GEST1, GEST2, GEST3, and GEST4 in the **Graph Variables** text box. Click **OK**. The resulting four graphs are shown below.

Copyright © 2012 Pearson Education, Inc. Publishing as Addison-Wesley.

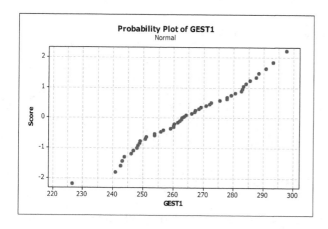

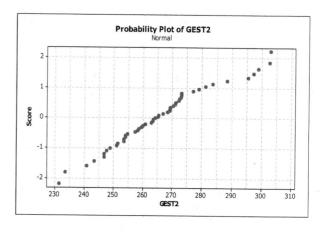

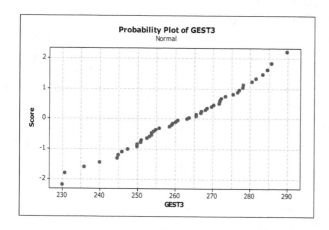

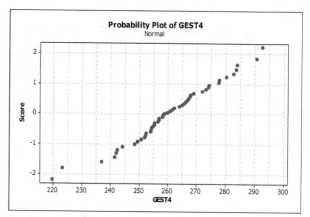

(c) Yes. Since the data were generated from a normal distribution, we would expect the four plots to be roughly linear, as they are. Your simulation will likely result in data and graphs that differ from ours.

Review Problems for Chapter 6

1. A density curve is a smooth curve that identifies the shape of a distribution for a variable. They are used to approximate percentages of all possible observations that lie within a range by finding corresponding area under its curve.

2. The percentage of all possible observations of a variable that lie between 25 and 50 equals the area under its density curve between <u>25 and 50</u>, expressed as a percentage.

3. (a) Since the area that lies to the left of 60 is 0.364, 36.4% of all possible observations of the variable are less than 60.

    (b) Since the area that lies to the left of 60 is 0.364, the area that lies to the right of 60 is 1.000 - 0.364 = 0.636. Thus, 63.6% of all possible observations of the variable are at least 60.

4. The area between 5 and 6 is 0.728. Therefore, the area outside of 5 and 6 (i.e. the area to the left of 5 or to the right of 6) is 1.000 - 0.728 = 0.272. Thus, 27.2% of the possible observations are either less than 5 or greater than 6.

5. Two primary reasons for studying the normal distribution are that:

    (a) It is often appropriate to use the normal distribution as the

Copyright © 2012 Pearson Education, Inc. Publishing as Addison-Wesley.

distribution of a population or random variable.

(b) The normal distribution is frequently employed in inferential statistics.

6. (a) A variable is normally distributed if its distribution has the shape of a normal curve.

(b) A population is normally distributed if a variable of the population is normally distributed and it is the only variable under consideration.

(c) The parameters for a normal curve are the mean μ and the standard deviation σ.

7. (a) False. There are many types of distributions that could have the same mean and standard deviation.

(b) True. The mean and standard deviation completely determine a normal distribution, so if two normal distributions have the same mean and standard deviation, then those two distributions are identical.

8. The percentages for a normally distributed variable and areas under the corresponding normal curve (expressed as a percentage) are identical.

9. The distribution of the standardized version of a normally distributed variable is the standard normal distribution, that is, a normal distribution with a mean of 0 and standard deviation of 1.

10. (a) True. The mean completely determines the center of the distribution.

(b) True. The standard deviation completely determines the shape of the distribution.

11. (a) The (second) curve with $\sigma = 6.2$ has the largest spread.

(b) The first and second curves are centered at $\mu = 1.5$.

(c) The first and third curves have the same shape because $\sigma = 3$ for both.

(d) The third curve is centered farthest to the left because it has the smallest value of $\mu = -2.7$.

(e) The fourth curve is the standard normal curve because $\mu = 0$ and $\sigma = 1$.

12. Key Fact 6.4.

13. (a) The table entry corresponding to the specified z-score is the area to the left of that z-score.

(b) The area to the right of a specified z-score is found by subtracting the table entry from 1.

(c) The area between two specified z-scores is found by subtracting the table entry for the smaller z-score from the table entry for the larger z-score.

14. (a) Find the table entry that is closest to the specified area. The z-score determined by locating the corresponding marginal values is the z-score that has the specified area to its left. If the specified area is exactly between two entries, then the two z-scores should be averaged.

(b) Subtract the specified area from 1. Find the entry in the table that is closest to the result of the subtraction, averaging if the area is exactly between two table entries. The z-score determined by locating the corresponding marginal values is the z-score that has the specified area to its right.

15. The value z_α is the z-score that has area α to its right under the standard normal curve.

Copyright © 2012 Pearson Education, Inc. Publishing as Addison-Wesley.

16. The 68.26-95.44-99.74 rule states that for a normally distributed variable: 68.26% of all possible observations lie within one standard deviation to either side of the mean; 95.44% of all possible observations lie within two standard deviations to either side of the mean; 99.74% of all possible observations lie within three standard deviation to either side of the mean.

17. The normal scores for a sample of observations are the observations we would expect to get for a sample of the same size for a variable having the standard normal distribution.

18. If we observe the values of a normally distributed variable for a sample, then a normal probability plot should be roughly <u>linear</u>.

19.

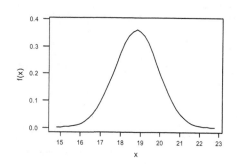

20. (a)

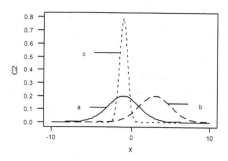

    (b) $z = (x - 18.8)/1.1$

    (c)

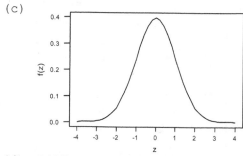

    (d) $P(17 \leq x \leq 20) = 0.8115$

    (e) The percentage of men who have forearm lengths less than 16 inches equals the area under the standard normal curve that lies to the <u>left</u> of <u>-2.55</u>.

21.   (a) The area to the right of 1.05 is 1.0000 - 0.8531 = 0.1469.
        (b) The area to the left of -1.05 is 0.1469 (by symmetry).
        (c) The area between -1.05 and 1.05 is 0.8531 - 0.1469 = 0.7062.

Copyright © 2012 Pearson Education, Inc. Publishing as Addison-Wesley.

22.   (a)   Area = 0.0013   (b)   Area = 1 - 0.7291 = 0.2709

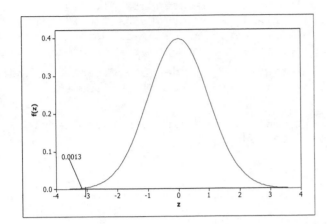 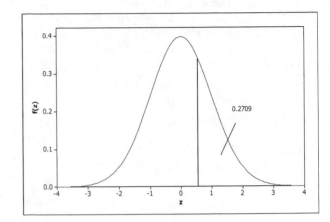

    (c)   Area = 0.9970 - 0.8665 = 0.1305   (d)   Area = 1.000 - 0.0197 = 0.9803

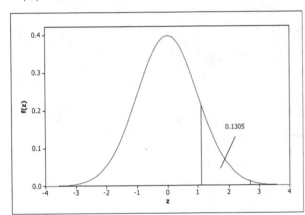 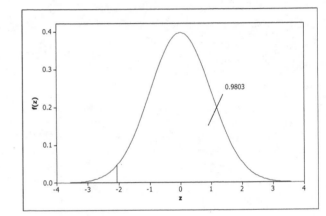

 (e)   Area = 0.0668 - 0.0000 = 0.0668   (f) Area = 0.8413 + (1 - 0.9987) = 0.8426

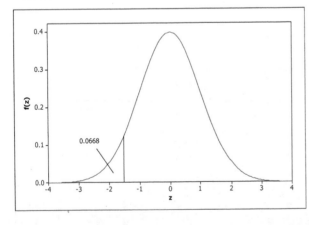 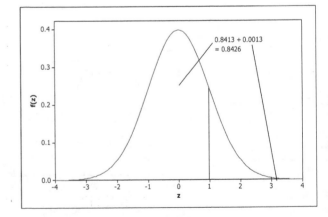

23.   (a) $z = -0.52$

       (b) z will have 0.9 to its left: $z = 1.28$

       (c) $z_{0.025} = 1.96$; $z_{0.05} = 1.645$; $z_{0.01} = 2.33$; $z_{0.005} = 2.575$

       (d) -2.575 and $+2.575$

Copyright © 2012 Pearson Education, Inc. Publishing as Addison-Wesley.

24.   (a)

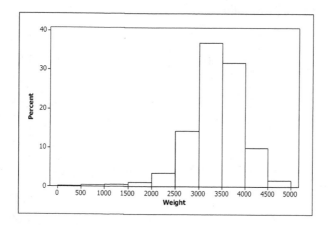

(b)   No.   The histogram indicates that the data are left skewed.

25.   (a)   The z-score for x = 1.4 is (1.4 - 1.3)/0.4 = 0.25.   The area to the
left of z = 0.25 is 0.5987, so the percentage of patients receiving
total knee arthroplasty who have a knee hyaluronic acid concentration
below 1.4 mg/ml is 0.5987.

(b)   The z-score for x = 1.0 is (1.0 - 1.3)/0.4 = -0.75 and for x = 2.0,
z = (2.0 - 1.3)/0.4 = 1.75.   The area to the left of z = -0.75 is
0.2266 and the area to the left of z = 1.75 is 0.9599.   Therefore, the
probability that the percentage of patients receiving total knee
arthroplasty who have a knee hyaluronic acid concentration between 1
and 2 mg/ml is 0.9599 - 0.2266 = 0.7333.

(c)   The z-score for x = 2.1 is (2.1 - 1.3)/0.4 = 2.00.   The area to the
left of z = 2.00 is 0.9772, so the percentage of patients receiving
total knee arthroplasty who have a knee hyaluronic acid concentration
above 2.1 mg/ml is 1.0000 - 0.9772 = 0.0228.

26.   (a)   An area of 0.2500 lies to the left of z = -0.67; an area of 0.5000 lies
to the left of z = 0.00; and an area of 0.7500 lies to the left of z =
0.67.   To convert these z-scores to GRE scores (x), we use $x = \mu + z\sigma$.
Thus Q_1 = 462 + (-0.67)(119) = 382.27; the median (Q_2) = 462 +
(0.00)(119) = 462; and Q_3 = 462 + (0.67)(119) = 541.73.   25% of the GRE
scores will be less than 382.27; 50% of the scores will be less than
462; and 75% of the scores will be less than 541.73.

(b)   An area of 0.9900 lies to the left of z = 2.33.   To convert this z-
scores to a GRE score (x), we use $x = \mu + z\sigma$.   Thus the 99th percentile
is 462 + (2.33)(119) = 739.27.   99% of the GRE scores will be less than
739.27.

27.   (a)   68.26% of students who took the verbal portion of the GRE scored
between <u>343</u> and <u>581</u>.

(b)   95.44% of students who took the verbal portion of the GRE scored
between <u>224</u> and <u>700</u>.

(c)   99.74% of students who took the verbal portion of the GRE scored
between <u>105</u> and <u>819</u>.

28.   (a)   Order the prices from smallest to largest.   Obtain the normal scores
for 12 observations from Table III.   The result is the table below.
The normal probability plot is to the left of the table.

Copyright © 2012 Pearson Education, Inc. Publishing as Addison-Wesley.

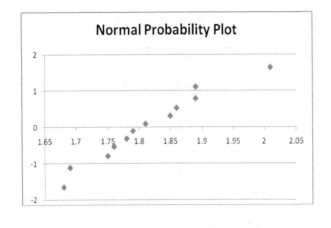

Price	Normal score
1.68	-1.64
1.69	-1.11
1.75	-0.79
1.76	-0.53
1.78	-0.31
1.79	-0.1
1.81	0.1
1.85	0.31
1.86	0.53
1.89	0.79
1.89	1.11
2.01	1.64

(b) There do not appear to be any outliers.

(c) The plot is roughly linear, so it seems reasonable to assume that the data come from a normal distribution.

(d) Using Minitab, enter the data in a column named PRICE. Select **Graph ▶ Probability Plot**, select **Single** and click **OK**. Enter PRICE in the **Graph variables,** click on the **Distributions** button and the **Data Display** tab. Check the **Symbols only** option and click OK. Then click on the **Scale** button and the **Y-Scale Type** tab and select the **Score** option. Click OK twice. The resulting graph is

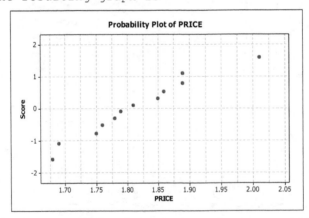

29. (a) Using Minitab, enter the data in a column named EMPLOYEES. Select **Graph ▶ Probability Plot**, select **Single** and click **OK**. Enter EMPLOYEES in the **Graph variables,** click on the **Distributions** button and the **Data Display** tab. Check the **Symbols only** option and click **OK**. Then click on the **Scale** button and the **Y-Scale Type** tab and select the **Score** option. Click OK twice. The resulting graph is

Copyright © 2012 Pearson Education, Inc. Publishing as Addison-Wesley.

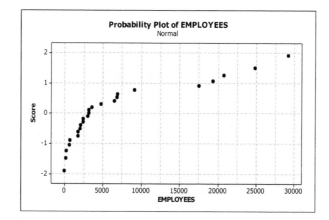

(b) There are five observations that are considerably greater than the other 20. If the distribution were otherwise normally distributed, these might be possible outliers.

(c) The graph is not close to being roughly linear. The data are very right skewed and hence not normally distributed, so the five possible outliers may be just part of the pattern of right skewness.

Copyright © 2012 Pearson Education, Inc. Publishing as Addison-Wesley.

Exercises 7.1

7.1 Sampling is often preferable to conducting a census because it is quicker, less costly, and sometimes it is the only practical way to get information.

7.3 (a) $\mu = \Sigma x/N = 6/3 = 2$

(b) for $n = 1$

Sample	$\overline{x}$
1	1.0
2	2.0
3	3.0

for $n = 2$

Sample	$\overline{x}$
1,2	1.2
1,3	2.0
2,3	2.5

for $n = 3$

Sample	$\overline{x}$
1,2,3	3.0

See the dotplots in part (c).

(c)

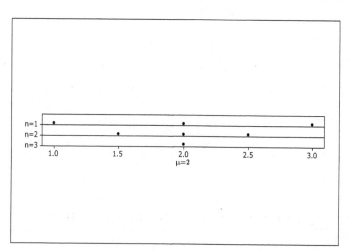

The sample means cluster more closely around the population mean as the sample size increases. Sampling error is smaller for large samples than for small samples.

(d) For $n = 1$, one out of the three sample means equaled the population mean. Thus, the probability that the sample mean equals the population mean for $n = 1$ is 1/3.

For $n = 2$, one out of the three sample means equaled the population

Copyright © 2012 Pearson Education, Inc. Publishing as Addison-Wesley.

mean. Thus, the probability that the sample mean equals the population mean for $n = 2$ is 1/3.

For $n = 3$, the one sample mean equals the population mean, Thus, the probability that the sample mean equals the population mean for $n = 3$ is 1.0.

(e) If the absolute value of the difference between the sample mean and the population mean is at most 0.5, that means that we are looking for sample means within the range 1.5 to 2.5, inclusive.

For $n = 1$, one out of the three sample means is in this range. Thus, the probability that the sampling error will be 0.5 or less for $n = 1$ is 1/3.

For $n = 2$, all of the three sample means are in this range. Thus, the probability that the sampling error will be 0.5 or less for $n = 2$ is 1.0.

For $n = 3$, the one sample mean is in this range. Thus, the probability that the sampling error will be 0.5 or less for $n = 3$ is 1.0.

7.5 (a) $\mu = \Sigma x/N = 10/4 = 2.5$

(b) for $n = 1$

Sample	$\bar{}$
1	1.0
2	2.0
3	3.0
4	4.0

for $n = 2$

Sample	$\bar{}$
1,2	1.5
1,3	2.0
1,4	2.5
2,3	2.5
2,4	3.0
3,4	3.5

for $n = 3$

Sample	$\bar{}$
1,2,3	2.0
1,2,4	2.3
1,3,4	2.7
2,3,4	3.0

for $n = 4$

Sample	$\bar{}$
1,2,3,4	2.5

See the dotplots in part (c).

(c)

Copyright © 2012 Pearson Education, Inc. Publishing as Addison-Wesley.

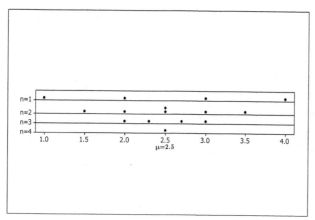

The sample means cluster more closely around the population mean as the sample size increases. Sampling error is smaller for large samples than for small samples.

(d) For $n = 1$, none of the four sample means equaled the population mean. Thus, the probability that the sample mean equals the population mean for $n = 1$ is 0.

For $n = 2$, two out of the six sample means equaled the population mean. Thus, the probability that the sample mean equals the population mean for $n = 2$ is $2/6 = 1/3$.

For $n = 3$, none of the four sample means equaled the population mean. Thus, the probability that the sample mean equals the population mean for $n = 3$ is 0.

For $n = 4$, the one sample mean equals the population mean, Thus, the probability that the sample mean equals the population mean for $n = 4$ is 1.

(e) If the absolute value of the difference between the sample mean and the population mean is at most 0.5, that means that we are looking for sample means within the range 2.0 to 3.0, inclusive.

For $n = 1$, two out of the four sample means is in this range. Thus, the probability that the sampling error will be 0.5 or less for $n = 1$ is $2/4 = 1/2$.

For $n = 2$, four of the six sample means are in this range. Thus, the probability that the sampling error will be 0.5 or less for $n = 2$ is $4/6 = 2/3$.

For $n = 3$, all four of the sample means are in this range. Thus, the probability that the sampling error will be 0.5 or less for $n = 3$ is 1.

For $n = 4$, the one sample mean is in this range. Thus, the probability that the sampling error will be 0.5 or less for $n = 4$ is 1.0.

7.7 (a) $\mu = \Sigma x/N = 15/5 = 3.0$

(b) for $n = 1$

Sample	$\overline{}$
1	1.0
2	2.0
3	3.0
4	4.0
5	5.0

Copyright © 2012 Pearson Education, Inc. Publishing as Addison-Wesley.

for $n = 2$

Sample	$\bar{x}$
1,2	1.5
1,3	2.0
1,4	2.5
1,5	3.0
2,3	2.5
2,4	3.0
2,5	3.5
3,4	3.5
3,5	4.0
4,5	4.5

for $n = 3$

Sample	$\bar{x}$
1,2,3	2.0
1,2,4	2.3
1,2,5	2.7
1,3,4	2.7
1,3,5	3.0
1,4,5	3.3
2,3,4	3.0
2,3,5	3.3
2,4,5	3.7
3,4,5	4.0

for $n = 4$

Sample	$\bar{x}$
1,2,3,4	2.50
1,2,3,5	2.75
1,2,4,5	3.00
1,3,4,5	3.25
2,3,4,5	3.50

for $n = 5$

Sample	$\bar{x}$
1,2,3,4,5	3.0

See the dotplots in part (c).

(c)

Copyright © 2012 Pearson Education, Inc. Publishing as Addison-Wesley.

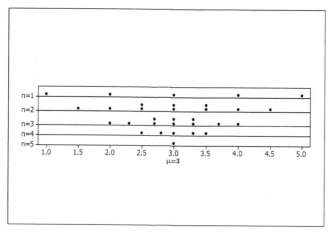

The sample means cluster more closely around the population mean as the sample size increases. Sampling error is smaller for large samples than for small samples.

(d) For $n = 1$, one of the five sample means equaled the population mean. Thus, the probability that the sample mean equals the population mean for $n = 1$ is 1/5.

For $n = 2$, two out of the ten sample means equaled the population mean. Thus, the probability that the sample mean equals the population mean for $n = 2$ is 2/10 = 1/5.

For $n = 3$, two out of the ten sample means equaled the population mean. Thus, the probability that the sample mean equals the population mean for $n = 3$ is 2/10 = 1/5.

For $n = 4$, one out of the five sample means equaled the population mean. Thus, the probability that the sample mean equals the population mean for $n = 4$ is 1/5.

For $n = 5$, the one sample mean equals the population mean, Thus, the probability that the sample mean equals the population mean for $n = 5$ is 1.

(e) If the absolute value of the difference between the sample mean and the population mean is at most 0.5, that means that we are looking for sample means within the range 2.5 to 3.5, inclusive.

For $n = 1$, one of the five sample means is in this range. Thus, the probability that the sampling error will be 0.5 or less for $n = 1$ is 1/5.

For $n = 2$, six of the ten sample means are in this range. Thus, the probability that the sampling error will be 0.5 or less for $n = 2$ is 6/10 = 3/5.

For $n = 3$, six of the ten sample means are in this range. Thus, the probability that the sampling error will be 0.5 or less for $n = 3$ is 6/10 = 3/5.

For $n = 4$, all of the five sample means are in this range. Thus, the probability that the sampling error will be 0.5 or less for $n = 4$ is 1.

For $n = 5$, the one sample mean is in this range. Thus, the probability that the sampling error will be 0.5 or less for $n = 5$ is 1.0.

7.9 (a) $\mu = \Sigma x/N = 21/6 = 3.5$

(b) for $n = 1$

Copyright © 2012 Pearson Education, Inc. Publishing as Addison-Wesley.

Sample	$\overline{x}$
1	1.0
2	2.0
3	3.0
4	4.0
5	5.0
6	6.0

for $n = 2$

Sample	$\overline{x}$
1,2	1.5
1,3	2.0
1,4	2.5
1,5	3.0
1,6	3.5
2,3	2.5
2,4	3.0
2,5	3.5
2,6	4.0
3,4	3.5
3,5	4.0
3,6	4.5
4,5	4.5
4,6	5.0
5,6	5.5

for $n = 3$

Sample	$\overline{x}$
1,2,3	2.0
1,2,4	2.3
1,2,5	2.7
1,2,6	3.0
1,3,4	2.7
1,3,5	3.0
1,3,6	3.3
1,4,5	3.3
1,4,6	3.7
1,5,6	4.0
2,3,4	3.0
2,3,5	3.3
2,3,6	3.7
2,4,5	3.7
2,4,6	4.0
2,5,6	4.3
3,4,5	4.0
3,4,6	4.3
3,5,6	4.7
4,5,6	5.0

Copyright © 2012 Pearson Education, Inc. Publishing as Addison-Wesley.

for *n* = 4

Sample	$\bar{}$
1,2,3,4	2.50
1,2,3,5	2.75
1,2,3,6	3.00
1,2,4,5	3.00
1,2,4,6	3.25
1,2,5,6	3.50
1,3,4,5	3.25
1,3,4,6	3.50
1,3,5,6	3.75
1,4,5,6	4.00
2,3,4,5	3.50
2,3,4,6	3.75
2,3,5,6	4.00
2,4,5,6	4.25
3,4,5,6	4.50

for *n* = 5

Sample	$\bar{}$
1,2,3,4,5	3.0
1,2,3,4,6	3.2
1,2,3,5,6	3.4
1,2,4,5,6	3.6
1,3,4,5,6	3.8
2,3,4,5,6	4.0

for *n* = 6

Sample	$\bar{}$
1,2,3,4,5,6	3.5

See the dotplots in part (c).

(c)

Copyright © 2012 Pearson Education, Inc. Publishing as Addison-Wesley.

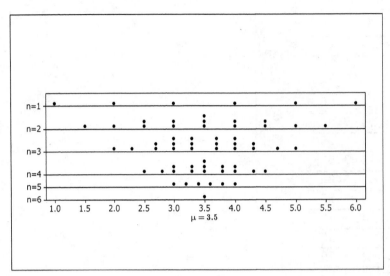

The sample means cluster more closely around the population mean as the sample size increases. Sampling error is smaller for large samples than for small samples.

(d) For $n = 1$, none of the six sample means equaled the population mean. Thus, the probability that the sample mean equals the population mean for $n = 1$ is 0.

For $n = 2$, three out of the fifteen sample means equaled the population mean. Thus, the probability that the sample mean equals the population mean for $n = 2$ is 3/15 = 1/5.

For $n = 3$, none of the twenty sample means equaled the population mean. Thus, the probability that the sample mean equals the population mean for $n = 3$ is 0.

For $n = 4$, three out of the fifteen sample means equaled the population mean. Thus, the probability that the sample mean equals the population mean for $n = 4$ is 3/15 = 1/5.

For $n = 5$, none of the six sample means equaled the population mean. Thus, the probability that the sample mean equals the population mean for $n = 5$ is 0.

For $n = 6$, the one sample mean equals the population mean, Thus, the probability that the sample mean equals the population mean for $n = 6$ is 1.

(e) If the absolute value of the difference between the sample mean and the population mean is at most 0.5, that means that we are looking for sample means within the range 3.0 to 4.0, inclusive.

For $n = 1$, two of the six sample means is in this range. Thus, the probability that the sampling error will be 0.5 or less for $n = 1$ is 2/6 = 1/3.

For $n = 2$, seven of the fifteen sample means are in this range. Thus, the probability that the sampling error will be 0.5 or less for $n = 2$ is 7/15.

For $n = 3$, twelve of the twenty sample means are in this range. Thus, the probability that the sampling error will be 0.5 or less for $n = 3$ is 12/20 = 3/5.

For $n = 4$, eleven of the fifteen sample means are in this range. Thus, the probability that the sampling error will be 0.5 or less for $n = 4$ is 11/15.

Copyright © 2012 Pearson Education, Inc. Publishing as Addison-Wesley.

For n = 5, all of the six sample means are in this range. Thus, the probability that the sampling error will be 0.5 or less for n = 5 is 1.

For n = 6, the one sample mean is in this range. Thus, the probability that the sampling error will be 0.5 or less for n = 6 is 1.0.

7.11   (a) $\mu = \Sigma x/N = 399/5 = 79.8$ inches.

(b)

Sample	Heights	$\overline{}$
T,K	80, 78	79.0
T,A	80, 84	82.0
T,D	80, 73	76.5
T,P	80, 84	82.0
K,A	78, 84	81.0
K,D	78, 73	75.5
K,P	78, 84	81.0
A,D	84, 73	78.5
A,P	84, 84	84.0
K,P	73, 84	78.5

(c)

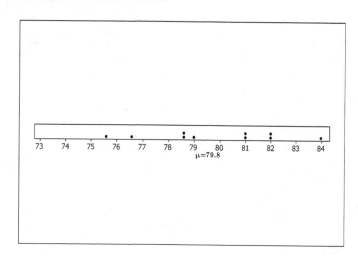

(d) $P(\overline{x} = \mu) = P(\overline{x} = 79.8) = 0.0$

(e) $P(79.8 - 1 \leq \overline{x} \leq 79.8 + 1) = P(78.8 \leq \overline{x} \leq 80.8)$

$$= 1/10 = 0.1$$

If we take a random sample of two heights, there is a 10% chance that the mean of the sample selected will be within one inch of the population mean.

Copyright © 2012 Pearson Education, Inc. Publishing as Addison-Wesley.

7.13 (b)

Sample	Heights	$\bar{\quad}$
T,K,A	80,78,84	80.67
T,K,D	80,78,73	77.00
T,K,P	80,78,84	80.67
T,A,D	80,84,73	79.00
T,A,P	80,84,84	82.67
T,D,P	80,73,84	79.00
K,A,D	78,84,73	78.33
K,A,P	78,84,84	82.00
K,D,P	78,73,84	78.33
A,D,P	84,73,84	80.33

(c)

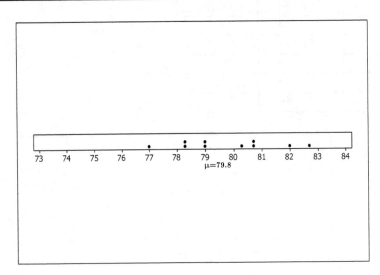

(d) $P(\bar{x} = \mu) = P(\bar{x} = 79.8) = 0.0$

(e) $P(79.8 - 1 \le \bar{x} \le 79.8 + 1) = P(78.8 \le \bar{x} \le 80.8)$

$$= 5/10 = 0.5$$

If we take a random sample of three heights, there is a 50% chance that the mean of the sample selected will be within one inch of the population mean.

7.15 (b)

Sample	Heights	
T,K,A,D,P	80,78,84,73,84	79.80

(c)

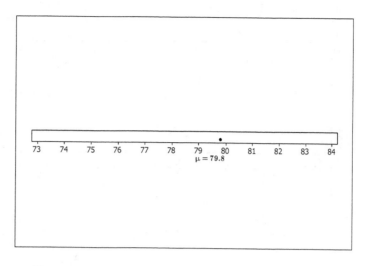

(d) $P(\overline{x} = \mu) = P(\overline{x} = 79.8) = 1.0$

(e) $P(79.8 - 1 \leq \overline{x} \leq 79.8 + 1) = P(78.8 \leq \overline{x} \leq 80.8) = 1.0$

   If we take a random sample of five heights, there is a 100% chance that
   the mean of the sample selected will be within one inch of the
   population mean.

7.17 (a) $\mu = \Sigma x/N = 180/6 = 30.0$ billion

   (b)

Sample	Wealth	$\overline{}$
G, B	40, 38	39.0
G, H	40, 35	37.5
G, E	40, 23	31.5
G, K	40, 22	31.0
G, A	40, 22	31.0
B, H	38, 35	36.5
B, E	38, 23	30.5
B, K	38, 22	30.0
B, A	38, 22	30.0
H, E	35, 23	29.0
H, K	35, 22	28.5
H, A	35, 22	28.5
E, K	23, 22	22.5
E, A	23, 22	22.5
K, A	22, 22	22.0

Copyright © 2012 Pearson Education, Inc. Publishing as Addison-Wesley.

(c)

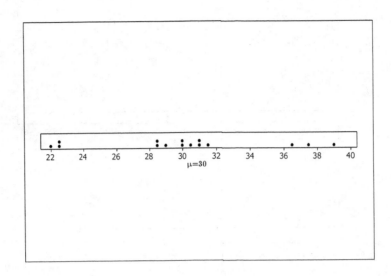

(d) $P(\bar{x} = \mu) = P(\bar{x} = 30.0) = 2/15 = 0.133$

(e) $P(30.0 - 2 \leq \bar{x} \leq 30.0 + 2) = P(28.0 \leq \bar{x} \leq 32.0)$

$$= 9/15$$

$$= 0.60$$

If we take a random sample of two of these six rich people, there is a 60% chance that their mean wealth will be within two billion of the population mean wealth.

7.19 (b)

Sample	Wealth	$\bar{x}$
G,B,H	40, 38, 35	37.7
G,B,E	40, 38, 23	33.7
G,B,K	40, 38, 22	33.3
G,B,A	40, 38, 22	33.3
G,H,E	40, 35, 23	32.7
G,H,K	40, 35, 22	32.3
G,H,A	40, 35, 22	32.3
G,E,K	40, 23, 22	28.3
G,E,A	40, 23, 22	28.3
G,K,A	40, 22, 22	28.0
B,H,E	38, 35, 23	32.0
B,H,K	38, 35, 22	31.7
B,H,A	38, 35, 22	31.7
B,E,K	38, 23, 22	27.7
B,E,A	38, 23, 22	27.7
B,K,A	38, 22, 22	27.3
H,E,K	35, 23, 22	26.7
H,E,A	35, 23, 22	26.7
H,K,A	35, 22, 22	26.3
E,K,A	23, 22, 22	22.3

Copyright © 2012 Pearson Education, Inc. Publishing as Addison-Wesley.

(c)

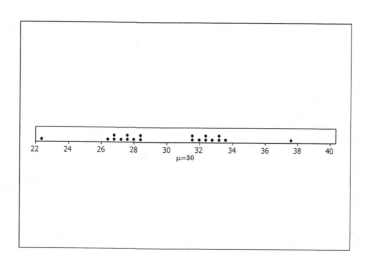

(e)   P(30.0 - 2 ≤ x̄ ≤ 30.0 + 2) = P(28.0 ≤ x̄ ≤ 32.0)

= 6/20 = 0.30

If we take a random sample of three wealthy people, there is a 30% chance that their mean wealth will be within two billion of the population mean wealth.

7.21   (b)

Sample	Wealth	x̄
G,B,H,E,K	40,38,35,23,22	31.6
G,B,H,E,A	40,38,35,23,22	31.6
G,B,H,K,A	40,38,35,22,22	31.4
G,B,E,K,A	40,38,23,22,22	29.0
G,H,E,K,A	40,35,23,22,22	28.4
B,H,E,K,A	38,35,23,22,22	28.0

(c)

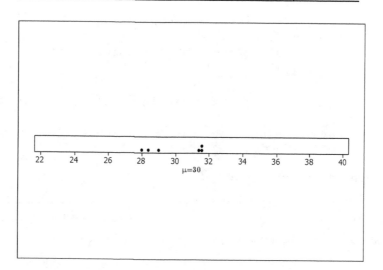

(d)   P(x̄ = μ) = P(x̄ = 30.0) = 0/6 = 0.000

(e)   P(30.0 - 2 ≤ x̄ ≤ 30.0 + 2) = P(28.0 ≤ x̄ ≤ 32.0)

Copyright © 2012 Pearson Education, Inc. Publishing as Addison-Wesley.

= 6/6 = 1

If we take a random sample of five wealthy people, there is an 100% chance that their mean wealth will be within two billion of the population mean wealth.

7.23 Increasing the sample size tends to reduce the sampling error.

7.25 (a) If a sample of size n = 1 is taken from a population of size N, there are N possible samples.

(b) Since each sample mean is based upon a single observation, the possible values of the sample mean and the population values are the same.

(c) There is no difference between taking a random sample of size n = 1 from a population and selecting a member at random from the population.

Exercises 7.2

7.27 Obtaining the mean and standard deviation of $\bar{x}$ is a first step in approximating the sampling distribution of the mean by a normal distribution because the normal distribution is completely determined by its mean and standard deviation.

7.29 Yes. The spread of the distribution of sample means becomes smaller as the sample size increases. Since that spread is measured by the standard deviation of all possible sample means, the standard deviation also gets smaller.

7.31 Standard error of the sample mean. The standard deviation of $\bar{x}$ determines the amount of sampling error to be expected when a population mean is estimated by a sample mean.

7.33 (a) For *n* = 1, the mean of the variable $\bar{x}$, $\mu_{\bar{x}}$, is found by summing up all

of the possible values for $\bar{x}$ and dividing by the number of possible samples. We get (1 + 2 + 3)/3 = 2.0.

For *n* = 2, the mean of the variable $\bar{x}$, $\mu_{\bar{x}}$, is found by summing up all

of the possible values for $\bar{x}$ and dividing by the number of possible samples. We get (1.5 + 2 + 2.5)/3 = 2.0.

For *n* = 3, the mean of the variable $\bar{x}$, $\mu_{\bar{x}}$, is found by summing up all

of the possible values for $\bar{x}$ and dividing by the number of possible samples. We get (2)/1 = 2.0.

For each sample size, $\mu_{\bar{x}} = 2$.

(b) From part (a), in Exercise 7.3, we calculated $\mu = \Sigma x/N = 6/3 = 2$. Formula 7.1 states that $\mu_{\bar{x}} = \mu$. Thus, $\mu_{\bar{x}} = 2$.

7.35 (a) For *n* = 1, the mean of the variable $\bar{x}$, $\mu_{\bar{x}}$, is found by summing up all

of the possible values for $\bar{x}$ and dividing by the number of possible samples. We get (1 + 2 + 3 + 4)/4 = 2.5.

Copyright © 2012 Pearson Education, Inc. Publishing as Addison-Wesley.

For $n = 2$, the mean of the variable $\bar{x}$, $\mu_{\bar{x}}$, is found by summing up all

of the possible values for $\bar{x}$ and dividing by the number of possible samples. We get $(1.5 + 2 + 2.5 + 2.5 + 3 + 3.5)/6 = 2.5$.

For $n = 3$, the mean of the variable $\bar{x}$, $\mu_{\bar{x}}$, is found by summing up all

of the possible values for $\bar{x}$ and dividing by the number of possible samples. We get $(2 + 2.3 + 2.7 + 3)/4 = 2.5$.

For $n = 4$, the mean of the variable $\bar{x}$, $\mu_{\bar{x}}$, is found by summing up all

of the possible values for $\bar{x}$ and dividing by the number of possible samples. We get $(2.5)/1 = 2.5$.

For each sample size, $\mu_{\bar{x}} = 2.5$.

(b) From part (a), in Exercise 7.5, we calculated $\mu = \Sigma x/N = 10/4 = 2.5$. Formula 7.1 states that $\mu_{\bar{x}} = \mu$. Thus, $\mu_{\bar{x}} = 2.5$.

7.37 (a) For $n = 1$, the mean of the variable $\bar{x}$, $\mu_{\bar{x}}$, is found by summing up all

of the possible values for $\bar{x}$ and dividing by the number of possible samples. We get $(1 + 2 + 3 + 4 + 5)/5 = 3.0$.

For $n = 2$, the mean of the variable $\bar{x}$, $\mu_{\bar{x}}$, is found by summing up all

of the possible values for $\bar{x}$ and dividing by the number of possible samples. We get $(1.5 + 2 + 2.5 + 3 + 2.5 + 3 + 3.5 + 3.5 + 4 + 4.5)/10 = 3.0$.

For $n = 3$, the mean of the variable $\bar{x}$, $\mu_{\bar{x}}$, is found by summing up all

of the possible values for $\bar{x}$ and dividing by the number of possible samples. We get $(2 + 2.3 + 2.7 + 2.7 + 3 + 3.3 + 3 + 3.3 + 3.7 + 4)/10 = 3.0$.

For $n = 4$, the mean of the variable $\bar{x}$, $\mu_{\bar{x}}$, is found by summing up all

of the possible values for $\bar{x}$ and dividing by the number of possible samples. We get $(2.5 + 2.75 + 3 + 3.25 + 3.5)/5 = 3.0$.

For $n = 5$, the mean of the variable $\bar{x}$, $\mu_{\bar{x}}$, is found by summing up all

of the possible values for $\bar{x}$ and dividing by the number of possible samples. We get $(3)/1 = 3.0$.

For each sample size, $\mu_{\bar{x}} = 3.0$.

(b) From part (a), in Exercise 7.7, we calculated $\mu = \Sigma x/N = 15/5 = 3$. Formula 7.1 states that $\mu_{\bar{x}} = \mu$. Thus, $\mu_{\bar{x}} = 3$.

Copyright © 2012 Pearson Education, Inc. Publishing as Addison-Wesley.

7.39 (a) For $n = 1$, the mean of the variable $\overline{x}$, $\mu_{\overline{x}}$, is found by summing up all

of the possible values for $\overline{x}$ and dividing by the number of possible samples. We get $(1 + 2 + 3 + 4 + 5 + 6)/6 = 3.5$.

For $n = 2$, the mean of the variable $\overline{x}$, $\mu_{\overline{x}}$, is found by summing up all

of the possible values for $\overline{x}$ and dividing by the number of possible samples. We get $(1.5 + 2 + 2.5 + 3 + 3.5 + 2.5 + 3 + 3.5 + 4 + 3.5 + 4 + 4.5 + 4.5 + 5 + 5.5)/15 = 3.5$.

For $n = 3$, the mean of the variable $\overline{x}$, $\mu_{\overline{x}}$, is found by summing up all

of the possible values for $\overline{x}$ and dividing by the number of possible samples. We get $(2 + 2.3 + 2.7 + 3 + 2.7 + 3 + 3.3 + 3.3 + 3.7 + 4 + 3 + 3.3 + 3.7 + 3.7 + 4 + 4.3 + 4 + 4.3 + 4.7 + 5)/20 = 3.5$.

For $n = 4$, the mean of the variable $\overline{x}$, $\mu_{\overline{x}}$, is found by summing up all

of the possible values for $\overline{x}$ and dividing by the number of possible samples. We get $(2.5 + 2.75 + 3 + 3 + 3.25 + 3.5 + 3.25 + 3.5 + 3.75 + 4 + 3.5 + 3.75 + 4 + 4.25 + 4.5)/15 = 3.5$.

For $n = 5$, the mean of the variable $\overline{x}$, $\mu_{\overline{x}}$, is found by summing up all

of the possible values for $\overline{x}$ and dividing by the number of possible samples. We get $(3 + 3.2 + 3.4 + 3.6 + 3.8 + 4)/6 = 3.5$.

For each sample size, $\mu_{\overline{x}} = 3.5$.

(b) From part (a), in Exercise 7.9, we calculated $\mu = \Sigma x/N = 21/6 = 3.5$. Formula 7.1 states that $\mu_{\overline{x}} = \mu$. Thus, $\mu_{\overline{x}} = 3.5$.

7.41 (a) $\mu_{\overline{x}} = \dfrac{\sum x}{N} = \dfrac{399}{5} = 79.8$

(b)

$$\mu_{\overline{x}} = \frac{\sum \overline{x}}{N} = \frac{79.0 + 82.0 + 76.5 + 82.0 + 81.0 + 75.5 + 81.0 + 78.5 + 84.0 + 78.5}{10}$$

$$= \frac{798}{10} = 79.8$$

(c) $\mu_{\overline{x}} = \mu = 79.8$

7.43 (b)

$$\mu_{\overline{x}} = \frac{\sum \overline{x}}{N} = \frac{80.7 + 77.0 + 80.7 + 79.0 + 82.7 + 79.0 + 78.3 + 82.0 + 78.3 + 80.3}{10}$$

$$= \frac{798}{10} = 79.8$$

Copyright © 2012 Pearson Education, Inc. Publishing as Addison-Wesley.

(c) $\mu_{\bar{x}} = \mu = 79.8$

7.45   (b)

$$\mu_{\bar{x}} = \frac{\sum \bar{x}}{N} = \frac{79.8}{1} = 79.8$$

(c) $\mu_{\bar{x}} = \mu = 79.8$

7.47   (a)   The population consists of all babies born in 1991.   The variable is the birth weight of the baby.

(b) $\mu_{\bar{x}} = \mu = 3369$ grams; $\sigma_{\bar{x}} = \sigma / \sqrt{n} = 581 / \sqrt{200} = 41.08$ grams

(c) $\mu_{\bar{x}} = \mu = 3369$ grams; $\sigma_{\bar{x}} = \sigma / \sqrt{n} = 10 / \sqrt{400} = 29.05$ grams

7.49   (a) $\mu_{\bar{x}} = \mu = \$65,100$; $\sigma_{\bar{x}} = \sigma / \sqrt{n} = 7200 / \sqrt{50} = \$1018.23$

Thus, for samples of size 50, the mean and standard deviation of all possible sample means are respectively, $65,100 and $1018.23.

(b) $\mu_{\bar{x}} = \mu = \$65,100$; $\sigma_{\bar{x}} = \sigma / \sqrt{n} = 7200 / \sqrt{100} = \$720.00$

Thus, for samples of size 100, the mean and standard deviation of all possible sample means are respectively, $65,100 and $720.

7.51   (a)   On the average, we would expect the sample mean of the four times to be the same as the population mean of 437 days.

(b)   The standard deviation of the sample mean is

$\sigma_{\bar{x}} = \sigma / \sqrt{n} = 399 / \sqrt{4} = 199.5$ days , so 99.74% of the time we would expect the sample mean to fall within 3 standard deviations of 437 days, i.e., between 437 - 3(199.5) = -161.5 and 437 + 3(199.5) = 1035.5.   Since the lower limit is clearly below zero and the time between earthquakes cannot be negative, we would expect the sample mean to fall between 0 and 1035.5 days 99.74% of the time.

7.53   (a)   Enter the data into cells A1 to A50 of Excel.   In cell B1, enter the expression =STDEVP(A1:A50) to obtain the population standard deviation of the data in B1.   The result is 1362.452   Note: Using DDXL procedures will yield a SAMPLE standard deviation of 1376.284, which you could then convert to a POPULATION standard deviation by multiplying by the square root of 49/50 to obtain the same result.

(b)   Equation 7.1 is the correct formula for obtaining the standard deviation of the sample mean since we are sampling without replacement from a finite population of 50 data values.

(c)   From Equation 7.1, $\sigma_{\bar{x}} = \sqrt{\dfrac{N-n}{N-1}} \dfrac{\sigma}{\sqrt{n}} = \sqrt{\dfrac{50-30}{50-1}} \dfrac{1362.452}{\sqrt{30}} = 158.920$.

From Equation 7.2, $\sigma_{\bar{x}} = \dfrac{\sigma}{\sqrt{n}} = \dfrac{1362.452}{\sqrt{30}} = 248.749$.

We expect the two results to be close when the sample size is small compared to the population size, that is, when it is 5% or less of the population size.   In this instance, the sample size is 30 and the population size is 50, so the sample size is 60% of the population size.   This causes the first square root in Equation 7.1 to be

Copyright © 2012 Pearson Education, Inc. Publishing as Addison-Wesley.

considerably less than 1, resulting in a large discrepancy between the two results.

(d) From Equation 7.1, $\sigma_{\bar{x}} = \sqrt{\dfrac{N-n}{N-1}}\dfrac{\sigma}{\sqrt{n}} = \sqrt{\dfrac{50-2}{50-1}}\dfrac{1362.452}{\sqrt{2}} = 953.518$.

From Equation 7.2, $\sigma_{\bar{x}} = \dfrac{\sigma}{\sqrt{n}} = \dfrac{1362.452}{\sqrt{2}} = 963.399$.

We expect the two results to be close when the sample size is small compared to the population size, that is, when it is 5% or less of the population size. In this instance, the sample size is 2 and the population size is 50, so the sample size is 4% of the population size. This causes the first square root in Equation 7.1 to be quite close to 1, resulting in only a small discrepancy between the two results.

(e) In Excel, in cells C1 to C50, enter the numbers 1 to 50. In cell D1, enter the expression =SQRT((50-C1)/(50-1))*1362.452/SQRT(C1). In Cell E1, enter the expression =1362.452/SQRT(C1). Copy cells D1 and E1 to the clipboard and then paste them into cells D2 through D50. This will produce your table in three columns. To save space, we show the table below using six columns.

n	Eq. 7.1	Eq. 7.2	n	Eq. 7.1	Eq. 7.2
1	1362.45	1362.45	26	187.00	267.20
2	953.52	963.40	27	179.64	262.20
3	770.39	786.61	28	172.53	257.48
4	660.04	681.23	29	165.63	253.00
5	583.91	609.31	30	158.92	248.75
6	527.08	556.22	31	152.38	244.70
7	482.40	514.96	32	145.98	240.85
8	445.97	481.70	33	139.70	237.17
9	415.43	454.15	34	133.52	233.66
10	389.27	430.85	35	127.42	230.30
11	366.49	410.79	36	121.38	227.08
12	346.36	393.31	37	115.37	223.99
13	328.36	377.88	38	109.38	221.02
14	312.11	364.13	39	103.37	218.17
15	297.31	351.78	40	97.32	215.42
16	283.73	340.61	41	91.19	212.78
17	271.18	330.44	42	84.95	210.23
18	259.51	321.13	43	78.53	207.77
19	248.61	312.57	44	71.87	205.40
20	238.38	304.65	45	64.88	203.10
21	228.72	297.31	46	57.40	200.88
22	219.58	290.48	47	49.17	198.73
23	210.88	284.09	48	39.73	196.65
24	202.58	278.11	49	27.81	194.64
25	194.64	272.49	50	0.00	192.68

When n = 2, n is less than 5% of N and the error is about 1%. When n is 10 (20% of N), the error is over 10%. When n is 25 (50% of N), the error is about 40%. We see that the larger n is relative to N, the greater the difference between the correct Equation 7.1 and the approximating Equation 7.2.

Copyright © 2012 Pearson Education, Inc. Publishing as Addison-Wesley.

7.55 (a) Yes. The mean of the sampling distribution of $\bar{x}$ is always μ. A demonstration of this is given in Example 7.4 and in Exercises 7.33 through 7.40.

  (b) No. For example, in Example 7.4, the population consists of five observations (76, 78, 79, 81, and 86). The population median is 79. For samples of size 2, the median and mean are identical and therefore have the same sampling distribution. The mean of the sampling distribution of the median equals the mean of the sampling distribution of the mean, which is shown to be 80 in the example. Since this is not equal to the population median, it is clear that, in general, the sample median is not an unbiased estimator of the population median.

7.57 (a) $\sigma_{\bar{x}} = \sqrt{\dfrac{N-n}{N-1}}\dfrac{\sigma}{\sqrt{n}} = \sqrt{\dfrac{5-1}{5-1}}\dfrac{3.41}{\sqrt{1}} = 3.41$ $\qquad$ $\sigma_{\bar{x}} = \dfrac{\sigma}{\sqrt{n}} = \dfrac{3.41}{\sqrt{1}} = 3.41$ $\qquad$ 20%

$\sigma_{\bar{x}} = \sqrt{\dfrac{N-n}{N-1}}\dfrac{\sigma}{\sqrt{n}} = \sqrt{\dfrac{5-2}{5-1}}\dfrac{3.41}{\sqrt{2}} = 2.09$ $\qquad$ $\sigma_{\bar{x}} = \dfrac{\sigma}{\sqrt{n}} = \dfrac{3.41}{\sqrt{2}} = 2.41$ $\qquad$ 40%

$\sigma_{\bar{x}} = \sqrt{\dfrac{N-n}{N-1}}\dfrac{\sigma}{\sqrt{n}} = \sqrt{\dfrac{5-3}{5-1}}\dfrac{3.41}{\sqrt{3}} = 1.39$ $\qquad$ $\sigma_{\bar{x}} = \dfrac{\sigma}{\sqrt{n}} = \dfrac{3.41}{\sqrt{3}} = 1.97$ $\qquad$ 60%

$\sigma_{\bar{x}} = \sqrt{\dfrac{N-n}{N-1}}\dfrac{\sigma}{\sqrt{n}} = \sqrt{\dfrac{5-4}{5-1}}\dfrac{3.41}{\sqrt{4}} = 0.85$ $\qquad$ $\sigma_{\bar{x}} = \dfrac{\sigma}{\sqrt{n}} = \dfrac{3.41}{\sqrt{4}} = 1.70$ $\qquad$ 80%

$\sigma_{\bar{x}} = \sqrt{\dfrac{N-n}{N-1}}\dfrac{\sigma}{\sqrt{n}} = \sqrt{\dfrac{5-5}{5-1}}\dfrac{3.41}{\sqrt{5}} = 0.00$ $\qquad$ $\sigma_{\bar{x}} = \dfrac{\sigma}{\sqrt{n}} = \dfrac{3.41}{\sqrt{5}} = 1.52$ $\qquad$ 100%

  (b) Equation 7.2 yields poor results when n = 2, 3, 4, or 5 because n is much larger than 5% of N. Equation 7.2 always yields the same result as Equation 7.1 when n = 1 [See Exercise 7.56 (b)], even though n is 20% of N in this example.

7.59 (a) Your answer will very likely be similar, but different from the one below, which we present for illustration. A histogram of the sampling

distribution of $\bar{x}$ for our samples of size 4 is shown.

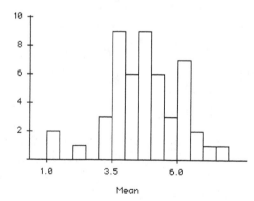

  (b) The mean of the population is

$\mu = \Sigma x / N = (0 + 1 + 2 + 3 + 4 + 5 + 6 + 7 + 8 + 9)/10 = 4.5$

The standard deviation of the population is

Copyright © 2012 Pearson Education, Inc. Publishing as Addison-Wesley.

$$\sigma = \sqrt{\frac{\Sigma(x_i - \mu)^2}{N}} = \sqrt{\frac{82.5}{10}} = 2.87 \;.$$

Thus, for samples of size 4 with replacement, the mean of $\bar{x}$ is 4.5 and

the standard deviation of $\bar{x}$ is $2.87/\sqrt{4} = 1.435$. We would expect the mean and standard deviation of the 50 sample means obtained in part (a) to be roughly 4.5 and 1.43 respectively.

(b) For the 50 samples we obtained, the mean was 4.62 and the standard deviation was 1.355.

(c) The answers in part (d) differ from the theoretical values given in part (c) as a result of random error or sampling variability.

7.61 (a) Theoretically, the mean of all possible sample means is the mean of the population, 8.7, and the standard deviation is $8.7/\sqrt{10} = 2.75$.

(b) We have Minitab take 1000 random samples of size n = 10 from an exponentially distributed population with mean 8.7 by choosing **Calc ▶**

Random Data ▶ Exponential..., entering <u>1000</u> in the **Generate rows of data** text box, entering <u>C1-C10</u> in the **Store in column(s)** text box, entering <u>8.7</u> in the **Mean** text box, and clicking **OK**. These commands tell Minitab to place a total of 10000 observations from the population into columns C1-C10, with 1000 observations in each column. Then, our first random sample of size n = 10 is the first row of columns C1-C10, our second random sample of size n = 10 is the second row of columns C1-C10, and so on.

(b) We compute the sample mean of each of the 1000 samples by choosing **Calc**

▶ Row statistics..., clicking on the **Mean** button, typing <u>C1-C10</u> in the **Input variables** text box and <u>XBAR</u> in the **Store result in** text box, and clicking **OK**. This command instructs Minitab to compute the means of the 1000 rows of C1-C10 and to place those means in a column named XBAR. Thus, the 2000 -values are now in XBAR. (We will not print these values.)

(c) We would expect the mean of the 1000 sample means to be roughly 8.7 and the standard deviation of the sample means to be about 2.75 since we are taking a sample of 1000 means from a theoretical sampling distribution with mean and standard deviation given in part (a).

(e) The mean is obtained by choosing **Calc ▶ Column statistics...**, clicking on the **Mean** button, selecting XBAR in the **Input variable** text box, and clicking **OK**. The result is

Mean of XBAR = 8.7252

Similarly, the standard deviation is obtained by choosing **Calc ▶ Column statistics...**, clicking on the **Standard deviation** button, selecting XBAR in the **Input variable** text box, and clicking **OK**. The result is

Standard deviation of XBAR = 2.7595.

Thus, the mean of our 1000 means was 8.7252 and the standard deviation was 2.7595.

(f) The answers in part (e) differ from the theoretical values given in part (d) as a result of random error or sampling variability.

Copyright © 2012 Pearson Education, Inc. Publishing as Addison-Wesley.

Exercises 7.3

7.63 (a) The sampling distribution of the mean is approximately normally distributed with mean

μ = 100 and standard deviation $\sigma_{\bar{x}} = 28/\sqrt{49} = 4$.

(b) No assumptions were made about the distribution of the population.

(c) Part (a) cannot be answered if the sample size is n = 16. Since the distribution of the population is not specified, we need a sample size of at least 30 to apply Key Fact 7.6.

7.65 (a) The probability distribution of $\bar{x}$ is normal with mean μ and standard deviation $\sigma_{\bar{x}} = \sigma/\sqrt{n}$.

(b) The answer to part (a) does not depend on how large the sample size is because the population being sampled is normally distributed.

(c) The mean of $\bar{x}$ is μ; its standard deviation is $\sigma_{\bar{x}} = \sigma/\sqrt{n}$.

(d) No. The mean and standard deviation of the sampling distribution of the sample mean are always as given in part (c) regardless of the distribution of the variable under consideration.

7.67 (a) All four graphs are centered at the same place because the population mean of $\bar{x}$ is μ and because normal curves are centered at their μ-parameter.

(b) Since $\sigma_{\bar{x}} = \sigma/\sqrt{n}$, we see that $\sigma_{\bar{x}}$ decreases as *n* increases. This results in a diminishing of the spread because the spread of a distribution is determined by its σ-parameter. As a consequence, we see that the larger the sample size, the greater the likelihood for small sampling error.

(c) The graphs in Figure 7.6(a) are bell-shaped because, for normally distributed populations, the random variable $\bar{x}$ is always normally distributed regardless of the sample size.

(d) The graphs in Figures 7.6(b) and 7.6(c) become bell-shaped as the sample size increases because of the central limit theorem; the probability distribution of $\bar{x}$ tends to a normal distribution as the sample size increases.

7.69 (a) Because the weights themselves are normally distributed, the sampling distribution for means of samples of size 3 will also be normal and will have mean $\mu_{\bar{x}} = \mu = 1.40\,kg$ and standard deviation

$\sigma_{\bar{x}} = 0.11/\sqrt{3} = 0.064$. Thus, the distribution of the possible sample means for samples of three brain weights will be normal with mean 1.40 kg and standard deviation 0.064 kg.

(b) Because the weights themselves are normally distributed, the sampling distribution for means of samples of size 12 will also be normal and will have mean $\mu_{\bar{x}} = \mu = 1.40\,kg$ and standard deviation

$\sigma_{\bar{x}} = 0.11/\sqrt{12} = 0.032$. Thus, the distribution of the possible sample means for samples of twelve brain weights will be normal with mean 1.40 kg and standard deviation 0.032 kg.

Copyright © 2012 Pearson Education, Inc. Publishing as Addison-Wesley.

(c) To facilitate the comparison of the three graphs, we have overlaid them on one set of axes.

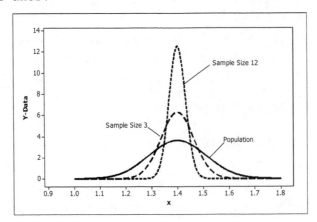

(d) $\sigma_{\bar{x}} = 0.11/\sqrt{3} = 0.064$; we want $P(1.30 \leq \bar{x} \leq 1.50)$.

z-score computations: Area less than z:

$$\bar{x} = 1.30 \rightarrow z = \frac{1.30 - 1.40}{0.064} = -1.56 \qquad\qquad 0.0594$$

$$\bar{x} = 1.50 \rightarrow z = \frac{1.50 - 1.40}{0.064} = 1.56 \qquad\qquad 0.9406$$

Total area = 0.9406 - 0.0594 = 0.8812

Note: If you round the standard deviation of the sample means differently, you may reach a slightly different answer.

Thus, 88.12% of samples of three Swedish men will have mean brain weights within 0.1 kg of the population mean brain weight of 1.40 kg.

(e) $\sigma_{\bar{x}} = 0.11/\sqrt{12} = 0.032$; we want $P(1.30 \leq \bar{x} \leq 1.50)$.

z-score computations: Area less than z:

$$\bar{x} = 1.30 \rightarrow z = \frac{1.30 - 1.40}{0.032} = -3.13 \qquad\qquad 0.0009$$

$$\bar{x} = 1.50 \rightarrow z = \frac{1.50 - 1.40}{0.032} = 3.13 \qquad\qquad 0.9991$$

Total area = 0.9991 - 0.0009 = 0.9982
Note: If you round the standard deviation of the sample means differently, you may reach a slightly different answer.

Thus, 99.82% of samples of twelve Swedish men will have mean brain weights within 0.1 kg of the population mean brain weight of 1.40 kg.

7.71 (a) The sampling distribution of the sample mean for samples of size 64 will be approximately normal with a mean of $49.0 thousand and a standard deviation of $\sigma_{\bar{x}} = 9.2/\sqrt{64} = 1.15$. Thus, if all possible sample means for samples of size 64 were found, their distribution would be approximately normal with mean $49.0 thousand and standard deviation $1,150.

Copyright © 2012 Pearson Education, Inc. Publishing as Addison-Wesley.

(b) The sampling distribution of the sample mean for samples of size 256 will be approximately normal with a mean of $49.0 thousand and a standard deviation of $\sigma_{\bar{x}} = 9.2/\sqrt{256} = 0.575$. Thus, if all possible sample means for samples of size 256 were found, their distribution would be approximately normal with mean $49.0 thousand and standard deviation $575.

(c) No. These results follow from the central limit theorem, which implies that for large samples (n ≥ 30), the distribution of the sample mean will be approximately normal regardless of the distribution of the original variable.

(d) $\sigma_{\bar{x}} = 9.2/\sqrt{64} = 1.15$; we want $P(48.0 \leq \bar{x} \leq 50.0)$.

z-score computations: Area less than z:

$$\bar{x} = 48.0 \rightarrow z = \frac{48.0 - 49.0}{1.15} = -0.87 \qquad 0.1922$$

$$\bar{x} = 50.0 \rightarrow z = \frac{50.0 - 49.0}{1.15} = 0.87 \qquad 0.8078$$

Total area = 0.8078 - 0.1922 = 0.6156

Thus, 61.56% of samples of size 64 will have a mean that is within $1000 of the population mean, i.e., 61.56% of the samples will have a sampling error less than $1000 for samples of size 64.

(e) $\sigma_{\bar{x}} = 9.2/\sqrt{256} = 0.575$; we want $P(48.0 \leq \bar{x} \leq 50.0)$.

z-score computations: Area less than z:

$$\bar{x} = 48.0 \rightarrow z = \frac{48.0 - 49.0}{0.575} = -1.74 \qquad 0.0409$$

$$\bar{x} = 50.0 \rightarrow z = \frac{50.0 - 49.0}{0.575} = 1.74 \qquad 0.9591$$

Total area = 0.9591 - 0.0409 = 0.9182

Thus, 91.82% of samples of size 256 will have a mean that is within $1000 of the population mean, i.e., 91.82% of the samples will have a sampling error less than $1000 for samples of size 256.

7.73 (a) The sampling distribution of the sample mean for n = 80 will be approximately normal with some mean μ and standard deviation $\sigma_{\bar{x}} = 8.3/\sqrt{80} = 0.928$.

(b) No. The sample size of 80 is large enough so that the Central Limit Theorem applies and the sampling distribution will be approximately normal.

(c) We want $P(\mu - 2 \leq \bar{x} \leq \mu + 2)$
z-score computations: Area less than z:

$$\bar{x} = \mu - 2 \rightarrow z = \frac{(\mu - 2) - \mu}{0.928} = -2.16 \qquad 0.0154$$

$$\bar{x} = \mu + 2 \rightarrow z = \frac{(\mu + 2) - \mu}{0.928} = 2.16 \qquad 0.9846$$

Total area = 0.9846 - 0.0154 = 0.9692

Thus, the probability is approximately 0.9692 that the sampling error

Copyright © 2012 Pearson Education, Inc. Publishing as Addison-Wesley.

made in estimating the population mean length of stay on the intervention ward by the mean length of stay of a sample of 80 patients will be at most 2 days.

7.75 $\sigma_{\overline{x}} = 40/\sqrt{250} = 2.53; P(|\overline{x} - \mu| \le 5)$

z-score computations: Area less than z:

$$\overline{x} = \mu - 5 \rightarrow z = \frac{(\mu - 5) - \mu}{2.53} = -1.98 \qquad 0.0239$$

$$\overline{x} = \mu + 5 \rightarrow z = \frac{(\mu + 5) - \mu}{2.53} = 1.98 \qquad 0.9761$$

Required Area = 0.9761 - 0.0239 = 0.9522

Thus, there is a 0.9522 probability that the contractor's estimate will be within 5 months of the true mean.

7.77 In solving this problem, we have to assume that the distribution of calcium intakes of adults with incomes below the poverty level is approximately normally distributed. Then, the distribution of the sample means will be approximately normally distributed. The distribution of the sample means have a mean 1000 mg and standard deviation of $\sigma_{\overline{x}} = \sigma/\sqrt{n} = 188/\sqrt{18} = 44.312$.

We want $P(\overline{x} \le 947.4)$:
z-score computation: Area less than z:

$$\overline{x} = 947.4 \rightarrow z = \frac{947.4 - 1000}{44.312} = -1.19 \qquad 0.1170$$

Total area = 0.1170

11.70% of all samples of size 18 of adults with incomes below the poverty level will have mean calcium intakes of at most 947.4 mg.

7.79 In solving this problem, we have to assume that the distribution of post-work hear rate for casting workers is approximately normally distributed. Then, the distribution of the sample means will be approximately normally distributed. The distribution of the sample means have a mean 72 bpm and standard deviation of $\sigma_{\overline{x}} = \sigma/\sqrt{n} = 11.2/\sqrt{29} = 2.080$. We want $P(\overline{x} > 78.3)$:

z-score computation: Area less than z:

$$\overline{x} = 78.3 \rightarrow z = \frac{78.3 - 72}{2.080} = 3.03 \qquad 0.9988$$

Total area = 1 - 0.9988 = 0.0012.

The probability is 0.0012 that samples of 29 casting workers will have a mean post=work hear rate exceeding 78.3 bpm.

7.81 (a) 68.26   (b)   95.44   (c)   99.74   (d)   100(1 - α)

7.83 (a) No. The original population is quite skewed. A sample size of 4 is quite small, so we would expect that a histogram of the means of such a sample would still exhibit some degree of skewness.

(b) We'll use Excel to carry out the simulation. First enter the values 1 through 7 in cells A1 to A7, then enter the frequencies 19.4 through 1.6 from the table in Example 7.9 into cells C1 through C7. In C8, enter the expression =Sum(C1:C7). Then in cell B1, enter the expression =C1/C8, copy this cell to the clipboard and then paste it into cells B2 through B7. Columns A and B now contain the probability distribution of x, the household size. Now from the Tools menu, select

Copyright © 2012 Pearson Education, Inc. Publishing as Addison-Wesley.

Data Analysis ▶ Random Number Generation. Enter 4 in the **Number of Variables** text box, enter 1000 in the **Number of Random Numbers** text box, choose **Discrete** in the **Distribution** box, enter the range A1:B7 in the **Value and Probability Input Range** text box, and enter A11 in the **Output Range** box. Click on **OK.** You will get 4 columns of 100 numbers between 1 and 7 in columns A, B, C, and D, starting in row 11 and ending in row 1010. In cell E11, enter the expression =AVERAGE(A11:D11). Copy this cell to the clipboard and then paste it into the cells E12 through E1010. Type the label **Mean** in cell E10. Use the mouse to highlight cells E10 though E1010. Now from the **DDXL** menu, select **Charts and Plots.** In the **Function Type** box, select **Histogram,** Click on **Mean** in the **Names and Columns** box and enter it in the **Quantitative Variables** box. Then click **OK.** Our result follows. Yours is very likely to be different. Our histogram is slightly skewed to the right.

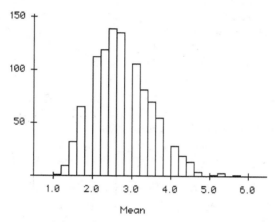

(c)   We would expect this histogram to be closer to bell-shaped than the one in part (a), but 10 is still a relatively small sample size, so there may still be a trace of skewness. Using the same procedure as in part (b), but with 10 variables, our result below is more symmetrical, but we definitely did not expect the one tall spike. Note that the scale is different from the first histogram. These means are more closely concentrated.

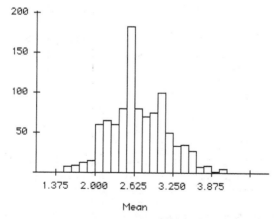

(d)   Since 100 is definitely a large sample size, we would expect the histogram of the sample means to be bell-shaped. Using the same procedure as in part (b), but with 100 variables, our result below is close to bell-shaped and again is more concentrated than the first two.

Copyright © 2012 Pearson Education, Inc. Publishing as Addison-Wesley.

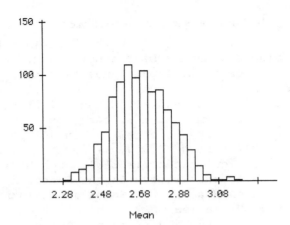

7.85 (a) We will describe a procedure using Minitab to sketch the exponential curve for this population, although the sketch could be done using any spreadsheet program as well. We begin by selecting typical values of x that will eventually be substituted into the exponential function itself to trace out values for y. Values of x selected are the integers 0 through 15. Name two columns in the Data window X and Y and enter the numbers 0, 1, 2, ...,15 in the X column. Then choose **Calc ▶**

Probability Distributions ▶ Exponential..., click on the **Probability density** button, click in the **Mean** text box and type 8.7. Then enter **X** in the **Input column** text box and **Y** in the **Optional storage** text box.

Click **OK**. Now choose **Graph ▶ Scatterplot...**, select the **With connect line** version and click **OK**. Select Y for row 1 of the **Y** column and X for row 1 of the **X** column. To plot a continuous curve, click on the **Data view** button and check only the **Connect line** box. Click **OK** twice. The result is shown below. The exponential distribution is reverse J-shaped.

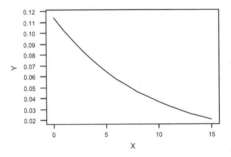

(b) To have Minitab take 1000 random samples of size n = 4 from an exponentially distributed population with mean μ = 8.7, we choose **Calc**

▶ Random Data ▶ Exponential..., enter 1000 in the **Generate rows of data** text box, enter C1-C4 in the **Store in column(s)** text box, enter 8.7 in the **Mean** text box, and click **OK**. These commands tell Minitab to place a total of 4000 observations from the population into columns C1-C4, with 1000 observations in each column. Then, our first random sample of size n = 4 is the first row of columns C1-C4, our second

Copyright © 2012 Pearson Education, Inc. Publishing as Addison-Wesley.

random sample of size n = 4 is the second row of columns C1-C4, and so on.

(c) We compute the sample mean of each of the 1000 samples by choosing **Calc ▶ Row statistics...**, clicking on the **Mean** button, entering <u>C1-C4</u> in the **Input variables** text box and <u>XBAR</u> in the **Store result in** text box, and clicking **OK**. This command instructs Minitab to compute the means of the 1000 rows of C1-C4 and to place those means in XBAR. Thus, the 1000 -values are now in XBAR. (We will not print these values.) Your results will differ from ours.

(d) The mean of the 1000 sample means is obtained by choosing **Calc ▶ Column statistics...**, clicking on the **Mean** button, selecting XBAR in the **Input variable** text box, and clicking **OK**. The result is

    Mean of XBAR = 8.4743

Similarly, the standard deviation of the 1000 sample means is obtained by choosing **Calc ▶ Column statistics...**, clicking on the **Standard deviation** button, selecting XBAR in the **Input variable** text box, and clicking **OK**. The result is

    Standard deviation of XBAR = 4.2498

(e) Theoretically, for all possible sample means for samples of size 4, the mean is the same as the population mean, 8.7, and the standard deviation is $\sigma/\sqrt{n} = 8.7/\sqrt{4} = 4.35$. Both of the simulated values obtained in part (d) are close to these values.

(f) To get a histogram of the 1000 sample means stored in XBAR, choose **Graph ▶ Histogram...**, selecting the **Simple** version and click **OK**. Enter XBAR in the **Graph variables** text box, and click **OK**. The result is shown below at the right. The histogram is not bell-shaped; given the right-skewness of the population, we would expect for samples of only size four, that the distribution of means would be somewhat skewed to the right.

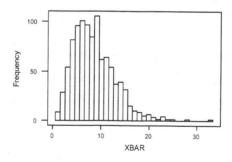

(g) Repeating the process above for samples of size 40 (use columns C1-C40 for the samples), we find that

    Mean of XBAR = 8.7367 and

    Standard deviation of XBAR = 1.4204

In theory, the mean of all possible samples of size 40 is 8.7 and the standard deviation is $\sigma/\sqrt{n} = 8.7/\sqrt{40} = 1.38$. The simulated values are very close to these.

Copyright © 2012 Pearson Education, Inc. Publishing as Addison-Wesley.

A histogram of the means of samples of size 40 is shown following, obtained as above. The simulated distribution of sample means is close to bell-shaped as we would expect for samples over size 30.

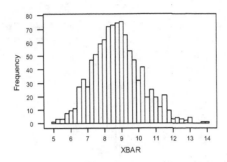

Review Problems for Chapter 7

1. Sampling errors are errors resulting from using a sample to estimate a population characteristic.

2. The sampling distribution of a statistic is the set of all possible observations of the statistic for a sample of a given size. It is important to know the sampling distribution in order to answer questions about the accuracy of estimating a population parameter using the sample statistic.

3. Two other terms are 'sampling distribution of the sample mean' and 'the

   distribution of the variable $\overline{x}$.'

4. The set of possible sample means exhibits less and less variability as the sample size increases, that is, the set becomes more and more clustered about the population mean. This means that as the sample size increases, there is a greater chance that the value of the sample mean from any sample is close to the value of the population mean.

5. (a) The sampling error results from using the sample mean income tax, $\overline{x}$, of the 292,966 tax returns sampled as an estimate of the mean income tax, μ, of all 2005 tax returns.

   (b) The sampling error is $10,319 - $10,407 = -$88.

   (c) No, not necessarily. However, increasing the sample size from 292,966 to 400,000 would increase the likelihood for a smaller sampling error.

   (d) Increase the sample size.

6. (a) $\mu = $108/6 = 18 (thousands)

Copyright © 2012 Pearson Education, Inc. Publishing as Addison-Wesley.

(b)

Sample	Salaries	
A, B, C, D	8, 12, 16, 20	14
A, B, C, E	8, 12, 16, 24	15
A, B, C, F	8, 12, 16, 28	16
A, B, D, E	8, 12, 20, 24	16
A, B, D, F	8, 12, 20, 28	17
A, B, E, F	8, 12, 24, 28	18
A, C, D, E	8, 16, 20, 24	17
A, C, D, F	8, 16, 20, 28	18
A, C, E, F	8, 16, 24, 28	19
A, D, E, F	8, 20, 24, 28	20
B, C, D, E	12, 16, 20, 24	18
B, C, D, F	12, 16, 20, 28	19
B, C, E, F	12, 16, 24, 28	20
B, D, E, F	12, 20, 24, 28	21
C, D, E, F	16, 20, 24, 28	22

(c)

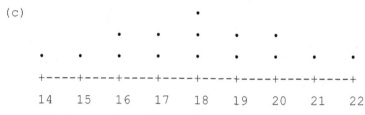

```
                        .
         .    .     .     .     .
  .   .    .     .     .     .     .     .     .
  +----+----+----+----+----+----+----+----+
  14   15   16   17   18   19   20   21   22
```

(d)
$$P(|\bar{x} - \mu| \le 1) = P(\bar{x} = 17) + P(\bar{x} = 18) + P(\bar{x} = 19)$$
$$= 2/15 + 3/15 + 2/15 = 7/15 = 0.4667$$

(e) $\mu_{\bar{x}} = \dfrac{\sum \bar{x}}{N} = \dfrac{270}{15} = 18.0$. The mean of the means of all of the possible samples of size 4 is $18.0 thousand.

(f) Yes. The mean of the sampling distribution of $\bar{x}$ is always the same as the mean of the population, which is $18 thousand in this case.

7. (a) The population consists of all new car sales in the U.S. in 2007. The variable under consideration is the amount spent for a new car.

(b) $\mu_{\bar{x}} = \$28,200; \sigma_{\bar{x}} = \$10200/\sqrt{50} = \$1442.50$

(c) $\mu_{\bar{x}} = \$28,200; \sigma_{\bar{x}} = \$10200/\sqrt{100} = \$1020.00$

(d) The value of $\sigma_{\bar{x}}$ will be smaller than $1020 because $\sigma_{\bar{x}} = \sigma/\sqrt{n}$. Thus, the larger the sample size, the smaller the value of $\sigma_{\bar{x}}$.

8. (a) False. By the central limit theorem, the random variable $\bar{x}$ is approximately normally distributed. Furthermore, $\mu_{\bar{x}} = \mu = 45$, and

$\sigma_{\bar{x}} = \sigma/\sqrt{n} = 7/\sqrt{196} = 0.5$. Thus, P($31 \le \bar{x} \le 59$) equals the area under

Copyright © 2012 Pearson Education, Inc. Publishing as Addison-Wesley.

the normal curve with parameters $\mu_{\bar{x}} = 45$ and $\sigma_{\bar{x}} = 0.5$ that lies between 31 and 59. Applying the usual techniques, we find that area to be 1.0000 to four decimal places. Hence, there is almost a 100% chance that the mean of the sample will be between 31 and 59.

(b) This is not possible to tell, since we do not know the distribution of the population.

(c) True. Referring to part (a), we see that $P(44 \le \bar{x} \le 46)$ equals the area under the normal curve with parameters $\mu_{\bar{x}} = 45$ and $\sigma_{\bar{x}} = 0.5$ that lies between 44 and 46. Applying the usual techniques, we find that area to be 0.9544. Hence, there is approximately a 95.44% chance that the mean of the sample will be between 44 and 46.

9. (a) False. Since the population is normally distributed, so is the random variable $\bar{x}$. Furthermore, $\mu_{\bar{x}} = \mu = 45$ and $\sigma_{\bar{x}} = \sigma / \sqrt{n} = 7 / \sqrt{196} = 0.5$. Hence, as in Problem 8(a), we find that there is almost a 100% chance that the mean of the sample will be between 31 and 59.

(b) True. Since the population is normally distributed, percentages for the population are equal to areas under the normal curve with parameters μ = 45 and σ = 7. Applying the usual techniques, we find that the area under that normal curve between 31 and 59 is 0.9544.

(c) True. From part (a), we see that the random variable $\bar{x}$ is normally distributed with $\mu_{\bar{x}} = 45$ and $\sigma_{\bar{x}} = 0.5$. Applying the usual techniques, we find that area to be 0.9544. Hence, we find that there is a 95.44% chance that the mean of the sample will be between 44 and 46.

10. (a) $\mu = 40 \ and \ \sigma = 12$

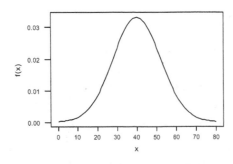

(b) normal distribution with $\mu_{\bar{x}} = 40 \ and \ \sigma_{\bar{x}} = 12 / \sqrt{4} = 6$

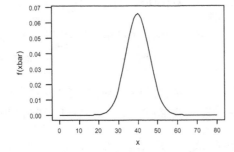

Copyright © 2012 Pearson Education, Inc. Publishing as Addison-Wesley.

(c) normal distribution with $\mu_{\bar{x}} = 40\ and\ \sigma_{\bar{x}} = 12/\sqrt{9} = 4$

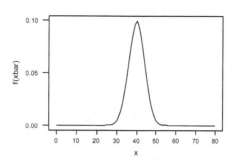

11.

$n = 4 : \mu_{\bar{x}} = 40\ and\ \sigma_{\bar{x}} = 12/\sqrt{4} = 6$

(a) P(31 ≤ x̄ ≤ 49):

z-score computations: Area less than z:

$\bar{x} = 31 \rightarrow z = \dfrac{31-40}{6} = -1.50$ 0.0668

$\bar{x} = 49 \rightarrow z = \dfrac{49-40}{6} = 1.50$ 0.9332

Total area = 0.9332 - 0.0668 = 0.8664, or 86.64%.

(b) The probability is 0.8664 that a sample of size four will have a mean within 9 mm of the population mean of 40 mm.

(c) There is an 86.64% chance that the sampling error when using the mean

krill length x̄ of the four krill will be less than 9 mm.

(d) $n = 9 : \mu_{\bar{x}} = 40\ and\ \sigma_{\bar{x}} = 12/\sqrt{9} = 4$

z-score computations: Area less than z:

$\bar{x} = 31 \rightarrow z = \dfrac{31-40}{4} = -2.25$ 0.0122

$\bar{x} = 49 \rightarrow z = \dfrac{49-40}{4} = 2.25$ 0.9878

Total area = 0.9878 - 0.0122 = 0.9756 or 97.56%

The probability is 0.9756 that a sample of size 9 will have a mean within 9 mm of the population mean of 40 mm.

There is an 97.56% chance that the sampling error when using the mean

krill length x̄ of the nine krill will be less than 9 mm.

12.   (a) For a normally distributed population, the random variable x̄ is normally distributed, regardless of the sample size. Also, we know that $\mu_{\bar{x}} = \mu$. Consequently, since the normal curve for a normally

Copyright © 2012 Pearson Education, Inc. Publishing as Addison-Wesley.

distributed population or random variable is centered at its μ-parameter, all three curves are centered at the same place.

(b) Curve B corresponds to the larger sample size. Since $\sigma_{\bar{x}} = \sigma/\sqrt{n}$, the larger the sample size, the smaller the value of σ and, hence, the smaller the spread of the normal curve for $\bar{x}$. Thus, Curve B, which has the smaller spread, corresponds to the larger sample size.

(c) The spread of each curve is different because $\sigma_{\bar{x}} = \sigma/\sqrt{n}$ and the spread of a normal curve is determined by $\sigma_{\bar{x}}$, Thus, different sample sizes result in normal curves with different spreads.

(d) Curve B corresponds to the sample size that will tend to produce less sampling error. The smaller the value of $\sigma_{\bar{x}}$, the smaller the sampling error tends to be.

(e) When x is normally distributed, $\bar{x}$ always has a normal distribution as well.

13. (a) Since 60 is a fairly large sample size, the distribution of the sample mean will be approximately normal with a mean of 4.60 mmol/l and a standard deviation of $\sigma_{\bar{x}} = \sigma/\sqrt{n} = 0.16/\sqrt{60} = 0.0207$.

(b) Since 120 is a large sample size, the distribution of the sample mean will be approximately normal with a mean of 4.60 mmol/l and a standard deviation of $\sigma_{\bar{x}} = \sigma/\sqrt{n} = 0.16/\sqrt{120} = 0.0146$.

(c) No. The Central Limit Theorem ensures that when n is large, the distribution of the sample mean will be approximately normal regardless of whether the blood glucose levels themselves are normally distributed.

14. (a) $n = 500: \mu_{\bar{x}} = \mu$ and $\sigma_{\bar{x}} = 50,900/\sqrt{500} = 2276.32$

P(μ - 2,000 ≤ $\bar{x}$ ≤ μ + 2,000):

z-score computations: Area less than z:

$$\bar{x} = \mu - 2000 \rightarrow z = \frac{(\mu - 2000) - \mu}{2276.32} = -0.88$$ 0.1894

$$\bar{x} = \mu + 2000 \rightarrow z = \frac{(\mu + 2000) - \mu}{2276.32} = 0.88$$ 0.8106

Total area = 0.8106 - 0.1894 = 0.6212

(b) To answer part (a), it is not necessary to assume that the population is normally distributed because the sample size is large and, therefore, $\bar{x}$ is approximately normally distributed, regardless of the distribution of the population of life-insurance amounts.

If the sample size were 20 instead of 500, it would be necessary to assume normality because the sample size is small.

(c) $n = 5000: \mu_{\bar{x}} = \mu$ and $\sigma_{\bar{x}} = 50,900/\sqrt{5000} = 719.83$

Copyright © 2012 Pearson Education, Inc. Publishing as Addison-Wesley.

z-score computations: Area less than z:

$$\bar{x} = \mu - 2000 \rightarrow z = \frac{(\mu - 2000) - \mu}{719.83} = -2.78$$ 0.0027

$$\bar{x} = \mu + 2000 \rightarrow z = \frac{(\mu + 2000) - \mu}{719.83} = 2.78$$ 0.9973

Total area = 0.9973 - 0.0027 = 0.9946

15.   (a) $P(\bar{x} \leq 4.5)$:

z-score computation: Area less than z:

$$\bar{x} = 4.5 \rightarrow z = \frac{4.5 - 5}{0.5} = -1.00$$ 0.1587

Total area = 0.1587

If the paint lasts 4.5 years, one would not consider this to be substantial evidence against the manufacturer's claim that the paint will last an average of five years.

Assuming the manufacturer's claim is correct, the probability is 0.1587 that the paint will last 4.5 years or less on a (randomly selected) house painted with the paint. In other words, there is a (fairly high) 15.87% chance that the paint would last 4.5 years or less, even if the manufacturer's claim is correct.

(b) $P(\bar{x} \leq 4.5)$:

$$n = 10 : \mu_{\bar{x}} = 5 \text{ and } \sigma_{\bar{x}} = 0.5 / \sqrt{10} = 0.158$$

z-score computation: Area less than z:

$$\bar{x} = 4.5 \rightarrow z = \frac{4.5 - 5}{0.158} = -3.16$$ 0.0008

Total area = 0.0008

For 10 houses, if the paint lasts an average of 4.5 years, I would consider this to be substantial evidence against the manufacturer's claim that the paint will last an average of five years.

Assuming the manufacturer's claim is correct, the probability is 0.0008 that the paint will last an average of 4.5 years or less for 10 (randomly selected) houses painted with the paint. In other words, there is less than a 0.1% chance that that would occur, if the manufacturer's claim is correct.

(c) $P(\bar{x} \leq 4.9)$:

z-score computation: Area less than z:

$$\bar{x} = 4.9 \rightarrow z = \frac{4.9 - 5}{0.158} = -0.63$$ 0.2643

Total area = 0.2643

For 10 houses, if the paint lasts an average of 4.9 years, I would not consider this to be substantial evidence against the manufacturer's

Copyright © 2012 Pearson Education, Inc. Publishing as Addison-Wesley.

claim that the paint will last an average of five years.
Assuming the manufacturer's claim is correct, the probability is 0.2643
that the paint will last an average of 4.9 years or less for 10
(randomly selected) houses painted with the paint. In other words,
there is a (fairly high) 26.43% chance that that would occur, even if
the manufacturer's claim is correct.

16. (a) The sample size is large enough that the distribution of the sample
means will be approximately normal with mean 6.83 and standard
deviation of $\sigma_{\bar{x}} = \sigma / \sqrt{n} = 4.28 / \sqrt{100} = 0.428$.

We want $P(\bar{x} > 7.5)$:

z-score computation: Area less than z:

$\bar{x} = 7.5 \rightarrow z = \dfrac{7.5 - 6.83}{0.428} = 1.57$ 0.9418

Total area = 1 – 0.9418 = 0.0582.

5.82% of all samples of size 100 days during the decade in question
will have a mean degree of cloudiness exceeding 7.5.

(b) For samples of size 5, it is not reasonable to use a normal
distribution to obtain the percentage required in part (a). From the
frequency distribution, we can see that the distribution of degree of
cloudiness is definitely not normally distributed. Therefore a sample
of at least 30 is required before we can assume that the distribution
of the sample means is approximately normally distributed.

17. (a) We have Minitab take 1000 random samples of size n = 4 from a normally
distributed population with mean 584 and standard deviation 151 by
choosing **Calc ▶ Random Data ▶ Normal...**, entering 1000 in the
Generate rows of data text box, entering C1–C4 in the **Store in
column(s)** text box, entering 584 in the **Mean** text box and 151 in the
Standard deviation text box, and clicking **OK**.

(b) We compute the sample mean of each of the 1000 samples by choosing **Calc
▶ Row statistics...**, clicking on the **Mean** button, typing C1–C4 in the
Input variables text box and XBAR in the **Store result in** text box, and
clicking **OK**. (We will not print these values.)

(c) The mean is obtained by choosing **Calc ▶ Column statistics...**, clicking
on the **Mean** button, selecting XBAR in the **Input variable** text box, and
clicking **OK**. Our result is
          Mean of XBAR = 580.266

Similarly, the standard deviation is obtained by choosing **Calc ▶ Column
statistics...**, clicking on the **Standard deviation** button, selecting
XBAR in the **Input variable** text box, and clicking **OK**. Our result is
          Standard deviation of XBAR = 76.8275
To get a histogram of the 1000 sample means stored in XBAR, choose
Graph ▶ Histogram..., select the **Simple** version and click **OK**. Enter
XBAR in the **Graph variables** text box, and click **OK**. The result
follows.

Copyright © 2012 Pearson Education, Inc. Publishing as Addison-Wesley.

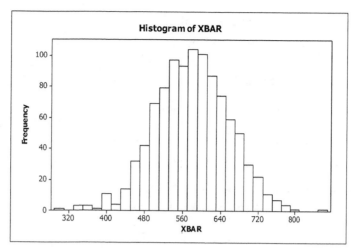

Histogram of XBAR

(d) Theoretically, the distribution of all possible sample means for samples of size four from this normal population should have mean 584, standard deviation $\sigma_{\bar{x}} = \sigma / \sqrt{n} = 151/\sqrt{4} = 75.5$, and a normal distribution.

(e) The histogram is close to bell-shaped, is centered near 584, and most of the data lies within three standard deviations (3 x 75.5 = 226.5) of 584, i.e. between 357.5 and 810.5. The mean of the 1000 sample means is 580.266, very close to 584, and the standard deviation of the sample means is 76.8275, very close to 75.5.

18. (a) A uniform distribution between 0 and 1:

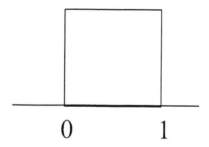

0   1

(b) We have Minitab take 2000 random samples of size n = 2 from a uniformly distributed population between 0 and 1 by choosing **Calc ▶ Random Data**

▶ **Uniform...**, entering <u>2000</u> in the **Generate rows of data** text box, entering <u>C1-C2</u> in the **Store in column(s)** text box, entering <u>0.0</u> in the **Lower endpoint** text box and <u>1.0</u> in the **Upper endpoint** text box, and clicking **OK**.

(c) We compute the sample mean of each of the 1000 samples by choosing **Calc ▶ Row statistics...**, clicking on the **Mean** button, typing <u>C1-C2</u> in the **Input variables** text box and <u>XBAR</u> in the **Store result in** text box, and clicking **OK**. (We will not print these values.)

(d) The mean of the sample means is obtained by choosing **Calc ▶ Column statistics...**, clicking on the **Mean** button, selecting XBAR in the **Input variable** text box, and clicking **OK**. Our result is

   Mean of XBAR = 0.50454

Similarly, the standard deviation is obtained by choosing **Calc ▶ Column statistics...**, clicking on the **Standard deviation** button, selecting

Copyright © 2012 Pearson Education, Inc. Publishing as Addison-Wesley.

XBAR in the **Input variable** text box, and clicking **OK**. Our result is

Standard deviation of XBAR = 0.20545

(e) Theoretically, the distribution of all possible sample means for samples of size two from this uniform normal population should have mean

$(0 + 1)/2 = .5$ and standard deviation $\sigma/\sqrt{n} = ((1-0)/\sqrt{12})/\sqrt{2} = 0.2041$. Minitab simulation results are very close to these theoretical values.

(f) To get a histogram of the 1000 sample means stored in XBAR, choose

Graph ▶ Histogram..., select the **Simple** version and click **OK**. Enter XBAR in the **Graph variables** text box, and click **OK**. The result is shown following.

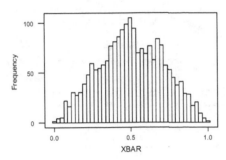

The histogram is more triangle-shaped than bell-shaped. Since the population is far from normal and the sample size is only two, we would not expect the sampling distribution to be bell-shaped. In fact, it can be shown using advanced methods that the sum of two uniformly distributed variables has a triangular distribution.

(g) Repeating the process above, but using C1-C35 for the data in each sample, we obtained

Mean of XBAR = 0.49932

Standard deviation of XBAR = 0.049240

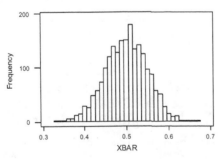

The theoretical distribution of the means has mean 0.5 and standard deviation $\sigma/\sqrt{n} = ((1-0)/\sqrt{12})/\sqrt{35} = 0.0488$.

The Minitab simulation values are very close to these. The histogram for the means of the samples of size 35 is shown. It is much more bell-shaped than for samples of size two, as we would expect since n > 30.

Copyright © 2012 Pearson Education, Inc. Publishing as Addison-Wesley.

Exercises 8.1

8.1 The value of a statistic that is used to estimate a parameter is called a <u>point estimate</u> of the parameter.

8.3 (a) $\bar{x} = \$526538/20 = \$26326.90$

(b) Since some sampling error is expected, it is unlikely that the sample mean will exactly equal the population mean μ.

8.5 (a) The confidence interval will be

$$\bar{x} - 2\sigma/\sqrt{n} \ \ to \ \ \bar{x} + 2\sigma/\sqrt{n}$$
$$26326.9 - 2(8100)/\sqrt{20} \ to \ 26326.9 + 2(8100)/\sqrt{20}$$
$$\$22,704.5 \ to \ \$29,949.3$$

(b) Since we know that 95.44% of all samples of 20 wedding costs have the property that the interval from $\bar{x}$ − \$3622.43 to $\bar{x}$ + \$3622.43 contains μ, we can be 95.44% confident that the interval from \$22707.47 to \$29952.33 contains μ.

(c) We can't be certain that the population mean lies in the interval, but we are 95.44% confident that it does.

8.7 (a) n = 35; a point estimate for the mean fuel tank capacity is

$\bar{x}$ = 664.9/35 = 19.00 gallons. This number is called a point estimate because it consists of a single value.

(b) The confidence interval will be

$$\bar{x} - 2\sigma/\sqrt{n} \ \ to \ \ \bar{x} + 2\sigma/\sqrt{n}$$
$$19.00 - 2(3.50)/\sqrt{35} \ to \ 19.00 + 2(3.50)/\sqrt{35}$$
$$17.82 \ to \ 20.18$$

(c) We could see if a histogram looked bell-shaped or if a normal probability plot produced a relatively straight line.

(d) It is not necessary that the fuel tank capacities be exactly normally distributed because for samples of size 35, the distribution of the sample means will be approximately normally distributed, regardless of the population distribution. Therefore, the confidence interval will be approximately correct.

8.9 (a) Using Minitab to retrieve the data from the WeissStats CD, we obtained the following normal probability plot by choosing **Stat ▶ Basic Statistics ▶ Normality Test**, entering LENGTHS in the **Variable** text box, and clicking **OK**.

Copyright © 2012 Pearson Education, Inc. Publishing as Addison-Wesley.

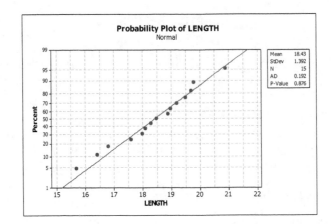

(b) Yes. There do not appear to be any outliers, and the data points fall very close to a straight line.

(c) The sample mean is 18.43 mm. Since $P(\bar{x} - 2\sigma/\sqrt{n} < \mu < \bar{x} + 2\sigma/\sqrt{n}) = 0.9544$, we can be 95.44% confident that the mean μ is somewhere between

$\bar{x} - 2\sigma/\sqrt{n}$ and $\bar{x} + 2\sigma/\sqrt{n}$. Since n = 15, $\bar{x}$ = 18.43, and σ = 1.76, the 95.44% confidence interval is

$$\bar{x} - 2\sigma/\sqrt{n} \text{ to } \bar{x} + 2\sigma/\sqrt{n}$$
$$18.43 - 2(1.76)/\sqrt{15} \text{ to } 18.43 + 2(1.76)/\sqrt{15}$$
$$17.52 \text{ to } 19.34$$

We can be 95.44% confident that the interval 17.52 mm to 19.34 mm contains the value of the mean μ.

(d) Yes. No. 18.14 lies between 17.52 and 19.34, so the interval we obtained in part (c) does contain μ. It is possible that our confidence interval will not contain μ. This should happen only about 4.56% of the time.

8.11 Since $P(\bar{x} - 3\sigma/\sqrt{n} < \mu < \bar{x} + 3\sigma/\sqrt{n}) = 0.9974$, we can be 99.74% confident that the mean μ is somewhere between $\bar{x} - 3\sigma/\sqrt{n}$ and $\bar{x} + 3\sigma/\sqrt{n}$. Since n = 36, $\bar{x}$ = 63.28, and σ = 7.2, the 99.74% confidence interval is (in thousands)

$$\bar{x} - 3\sigma/\sqrt{n} \text{ to } \bar{x} + 3\sigma/\sqrt{n}$$
$$63.28 - 3(7.2)/\sqrt{36} \text{ to } 63.28 + 3(7.2)/\sqrt{36}$$
$$59.68 \text{ to } 66.88$$

We can be 99.74% confident that the interval \$59,680 to \$66,880 contains the value of the mean μ.

Exercises 8.2

8.13 (a) Confidence level = 0.90; α = 0.10

(b) Confidence level = 0.99; α = 0.01

8.15 (a) By saying that a $1-\alpha$ confidence interval is exact, we mean that the true confidence level is equal to $1-\alpha$.

Copyright © 2012 Pearson Education, Inc. Publishing as Addison-Wesley.

(b) By saying that a $1-\alpha$ confidence interval is approximately correct, we mean that the true confidence level is only approximately equal to $1-\alpha$.

8.17 When we use the abbreviation "normal population," we mean that the variable under consideration is normally distributed on the population of interest.

8.19 A statistical procedure is robust if it is insensitive to departures from the assumptions on which it is based.

8.21 (a) The z-interval procedure is reasonable since the population is roughly normal and the sample size is over 15.

(b) The z-interval procedure is not reasonable since the sample size is too small for a population far from normal.

(c) The z-interval procedure is reasonable since the sampling distribution of $\bar{x}$ will be very close to normal for a sample size of 250 even though the population itself is far from normal.

8.23 A 95% confidence interval will give a more precise (shorter interval) estimate of μ than will a 99% confidence interval.

8.25 The 95% confidence interval for μ is

$$\bar{x} - z_{\alpha/2}\sigma/\sqrt{n} \quad to \quad \bar{x} + z_{\alpha/2}\sigma/\sqrt{n}$$
$$20 - 1.96(3)/\sqrt{36} \; to \; 20 + 1.96(3)/\sqrt{36}$$
$$19.02 \; to \; 20.98$$

8.27 The 90% confidence interval for μ is

$$\bar{x} - z_{\alpha/2}\sigma/\sqrt{n} \quad to \quad \bar{x} + z_{\alpha/2}\sigma/\sqrt{n}$$
$$30 - 1.645(4)/\sqrt{25} \; to \; 30 + 1.645(4)/\sqrt{25}$$
$$28.68 \; to \; 31.32$$

8.29 The 99% confidence interval for μ is

$$\bar{x} - z_{\alpha/2}\sigma/\sqrt{n} \quad to \quad \bar{x} + z_{\alpha/2}\sigma/\sqrt{n}$$
$$50 - 2.575(5)/\sqrt{16} \; to \; 50 + 2.575(5)/\sqrt{16}$$
$$46.78 \; to \; 53.22$$

8.31 (a) The sample mean is \$113.97 million / 18 = \$6.332 million. The 95% confidence interval for μ is

$$\bar{x} - z_{\alpha/2}\sigma/\sqrt{n} \quad to \quad \bar{x} + z_{\alpha/2}\sigma/\sqrt{n}$$
$$6.332 - 1.96(2.04)/\sqrt{18} \; to \; 6.332 + 1.96(2.04)/\sqrt{18}$$
$$\$5.389 \; to \; \$7.274 \; (million)$$

(b) We can be 95% confident that the interval from \$5.389 million to \$7.274 million contains the population mean venture capital investment in the fiber optics business sector.

8.33 The sample mean is 6.31 / 12 = 0.526 ppm. The 95% confidence interval for μ is

Copyright © 2012 Pearson Education, Inc. Publishing as Addison-Wesley.

$$\bar{x} - z_{\alpha/2}\sigma/\sqrt{n} \quad to \quad \bar{x} + z_{\alpha/2}\sigma/\sqrt{n}$$

$$0.526 - 2.575(0.37)/\sqrt{12} \quad to \quad 0.526 + 2.575(0.37)/\sqrt{12}$$

$$0.251 \; to \; 0.801 \;\; \text{ppm}$$

We can be 99% confident that the interval from 0.251 to 0.801 ppm contains the population mean cadmium level in *Boletus pinicola* mushrooms.

8.35 $n = 32$; $\bar{x} = 33.4$; $\sigma = 42$

Step 1: $\alpha = 0.05$; $z_{\alpha/2} = z_{0.025} = 1.96$

Step 2:

$$\bar{x} - z_{\alpha/2}\sigma/\sqrt{n} \; to \; \bar{x} + z_{\alpha/2}\sigma/\sqrt{n}$$

$$33.4 - 1.96(42)/\sqrt{32} \quad to \quad 33.4 + 1.96(42)/\sqrt{32}$$

$$18.8 \; to \; 48.0$$

We can be 95% confident that the mean duration of imprisonment, μ, of all East German political prisoners with chronic PTSD is somewhere between 18.8 and 48.0 months.

8.37 $n = 18$; $\bar{x} = 6.33$; $\sigma = 2.04$

(a) Step 1: $\alpha = 0.01$; $z_{\alpha/2} = z_{0.005} = 2.575$

Step 2:

$$\bar{x} - z_{\alpha/2}\sigma/\sqrt{n} \; to \; \bar{x} + z_{\alpha/2}\sigma/\sqrt{n}$$

$$6.33 - 2.575(2.04)/\sqrt{18} \quad to \quad 6.33 + 2.575(2.04)/\sqrt{18}$$

$$5.09 \; to \; 7.57$$

(b) The confidence interval in part (a) is longer than the one in Exercise 8.31 because we have changed the confidence level from 95% in Exercise 8.31 to 99% in this exercise. Notice that increasing the confidence level from 95% to 99% increases the $z_{\alpha/2}$-value from 1.96 to 2.575. The larger z-value, in turn, results in a longer interval. In order to achieve a higher level of confidence that the interval contains the population mean, we need a longer interval.

(c)

We can be 95% confident
that µ lies in here

$$\vdash\!\!\!-\!\!\!-\!\!\!-\!\!\!-\!\!\!-\!\!\!-\!\!\!-\!\!\dashv$$
5.39   7.27

We can be 99% confident
that µ lies in here

$$\vdash\!\!\!-\!\!\!-\!\!\!-\!\!\!-\!\!\!-\!\!\!-\!\!\!-\!\!\!-\!\!\dashv$$
5.09   7.57

(d) The 95% confidence interval is shorter and therefore provides a more precise estimate of µ.

Copyright © 2012 Pearson Education, Inc. Publishing as Addison-Wesley.

8.39 (a)   Using Minitab, retrieve the data from the WeissStats CD. Then choose

Stat ▶ Basic statistics ▶ 1-Sample z. Enter <u>TIME</u> in the **Samples in Columns** text box and enter <u>30</u> in the **Standard deviation** text box. Click on the **Graphs** button and check the boxes for **Histogram of data** and **Boxplot of data**, then click **OK** twice. The confidence interval appears in the Sessions Window as

```
Variable    N     Mean   StDev  SE Mean        95% CI
TIME       20   289.950  30.741   6.708  (276.802, 303.098)
```

The histogram and boxplot are shown following.

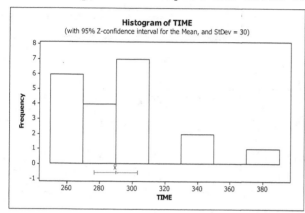

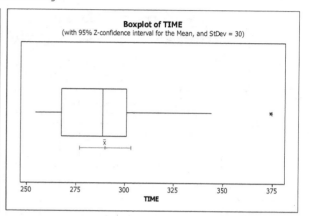

To get the probability plot and stem-and leaf diagram, choose **Stat ▶**

Basic statistics ▶ Normality test, enter <u>TIME</u> in the **Variable** text box and click **OK**. Then choose **Graph ▶ Stem-and-Leaf**, and enter <u>TIME</u> in the **Graph Variables** text box and click **OK**. The results are

```
Stem-and-leaf of TIME   N  = 20
Leaf Unit = 1.0


    2    25   45
    6    26   0779
    9    27   235
   10    28   7
   10    29   01257
    5    30   22
    3    31
    3    32
    3    33   3
    2    34   4

   HI 374
```

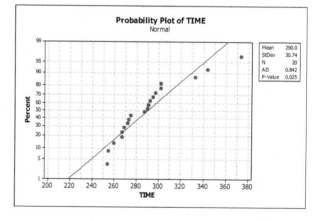

(c)   The value 374 is a potential outlier. Remove it from the data and repeat the process in part (a). The result is

```
Variable    N     Mean   StDev  SE Mean        95% CI
TIME       19   285.526  24.174   6.882  (272.037, 299.016)
```

(d)   The graphs (not shown here) that were produced with the confidence interval now show that 344 is an outlier in the sample of 19 observations. The data are also skewed to the right both before and after deleting the first outlier. The relatively small sample size

Copyright © 2012 Pearson Education, Inc. Publishing as Addison-Wesley.

makes it unwise to use the z-interval procedure with these data.

8.41 (a) We will get the confidence interval for part (c) along with the histogram and boxplot. Using Minitab, retrieve the data from the WeissStats CD. Then choose **Stat ▶ Basic statistics ▶ 1-Sample z.** Enter TEMP in the **Samples in Columns** text box and enter 0.63 in the **Standard deviation** text box. Click on the Options button and enter 99.0 in the **Confidence level** text box and click **OK**. Click on the **Graphs** button and check the boxes for **Histogram of data** and **Boxplot of data**, then click **OK** twice. To get the probability plot and stem-and leaf diagram, choose **Stat ▶ Basic statistics ▶ Normality test**, enter TEMP in the **Variable** text box and click **OK**. Then choose **Graph ▶ Stem-and-Leaf**, and enter TEMP in the **Graph Variables** text box and click **OK**. The results are

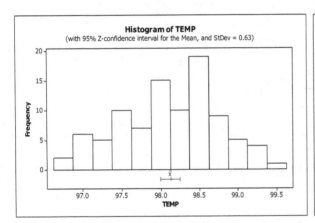

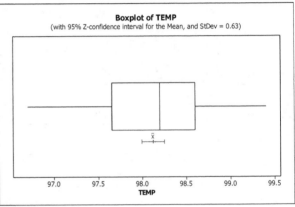

```
Stem-and-leaf of TEMP   N  = 93
Leaf Unit = 0.10

    1    96   7
    3    96   89
    8    97   00001
   13    97   22233
   19    97   444444
   26    97   6666777
   31    97   88889
   45    98   00000000000111
  (10)   98   2222222233
   38    98   4444445555
   28    98   66666666677
   17    98   8888888
   10    99   00001
    5    99   2233
    1    99   4
```

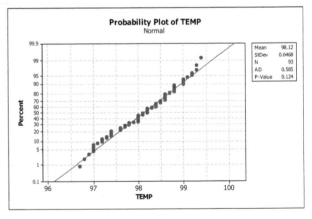

(b) Yes. There are no outliers, the sample size is large (93), and the distribution of the data appears to be approximately normal.

(c) The output from part (a) in the Sessions Window is

```
Variable   N    Mean    StDev  SE Mean       99% CI
TEMP       93  98.1237  0.6468  0.0653  (97.9554, 98.2919)
```

We can be 99% confident that the interval (97.9554, 98.2919) contains the mean body temperature μ of healthy humans. The output is

Copyright © 2012 Pearson Education, Inc. Publishing as Addison-Wesley.

surprising since the confidence interval does not inclue the generally accepted normal temperature of 98.6° F.

8.43 (a) We will get the confidence interval for part (c) along with the histogram and boxplot. Using Minitab, retrieve the data from the WeissStats CD. Then choose **Stat ▶ Basic statistics ▶ 1-Sample z**. Enter SPEED in the **Samples in Columns** text box and enter 3.2 in the **Standard deviation** text box. Click on the **Options** button and enter 95.0 in the **Confidence level** text box and click **OK**. Click on the **Graphs** button and check the boxes for **Histogram of data** and **Boxplot of data**, then click **OK** twice. To get the probability plot and stem-and leaf diagram, choose **Stat ▶ Basic statistics ▶ Normality test**, enter SPEED in the **Variable** text box and click **OK**. Then choose **Graph ▶ Stem-and-Leaf**, and enter SPEED in the **Graph Variables** text box and click **OK**. The results are

Variable	N	Mean	StDev	SE Mean	95% CI
SPEED	35	59.5257	4.2739	0.5409	(58.4656, 60.5859)

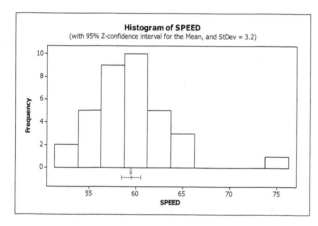

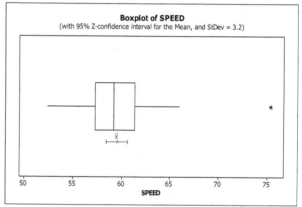

```
Stem-and-leaf of SPEED   N  = 35
Leaf Unit = 0.10

    2    52   46
    2    53
    4    54   78
    7    55   459
    8    56   5
   13    57   35688
   16    58   137
   (5)   59   02678
   14    60   12679
    9    61   36
    7    62   36
    5    63   4
    4    64
    4    65   02
    2    66   0

   HI 753
```

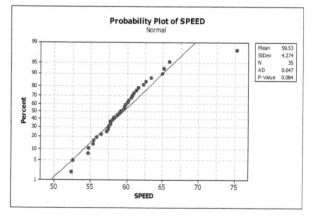

(c) After removing 75.3 from the data, the 95% confidence interval is

Variable	N	Mean	StDev	SE Mean	95% CI
SPEED	34	59.0618	3.3253	0.5488	(57.9861, 60.1374)

Copyright © 2012 Pearson Education, Inc. Publishing as Addison-Wesley.

(d) The outlier had an effect on the mean of about 0.46 mph and, therefore, about the same effect on the endpoints of the confidence interval. The movement of the interval is about one-fourth the length of the interval and is significant in that sense. On the other hand, the change in the mean represents less than a 1% error and is probably not very important, although this is a matter of judgment. The sample size of 35 appears to be large enough that one outlier of the size observed does not have a great effect.

8.45 (a) To increase the precision (shorten the confidence interval) without reducing the confidence level, we should increase the sample size n.

(b) To increase our level of confidence without reducing the precision, we should increase the sample size n.

8.47 We can be $100(1 - \alpha)\%$ confident that the mean μ is greater than the lower confidence bound and $100(1 - \alpha)\%$ confident that the mean μ is less than the upper confidence bound.

8.49 (a) The sample mean is 6.31 / 12 = 0.526 ppm. The 99% lower confidence bound for μ is

$$\bar{x} - z_\alpha \sigma / \sqrt{n}$$

$$0.526 - 2.33(0.37) / \sqrt{12}$$

$$0.277 \text{ ppm}$$

We can be 99% confident that the population mean cadmium level in *Boletus pinicola* mushrooms is greater than 0.277 ppm.

(b) This lower confidence bound is greater than the lower confidence limit of 0.251 found in Exercise 8.33. This is because the z-value used for a one-sided 99% confidence bound is 2.33, whereas, the z-value used for a two-sided 99% confidence interval is 2.575.

Exercises 8.3

8.51 The length of a confidence interval, and thus the precision with which $\bar{x}$ estimates μ, is determined by the margin of error.

8.53 (a) The length of the confidence interval is twice the margin of error; i.e., 2 x 3.4 = 6.8.

(b) The confidence interval is 52.8 $\pm$ 3.4 = (49.4, 56.2).

(c)

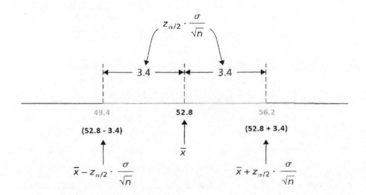

Copyright © 2012 Pearson Education, Inc. Publishing as Addison-Wesley.

8.55 (a) The margin of error is 1/2 the length of the confidence interval; i.e., (1/2) x 20 = 10.

(b) The confidence interval is 60 ± 10 = (50, 70).

(c)

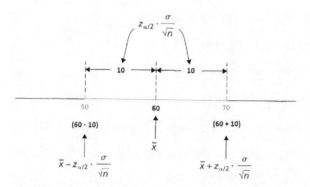

8.57 (a) True. The length of the confidence interval is twice the margin of error.

(b) True. The margin of error is one-half the length of the confidence interval.

(c) False. One must also know the sample mean.

(d) True. The confidence interval is mean ± margin of error.

8.59 (a) We want a whole number because the number of observations to be taken cannot be fractional. It must be an integer because we cannot sample a partial member of a population.

(b) The original number computed is the smallest value of n that will provide the required margin of error. If we were to round down, the actual sample size would be slightly too small.

8.61 n = 10; $\bar{x}$ = 34.2 cm; σ = 2.1 cm

(a) α = 0.10; $z_{\alpha/2}$ = $z_{0.05}$ = 1.645, so the 90% confidence interval for μ is

$$\bar{x} - 1.645(2.1)/\sqrt{10} \quad to \quad \bar{x} + 1.645(2.1)/\sqrt{10}$$
$$34.2 - 1.645(2.1)/\sqrt{10} \; to \; 34.2 + 1.645(2.1)/\sqrt{10}$$
$$33.1 \; to \; 35.3 \; cm$$

(b) The margin of error is $E = 1.645\sigma/\sqrt{n} = 1.645(2.1)/\sqrt{10} = 1.1 \; cm$.

(c) We are 90% confident that our error in estimating μ by $\bar{x}$ is at most 1.1 cm.

(d) α = 0.05; $z_{\alpha/2}$ = $z_{0.025}$ = 1.96, so $n = \left(\dfrac{z_{\alpha/2} \cdot \sigma}{E}\right)^2 = \left(\dfrac{1.96 \cdot 2.1}{0.5}\right)^2 = 67.76 \; or \; 68.$

8.63 (a) E = 0.5($7.247 million − $5.389 million) = 0.943 million

Copyright © 2012 Pearson Education, Inc. Publishing as Addison-Wesley.

(b) $E = 1.96\sigma/\sqrt{n} = 1.96(2.04)/\sqrt{18} = \$0.942$ million . Note: there is a small difference in the answers do to rounding.

8.65 (a) E = (48.0 - 18.8)/2 = 14.6 months

(b) We are 95% confident that the maximum error made in using $\overline{x}$ to estimate μ is 14.6 months.

(c) The margin of error of the estimate is specified to be E = 12.0 months. Also, for a 99% confidence interval, $z_{\alpha/2}$ = $z_{0.005}$ = 2.575.

$$n = \left[\frac{z_{\alpha/2}\sigma}{E}\right]^2 = \left[\frac{2.575(42)}{12}\right]^2 = 81.2 \rightarrow 82$$

(d)

$$\overline{x} - z_{\alpha/2}\sigma/\sqrt{n} \text{ to } \overline{x} + z_{\alpha/2}\sigma/\sqrt{n}$$
$$36.2 - 2.575(12)/\sqrt{82} \text{ to } 36.2 + 2.575(12)/\sqrt{82}$$
$$24.3 \text{ to } 48.1$$

8.67 n = 900; α = 0.05; $z_{\alpha/2}$ = $z_{0.025}$ = 1.96; σ = 12.1

$$E = 1.96(12.1)/\sqrt{900} = 0.791$$

8.69 (a) The margin of error of the estimate is specified to be E = 2 years.

$$n = \left[\frac{z_{\alpha/2}\sigma}{E}\right]^2 = \left[\frac{1.96(13.36)}{2.0}\right]^2 = 171.4 \rightarrow 172$$

(b) We used s in place of σ because σ was unknown. We can do this because the sample of size 36 is large enough to provide an estimate of σ and the variation is not likely to change much from one year to the next.

8.71 (a) α = 0.05; $z_{\alpha/2}$ = $z_{0.025}$ = 1.96; σ = 10

$$E = 1.96(10) / \sqrt{4} = 9.80$$

(b) α = 0.05; $z_{\alpha/2}$ = $z_{0.025}$ = 1.96; σ = 10

$$E = 1.96(10) / \sqrt{16} = 4.90$$

(c) It appears that quadrupling the sample size will halve the margin of error. Therefore, increasing n from 16 to 64 will decrease the margin of error from 4.90 to 2.45.

Exercises 8.4

8.73 The formula for the standardized version of $\overline{x}$ uses σ in the denominator, whereas the studentized version uses s.

8.75 (a) The standardized version of $\overline{x}$ is
$$z = (\overline{x} - \mu)/(\sigma/\sqrt{n}) = (108 - 100)/(16/\sqrt{4}) = 1.00 .$$

Copyright © 2012 Pearson Education, Inc. Publishing as Addison-Wesley.

(b) The studentized version of $\bar{x}$ is
$$t = (\bar{x} - \mu)/(s/\sqrt{n}) = (108 - 100)/(12/\sqrt{4}) = 1.333 \, .$$

8.77 (a) Standard normal

(b) t-distribution with 11 degrees of freedom

8.79 The variation in the possible values of the standardized version of $\bar{x}$ is due only to the variation in $\bar{x}$ while the variation in the studentized version results not only from the variation in $\bar{x}$, but also from the variation in the sample standard deviation.

8.81 For df = 6:

(a) $t_{0.10} = 1.440$ (b) $t_{0.025} = 2.447$ (c) $t_{0.01} = 3.143$

8.83 (a) $t_{0.10} = 1.323$ (b) $t_{0.01} = 2.518$

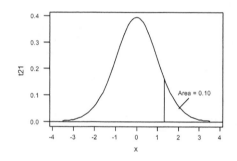

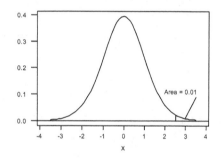

(c) $-t_{0.025} = -2.080$ (d) $\pm t_{0.05} = \pm 1.721$

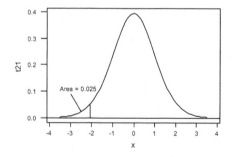

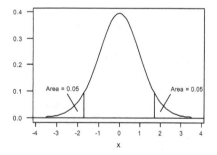

8.85 It is reasonable to use the t-interval procedure since the sample size is large and for large degrees of freedom (99), the t-distribution is very similar to the standard normal distribution. Another way of expressing this is that the sampling distribution of $\bar{x}$ is approximately normal when n is large, so the standardized and studentized versions of $\bar{x}$ are essentially the same.

Copyright © 2012 Pearson Education, Inc. Publishing as Addison-Wesley.

8.87 The 95% confidence interval for μ is

$$\bar{x} - t_{\alpha/2}s/\sqrt{n} \quad to \quad \bar{x} + t_{\alpha/2}s/\sqrt{n}$$
$$20 - 2.030(3)/\sqrt{36} \; to \; 20 + 2.030(3)/\sqrt{36}$$
$$18.98 \; to \; 21.02$$

8.89 The 90% confidence interval for μ is

$$\bar{x} - t_{\alpha/2}s/\sqrt{n} \quad to \quad \bar{x} + t_{\alpha/2}s/\sqrt{n}$$
$$30 - 1.711(4)/\sqrt{25} \; to \; 30 + 1.711(4)/\sqrt{25}$$
$$28.63 \; to \; 31.37$$

8.91 The 99% confidence interval for μ is

$$\bar{x} - t_{\alpha/2}s/\sqrt{n} \quad to \quad \bar{x} + t_{\alpha/2}s/\sqrt{n}$$
$$50 - 2.947(5)/\sqrt{16} \; to \; 50 + 2.947(5)/\sqrt{16}$$
$$46.32 \; to \; 53.68$$

8.93 n = 30; df = 29; $t_{\alpha/2}$ = $t_{0.05}$ = 1.699; $\bar{x}$ = 27.97 minutes; s = 10.04 minutes

(a)

$$\bar{x} - t_{\alpha/2}s/\sqrt{n} \; to \; \bar{x} + t_{\alpha/2}s/\sqrt{n}$$
$$27.97 - 1.699(10.04)/\sqrt{30} \quad to \quad 27.97 + 1.699(10.04)/\sqrt{30}$$
$$24.86 \; to \; 31.08$$

(b) We are 90% confident that the interval 24.86 to 31.08 minutes contains the mean, μ, of all commute times for working adults in the Washington, D.C. area.

8.95 n = 10; df = 9; $t_{\alpha/2}$ = $t_{0.025}$ = 2.262; $\bar{x}$ = 2.33 hours; s = 2.002 hours

(a)

$$\bar{x} - t_{\alpha/2}s/\sqrt{n} \; to \; \bar{x} + t_{\alpha/2}s/\sqrt{n}$$
$$2.33 - 2.262(2.002)/\sqrt{10} \quad to \quad 2.33 + 2.262(2.002)/\sqrt{10}$$
$$0.90 \; to \; 3.76$$

(b) We are 95% confident that the interval 0.90 to 3.76 hours contains the mean, μ, of addition sleep obtained by a patients using laevohysocyamine hydrobromide. Since this interval was entirely above zero, we could conclude that the treatment was effective.

8.97 n = 77; df = 76; $t_{\alpha/2}$ = $t_{0.025}$ = 1.992; $\bar{x}$ = 0.199 m/sec; s = 0.210 m/sec

(a)

$$\bar{x} - t_{\alpha/2}s/\sqrt{n} \; to \; \bar{x} + t_{\alpha/2}s/\sqrt{n}$$
$$0.199 - 1.992(0.210)/\sqrt{77} \quad to \quad 0.199 + 1.992(0.210)/\sqrt{77}$$
$$0.151 \; to \; 0.247$$

(b) We are 95% confident that the interval 0.151 m/sec to 0.247 m/sec contains the mean increase, μ, of aortic-jet velocity of patiens with calcific aortic stenosis who received 80 mg of atorvasatin daily.

Copyright © 2012 Pearson Education, Inc. Publishing as Addison-Wesley.

Since this interval lies entirely above zero, we conclude that there is an increase in aortic-jet velocity for such patients.

8.99 We used Minitab to obtain a boxplot and a normal probability plot of the data, which are shown below. The plots show that the values 6.7 and 7.6 are potential outliers and the distribution is skewed right. Since the sample size is only 22, it is not reasonable to use the t-interval procedure to obtain a confidence interval for the population mean.

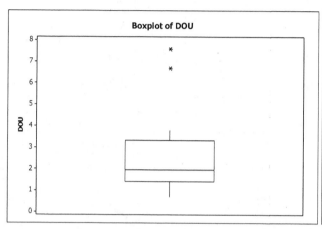
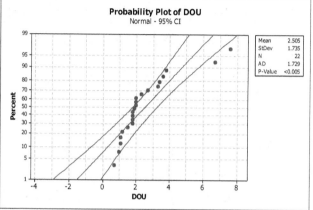

8.101 We used Minitab to obtain a boxplot and a normal probability plot of the data, which are shown below. The plots show no outliers and the distribution is close to symmetrical. Since the sample size is 20, it is reasonable to use the t-interval procedure to obtain a confidence interval for the population mean.

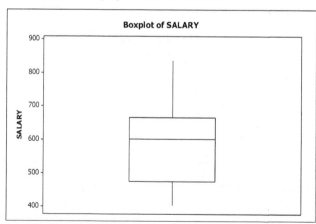
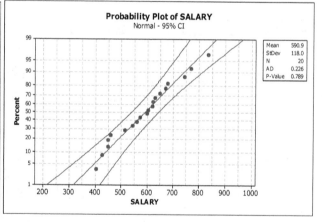

8.103 (a) We will get the confidence interval for part (c) along with the histogram and boxplot. Using Minitab, retrieve the data from the WeissStats CD. Then choose **Stat ▶ Basic statistics ▶ 1-Sample t.** Enter DEPTH in the **Samples in Columns** text box. Click on the **Options** button and enter 90.0 in the **Confidence level** text box and click **OK**. Click on the **Graphs** button and check the boxes for **Histogram of data** and **Boxplot of data**, then click **OK** twice. To get the probability plot and stem-and leaf diagram, choose **Stat ▶ Basic statistics ▶ Normality test**, enter DEPTH in the **Variable** text box and click **OK**. Then choose

Copyright © 2012 Pearson Education, Inc. Publishing as Addison-Wesley.

Graph ▶ Stem-and-Leaf, and enter <u>DEPTH</u> in the **Graph Variables** text box and click **OK**. The results are

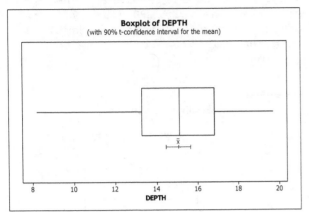

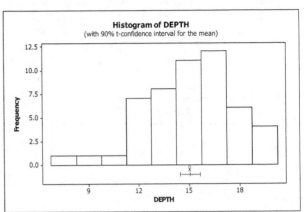

```
Stem-and-leaf of DEPTH   N  = 51
Leaf Unit = 0.10

     1    8   2
     2    9   7
     2   10
     4   11   08
    12   12   01123588
    16   13   3399
    23   14   0245799
    (8)  15   00134689
    20   16   002567789
    11   17   24459
     6   18   2389
     2   19   37
```

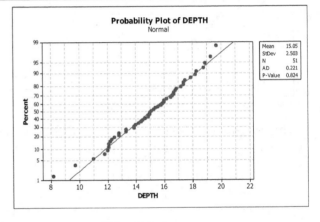

(b) Yes. The plots indicate no outliers, the sample size is quite large at 51, and the distribution of the data is only a little left skewed.

(c) The output for the confidence interval obtained in part (a) is

```
Variable   N     Mean   StDev  SE Mean       90% CI
DEPTH      51  15.0510  2.5028  0.3505  (14.4636, 15.6383)
```

We can be 90% confident that the mean, μ, for all burrow depths is somewhere between 14.46 cm and 15.64 cm.

8.105 (a) We will get the confidence interval for part (c) along with the histogram and boxplot. Using Minitab, retrieve the data from the WeissStats CD. Then choose **Stat ▶ Basic statistics ▶ 1-Sample t**. Enter <u>WITHOUT</u> in the **Samples in Columns** text box. Click on the **Options** button and enter <u>95.0</u> in the **Confidence level** text box and click **OK**. Click on the **Graphs** button and check the boxes for **Histogram of data** and **Boxplot of data**, then click **OK** twice. To get the probability plot and stem-and leaf diagram, choose **Stat ▶ Basic statistics ▶ Normality test**, enter <u>WITHOUT</u> in the **Variable** text box and click **OK**. Then choose **Graph ▶ Stem-and-Leaf**, and enter <u>WITHOUT</u> in the **Graph Variables** text box and click **OK**. The results are

Copyright © 2012 Pearson Education, Inc. Publishing as Addison-Wesley.

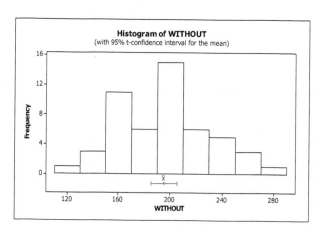

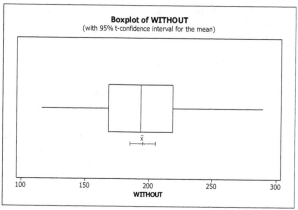

Stem-and-leaf of WITHOUT   N = 51
Leaf Unit = 10

```
   1    1  1
   2    1  3
   9    1  4455555
  19    1  6666667777
  (8)   1  88999999
  24    2  000000000111
  12    2  2223333
   5    2  45
   3    2  66
   1    2  8
```

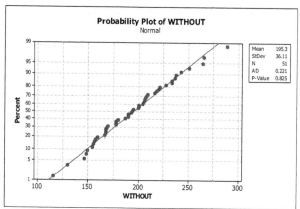

(b) Yes. The plots indicate no outliers, the sample size is quite large at 51, and the distribution of the data is quite symmetric.

(c) The output for the confidence interval obtained in part (a) is

```
Variable    N     Mean    StDev  SE Mean      95% CI
WITHOUT    51   195.275   36.110   5.056   (185.118, 205.431)
```

We can be 95% confident that the mean, μ, for all plasma cholesterol concentrations of patients without evidence of heart disease is somewhere between 185.118 mg/dl and 203.431 mg/dl.

(d) Repeating the process in part (a) using the WITH data, we obtain

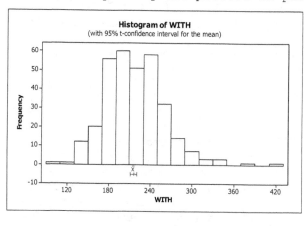

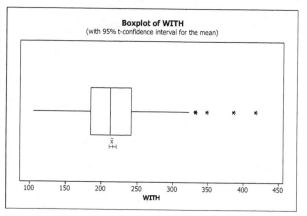

Copyright © 2012 Pearson Education, Inc. Publishing as Addison-Wesley.

```
Stem-and-leaf of WITH   N  = 320
Leaf Unit = 10

    2    1  01
    8    1  333333
   19    1  44444455555
   63    1  6666666666666667777777777777777777777777777777
  126    1  888888888888888888888888888889999999999999999999999999999999999
  (50)   2  0000000000000000000000001111111111111111111111111111
  144    2  2222222222222222222222222333333333333333333333333333333333
   87    2  44444444444444444444444444445555555555555555555
   44    2  66666666666666677777777
   21    2  8888889999
   11    3  00011
    6    3  2
```

HI 33, 33, 34, 38, 41

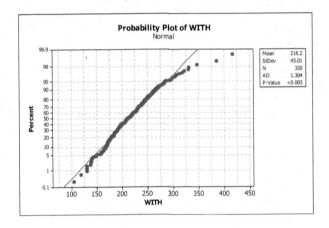

The data are quite symmetric except for five outliers on the high side. Nevertheless, the large sample size of 320 makes it reasonable to use the t-interval procedure with these data. The results of that procedure are

```
Variable   N    Mean    StDev  SE Mean      95% CI
WITH      320  216.191  43.015   2.405  (211.460, 220.921)
```

We can be 95% confident that the mean, μ, for all plasma cholesterol concentrations of patients with evidence of heart disease is somewhere between 211.460 mg/dl and 220.921 mg/dl.

8.107 (a) We could not provide entries for every possible degrees of freedom because the number of possibilities is infinite.

(b) As the degrees of freedom increase, the difference in consecutive entries becomes very small, making it unnecessary to list every possibility.

(c) Anytime the actual degrees of freedom lies between two consecutive entries in the table, we should use the table value associated with the lower number of degrees of freedom. Thus for $t_{0.05}$ with 87 df, we should use the value associated with 80 df, 1.664; for 125 df, use 1.660; for 650 df, use 1.660; and for 3000 df, use 1.645. This is a conservative approach, resulting in margins of error that are never smaller than we are entitled to have.

Copyright © 2012 Pearson Education, Inc. Publishing as Addison-Wesley.

8.109 The observed values of the studentized and standardized versions of $\bar{x}$ are the same for any sample size n whenever the sample standard deviation s is identical to the population standard deviation σ.

8.111 (a) Your results will vary. To obtain the 2000 samples using Minitab,

Choose **Calc ▶ Random Data ▶ Normal...**, enter 2000 in the **Generate rows of data** text box, enter C1-C5 in the **Store in Column(s):** text box, enter .270 in the **Mean** text box, and enter .031 in the **Standard deviation:** text box. Click **OK**.

(b) To find the mean and sample standard deviation in each row (sample), choose **Calc ▶ Row statistics...** and click on **Mean**. Enter C1-C5 in the **Input variable(s):** text box and enter C6 in the **Store result in:** text box. Repeat this last process, selecting **Standard Deviation** instead of **Mean** and put the results in C7.

(c) To obtain the Standardized version of each mean in the sample, choose **Calc ▶ Calculator...**, enter STANDARD in the **Store results in variable:** text box, enter (C6-.270)/(.031/SQRT(5)) in the **Expression:** text box and click **OK**.

(d) To facilitate a comparison in part (k),., do part (h) now. Choose **Graph ▶ Histogram...**, select the **Simple** version, and click **OK**. Enter STANDARD STUDENT in the **Graph variables** text box. Click the **Multiple graphs** button, click the **On separate graphs** button and check the box for **Same X, including same bins**. Click **OK** and click **OK**. Our first graph is shown below, but yours will differ, yet look similar.

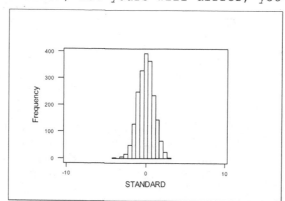

(e) In theory, the distribution of the standardized version of $\bar{x}$ is standard normal.

(f) The histogram in (d) appears to be very close to standard normal. Recall that the distribution is centered at zero and 99.74% of the data should be within 3 standard deviations of the mean. It does appear that this is so for this simulated data.

(g) To obtain the Studentized version of each $\bar{x}$ in the sample, choose **Calc ▶ Calculator...**, enter STUDENT in the **Store results in variable:** text box, enter (C6-.270)/(C7/SQRT(5) in the **Expression:** text box and click **OK**.

(h) This graph was produced in produced (d). Your graph should be similar to ours.

Copyright © 2012 Pearson Education, Inc. Publishing as Addison-Wesley.

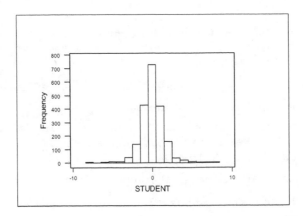

(i) In theory, the distribution of the studentized version of $\bar{x}$ is a t-distribution with 4 degrees of freedom.

(j) The distribution shown is symmetric about zero as is a t-distribution.

(k) The histogram of the studentized version in (h) is more spread out than that of the standardized version in (d). The reason is that there is more variability in the t-distribution due to the extra uncertainty arising from the use of s instead of σ.

8.113 We can be $100(1 - \alpha)\%$ confident that the mean μ is greater than the lower confidence bound and $100(1 - \alpha)\%$ confident that the mean μ is less than the upper confidence bound.

8.115 (a) n=20; $\bar{x}$ = 4.760; s = 2.297; df = 19; t_α = $t_{0.10}$ = 1.328

$$\bar{x} - t_\alpha s / \sqrt{n}$$

$$4.760 - 1.328(2.297) / \sqrt{20}$$

4.078 hours

We can be 90% confident that the population mean amount of television watched per day last year by the average person is greater than 4.078 hours.

(b) This lower confidence bound is greater than the lower confidence limit of 3.872 found in Exercise 8.94. This is because the t-value used for a one-sided 90% confidence bound is 1.328, whereas, the t-value used for the two-sided 90% confidence interval was 1.729.

8.117 (a) n=1009; $\bar{x}$= 639.00; s= 477.98; df =1008 (using df=1000); t_α = $t_{0.05}$ = 1.646

$$\bar{x} + t_\alpha s / \sqrt{n}$$

$$639.00 + 1.646(477.98) / \sqrt{1009}$$

$663.77

(b) We can be 95% confident that the population mean amount spent on Christmas gifts in 2008 is less than $663.77.

(c) According to our confidence interval in part (b), we are 95% confident that on average the amount spent on Christmas gifts in 2008 is less than the average amount spent on Christmas gifts in 2007. It seems like consumers, perhaps hurt by the recession, spent less money in 2008 than 2007, on average.

Copyright © 2012 Pearson Education, Inc. Publishing as Addison-Wesley.

Review Problems for Chapter 8

1. A point estimate of a parameter consists of a single value with no indication of the accuracy of the estimate. A confidence interval consists of an interval of numbers obtained from a point estimate of the parameter together with a percentage that specifies how confident we are that the parameter lies in the interval.

2. False. We are 95% confident that the mean lies in the interval from 33.8 to 39.0, but about 5% of the time, the procedure will produce an interval that does not contain the population mean. Therefore, we cannot say that the mean must lie in the interval.

3. No. The z-interval procedure can be used almost anytime with large samples

    because the sampling distribution of $\overline{x}$ is approximately normal for large n. The same is true for the t-interval procedure because when n is large, the t distribution is very similar to the normal distribution. However, when n is small, especially when n is 15 or less, the z-interval and t-interval procedures will not provide reliable estimates if the distribution of the underlying variable is not normal. For sample sizes in the range of 15 to 30, both procedures can be used if the data is roughly normal and has no outliers.

4. Approximately 950 of 1000 95% confidence intervals for a population mean would actually contain the true value of the mean.

5. Before applying a particular statistical inference procedure, we should look at graphical displays of the sample data to see if there appear to be any violations of the conditions required for the use of the procedure.

6. (a) Reducing the sample size from 100 to 50 will reduce the precision of the estimate (result in a longer confidence interval).

    (b) Reducing the confidence level from .95 to .90 while maintaining the sample size will increase the precision of the estimate (result in a shorter confidence interval).

7. (a) The length of the confidence interval is twice the margin of error or 2 x 10.7 = 21.4.

    (b) The confidence interval will be 75.2 $\pm$ 10.7 = (64.5, 85.9)

8. (a) $E = z_{\alpha/2}(\sigma/\sqrt{n}) = 1.645(12/\sqrt{9}) = 6.58$

    (b) To obtain the confidence interval, you also need to know $\overline{x}$.

9. (a) The standardized value of $\overline{x}$ is $z = \dfrac{\overline{x} - \mu}{\sigma/\sqrt{n}} = \dfrac{262.1 - 266}{16/\sqrt{10}} = -0.77$

    (b) The studentized value of $\overline{x}$ is $t = \dfrac{\overline{x} - \mu}{s/\sqrt{n}} = \dfrac{262.1 - 266}{20.4/\sqrt{10}} = -0.61$

10. (a) standard normal distribution

    (b) t distribution with 14 degrees of freedom

11. The curve that looks more like the standard normal curve has the larger degrees of freedom because, as the number of degrees of freedom gets larger, t-curves look increasingly like the standard normal curve.

12. (a) The t-interval procedure should be used.

    (b) The z-interval procedure should be used.

    (c) The z-interval procedure should be used.

Copyright © 2012 Pearson Education, Inc. Publishing as Addison-Wesley.

(d) Neither procedure should be used.

(e) The z-interval procedure should be used.

(f) Neither procedure should be used.

13. $n = 36$, $\bar{x} = 58.53$, $\sigma = 13.0$, $z_{\alpha/2} = z_{0.025} = 1.96$

$$\bar{x} - z_{\alpha/2}(\sigma/\sqrt{n}) \text{ to } \bar{x} + z_{\alpha/2}(\sigma/\sqrt{n})$$

$$58.53 - 1.96 \cdot (13.0/\sqrt{36}) \text{ to } 58.53 + 1.96 \cdot (13.0/\sqrt{36})$$

$$54.3 \text{ to } 62.8 \text{ years}$$

14. A confidence-interval estimate specifies how confident we are that a (unknown) parameter lies in the interval. This interpretation is presented correctly by (c). A *specific* confidence interval either will or will not contain the true value of the population mean μ; the *specific* interval is either sure to contain μ or sure not to contain μ. This interpretation is *not* presented correctly by (a), (b), and (d).

15. $n = 461$, $= 11.9$ mm, $\sigma = 2.5$ mm, $z_{\alpha/2} = z_{0.05} = 1.645$

(a)

$$\bar{x} - z_{\alpha/2}\sigma/\sqrt{n} \text{ to } \bar{x} + z_{\alpha/2}\sigma/\sqrt{n}$$

$$11.9 - 1.645(2.5)/\sqrt{461} \text{ to } 11.9 + 1.645(2.5)/\sqrt{461}$$

$$11.71 \text{ to } 12.09$$

(b) We can be 90% confident that the mean length, μ, of N. *trivittata* is somewhere between 11.71 and 12.09 mm.

(c) Since the sample size is very large, the distribution of sample means will be approximately normal regardless of the shape of the original distribution. It would be nice if the normal probability plot were roughly linear and did not indicate the presence of any extreme outliers, but some non-linearity and a few moderate outliers will not likely invalidate the use of the z-interval procedure.

16. (a) $E = z_{\alpha/2}\sigma/\sqrt{n} = 1.645(2.5)/\sqrt{461} = 0.19$

(b) We can be 90% confident that the maximum error made in using $\bar{x}$ to estimate μ is 0.19 mm.

(c) The margin of error of the estimate is specified to be E = 0.1 mm.

$$n = \left[\frac{z_{\alpha/2}\,\sigma}{E} \right]^2 = \left[\frac{1.645\,(2.5)}{0.1} \right]^2 = 1691.3 \rightarrow 1692$$

(d)

$$\bar{x} - z_{\alpha/2}\sigma/\sqrt{n} \text{ to } \bar{x} + z_{\alpha/2}\sigma/\sqrt{n}$$

$$12.0 - 1.645(2.5)/\sqrt{1692} \text{ to } 12.0 + 1.645(2.5)/\sqrt{1692}$$

$$11.90 \text{ to } 12.10$$

Copyright © 2012 Pearson Education, Inc. Publishing as Addison-Wesley.

17.   (a) $t_{0.025} = 2.101$ (b) $t_{0.05} = 1.734$

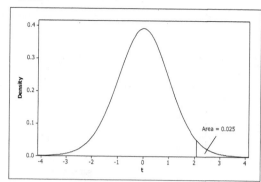 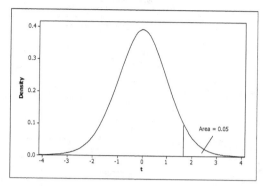

(c) $-t_{0.10} = -1.330$ (d) $\pm t_{0.005} = \pm 2.878$

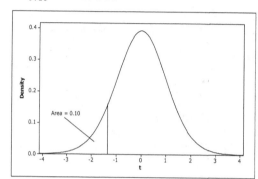 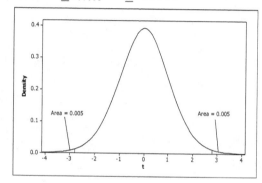

18. $n = 16$; $df = 15$; $t_{\alpha/2} = t_{0.025} = 2.131$; $\bar{x} = 85.99$ mm Hg; $s = 8.08$ mm Hg

(a)

$$\bar{x} - t_{\alpha/2}s/\sqrt{n} \text{ to } \bar{x} + t_{\alpha/2}s/\sqrt{n}$$

$$85.99 - 2.131(8.08)/\sqrt{16} \text{ to } 85.99 + 2.131(8.08)/\sqrt{16}$$

$$81.69 \text{ to } 90.29$$

We can be 95% confident that the mean arterial blood pressure μ of all children of diabetic mothers is somewhere between 81.69 and 90.29 mm Hg.

(b) Using Minitab, choose **Stat ▶ Basic statistics ▶ 1-Sample t...**, enter PRESSURE in the **Samples in columns** text box. Click the **options...** button, type 95 in the **Confidence interval** text box, click the **Graphs** and check the boxes for **Histogram of data** and **Boxplot of data**, and click **OK** twice. Then choose **Stat ▶ Basic statistics ▶ Normality test**, enter PRESSURE in the **Variable** text box and click **OK**. Finally, choose **Graph ▶ Stem-and-Leaf**, enter PRESSURE in the **Graph Variables** text box and click **OK**. The results are

Copyright © 2012 Pearson Education, Inc. Publishing as Addison-Wesley.

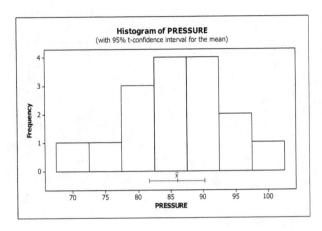

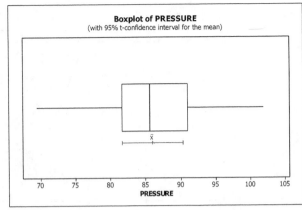

Stem-and-leaf of PRESSURE   N = 16
   Leaf Unit = 1.0

```
   1   6   9
   1   7
   3   7   58
   8   8   12244
   8   8   678
   5   9   014
   2   9   6
   1  10   1
```

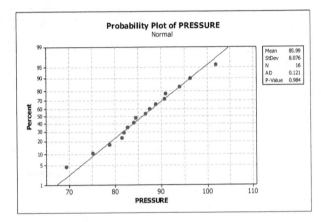

(c) Yes. There are no outliers and the distribution of the data is
     approximately normal.

19.   (a) Using Minitab, choose **Stat ▶ Basic statistics ▶ 1-Sample t...,** enter
            PRICE in the **Samples in columns** text box. Click the **options…** button,
            enter 90 in the **Confidence interval** text box and click **OK**, click the
            Graphs button and check the boxes for **Histogram of data** and **Boxplot of**

            data, and click **OK** twice. Then choose **Stat ▶ Basic statistics ▶**
            Normality test, enter PRICE in the **Variable** text box and click **OK**.

            Finally, choose **Graph ▶ Stem-and-Leaf,** enter PRICE in the **Graph**
            Variables text box and click **OK**. The confidence interval is

```
   Variable   N     Mean   StDev   SE Mean        90% CI
   PRICE     18  1964.72  206.45     48.66  (1880.07, 2049.37)
```

We can be 90% confident that the interval ($1880.07, $2049.37) contains
the mean diamond price for one-half carat diamonds.

Copyright © 2012 Pearson Education, Inc. Publishing as Addison-Wesley.

(b)   The graphs obtained by the procedures in part (a) are

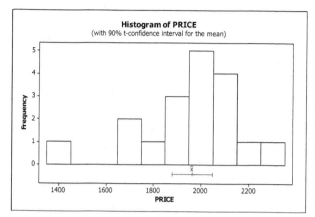

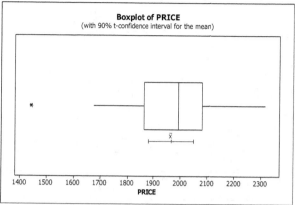

```
Stem-and-leaf of PRICE    N   = 18
Leaf Unit = 10

LO 144


   2   16   7
   3   17   1
   5   18   27
  (6)  19   448899
   7   20   377
   4   21   04
   2   22   3
   1   23   1
```

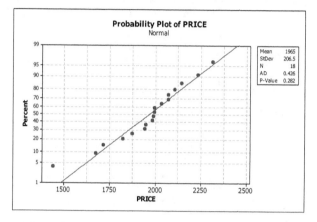

(c)   No. Since the sample size is small (18), and there is an outlier (1442), it is not reasonable to use the t-interval procedure.

20.   (a)   We will import the data into Minitab.

(b)   Choose **Stat ▶ Basic statistics ▶ 1-Sample t..**, enter <u>DURATION</u> in the **Samples in columns** text box. Click the **options…** button, enter <u>99</u> in the **Confidence interval** text box and click **OK**, click the **Graphs** button and check the boxes for **Histogram of data** and **Boxplot of data,** and click **OK** twice. Then choose **Stat ▶ Basic statistics ▶ Normality test**, enter <u>DURATION</u> in the **Variable** text box and click **OK**. Finally, choose **Graph ▶ Stem-and-Leaf**, enter <u>DURATION</u> in the **Graph Variables** text box and click **OK**. The graphs are

Copyright © 2012 Pearson Education, Inc. Publishing as Addison-Wesley.

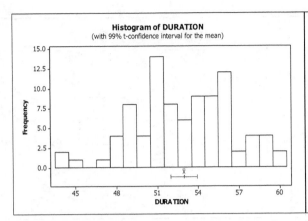

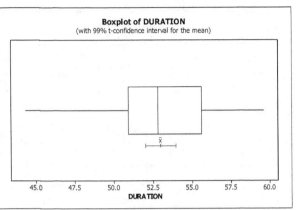

```
Stem-and-leaf of DURATION   N   = 90
Leaf Unit = 0.10

    2   44   33
    3   45   0
    3   46
    4   47   3
    9   48   00036
   16   49   0000033
   22   50   003366
   34   51   000000003333
   45   52   00000003666
   45   53   000
   42   54   000000033666
   30   55   0000336666
   20   56   000003336
   11   57   0
   10   58   00036
    5   59   00356
```

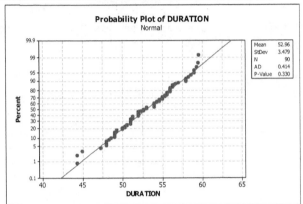

(c) Yes. The sample size is large (90) and there are no outliers.

(d) The procedure in part (a) resulted in the following confidence interval.

```
Variable   N     Mean    StDev   SE Mean        99% CI
DURATION   90   52.9633  3.4794   0.3668   (51.9979, 53.9287)
```

We can be 99% confident that the interval (51.9979, 53.9287) contains the population mean μ of the larval duration of convict surgeonfish in days.

21. (a) Using Minitab, choose **Calc ▶ Random data ▶ Sample from columns**, enter 35 in the **Sample ___ rows from column(s)** text box, enter <u>MILEAGE</u> in the first text box and enter <u>SAMPLE</u> in the **Store samples in** text box. Click **OK**.

   (b) Choose **Stat ▶ Basic statistics ▶ 1-Sample t..**, enter <u>SAMPLE</u> in the **Samples in columns** text box. Click the **options…** button, enter <u>95</u> in the **Confidence interval** text box and click **OK**. The result is

```
Variable   N     Mean    StDev   SE Mean        95% CI
SAMPLE     35   24.4857  4.0174   0.6791   (23.1057, 25.8657)
```

   (c) Choose **Calc ▶ Column statistics** and check the box for the mean. Enter <u>MILEAGE</u> in the **Input variable** text box and click **OK**. The population

Copyright © 2012 Pearson Education, Inc. Publishing as Addison-Wesley.

mean is 24.8234. This value lies in the 95% confidence interval found in part (b). It would not necessarily have to lie in the confidence interval. Since the sample was randomly selected, it is possible to select a sample for which the confidence interval does not contain the population mean.

22. (a) The population consists of all eruptions of the Old Faithful Geyser. The variable under consideration is the time between eruptions (in minutes).

(b) Using Minitab, choose **Stat ▶ Basic statistics ▶ 1-Sample t..**, enter TIME in the **Samples in columns** text box. Click the **options…** button, enter 99 in the **Confidence interval** text box and click **OK**. The resulting confidence interval is

```
Variable    N     Mean   StDev  SE Mean       99% CI
TIME      500  91.5780  9.2022   0.4115  (90.5139, 92.6421)
```

We can be 99% confident that the population mean time between eruptions of the Old Faithful Geyser lies in the interval from 90.5139 to 92.6421 minutes.

(c) Strictly speaking, the confidence interval applies to the population from which the sample was drawn. There were no observations in the sample from the future, so the interval is not relevant to the future. In geologic terms, however, five years is insignificant, so the confidence interval is probably still somewhat meaningful. One should consider that unforeseen events, such as earthquakes or volcanic eruptions in the region could have a substantial effect on the time between eruptions. In that case, the current sample would not be representative of the population in five years.

23. (a) Your results will vary. To obtain the 3000 samples using Minitab, Choose **Calc ▶ Random Data ▶ Normal...**, enter 3000 in the **Generate rows of data** text box, type C1-C4 in the **Store in Column(s):** text box, enter 4.66 in the **Mean** text box, and enter 0.75 in the **Standard deviation:** text box. Click **OK**.

(b) To find the mean and sample standard deviation in each row (sample), choose **Calc ▶ Row statistics...** and click on **Mean**. Enter C1-C4 in the **Input variable(s):** text box and enter XBAR in the **Store result in:** text box. Repeat this last process, selecting **Standard Deviation** instead of **Mean** and put the results in SD.

(c) To obtain the Standardized version of each $\bar{x}$ in the sample, choose **Calc ▶ Calculator...**, enter STANDARD in the **Store results in variable:** text box, enter ('XBAR'-4.66)/(0.75/SQRT(4)) in the **Expression:** text box and click **OK**.

(d) To facilitate a comparison in part (h), **do part (g) now**. Then choose **Graph ▶ Histogram...**, select the **Simple** version, and click **OK**. Enter STANDARD and STUDENT in the **Graph variables** text box. Click on the **Multiple graphs** button, then click the **On separate graphs** button and check the boxes for **Same X, including same bins** and **Same Y**. Click **OK** twice. The first graph is shown below. Yours will differ, yet look similar.

Copyright © 2012 Pearson Education, Inc. Publishing as Addison-Wesley.

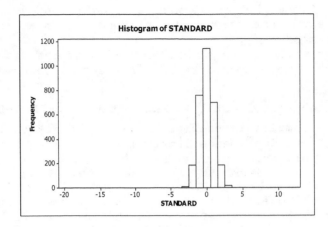

(e) In theory, the distribution of the standardized version of $\bar{x}$ is standard normal.

(f) The histogram in (e) appears to be very close to standard normal. Recall that the distribution is centered at zero and 99.74% of the data should be within 3 standard deviations of the mean. It does appear that this is so for this simulated data.

(g) To obtain the Studentized version of each $\bar{x}$ in the sample, choose **Calc** ▶ **Calculator...**, enter STUDENT in the **Store results in variable** text box, enter ('SBAR'-4.66)/('SD'/SQRT(4) in the **Expression:** text box and click **OK**.

(h) This graph was produced by the procedure in part (d). Our second graph follows, but yours will differ, yet look similar.

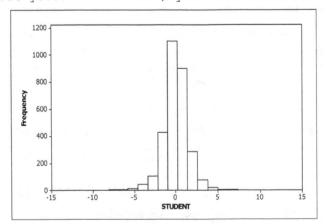

(i) In theory, the distribution of the studentized version of $\bar{x}$ is a t-distribution with 3 degrees of freedom.

(j) The distribution shown is symmetric about zero as is a t-distribution.

(k) The histogram of the studentized version in (h) is much more spread out than that of the standardized version in (d). The reason is that there is more variability in the t-distribution due to the extra uncertainty arising from the use of s instead of σ.

Copyright © 2012 Pearson Education, Inc. Publishing as Addison-Wesley.

CHAPTER 9 ANSWERS

Exercises 9.1

9.1 A hypothesis is a statement that something is true.

9.3 (a) The population mean μ is equal to some fixed amount; i.e., $\mu = \mu_0$.

   (b) The population mean μ is greater than μ_0; i.e., H_a: $\mu > \mu_0$.

   The population mean μ is less than μ_0; i.e., H_a: $\mu < \mu_0$.

   The population mean μ is unequal to μ_0; i.e., H_a: $\mu \neq \mu_0$.

9.5 Let μ denote the mean cadmium level in *Boletus pinicola* mushrooms.

   (a) H_0: $\mu = 0.5$ ppm   (b) H_a: $\mu > 0.5$ ppm   (c)   right-tailed test

9.7 Let μ denote the mean daily intake of iron by adult females under age 51.

   (a) H_0: $\mu = 18$ mg/day (b) H_a: $\mu < 18$ mg/day   (c)   left-tailed

9.9 Let μ denote the mean length of imprisonment for motor-vehicle theft offenders in Australia.

   (a) H_0: $\mu = 16.7$ months   (b) H_a: $\mu \neq 16.7$ months

   (c) two-tailed

9.11 Let μ denote the mean body temperature of healthy humans.

   (a) H_0: $\mu = 98.6°F$ (b) H_a: $\mu \neq 98.6°F$ (c)   two-tailed

9.13 Let μ denote the mean local monthly bill for cell phone users in the U.S. for last year.

   (a) H_0: $\mu = \$49.94$ (b) H_a: $\mu < \$49.94$ (c)   left-tailed

9.15 (a) If the null hypothesis is in fact false, you cannot possibly make a Type I error. You would either reject the null hypothesis when it is false, which is a correct decision, or you would not reject the null hypothesis when it is false, which is a Type II error.

   (b) If the null hypothesis is in fact false, you could possibly make a Type II error. You would either reject the null hypothesis when it is false, which is a correct decision, or you would not reject the null hypothesis when it is false, which is a Type II error.

9.17 True. The significance level is equal to the probability of a Type I error, which is the probability of rejecting a true null hypothesis. Therefore, the smaller the significance level, the smaller the probability of rejecting a true null hypothesis.

9.19 A Type I error is made when a true null hypothesis is rejected. The probability of making this error is denoted by α. A Type II error is made when a false null hypothesis is not rejected. We denote the probability of a Type II error by β.

9.21 (a) A Type I error would occur if, in fact, $\mu = 0.5$ ppm, but the results of the sampling lead to the conclusion that $\mu > 0.5$ ppm.

   (b) A Type II error would occur if, in fact, $\mu > 0.5$ ppm, but the results of the sampling fail to lead to that conclusion.

   (c) A correct decision would occur if, in fact, $\mu = 0.5$ ppm and the results of the sampling do not lead to the rejection of that fact; or

Copyright © 2012 Pearson Education, Inc. Publishing as Addison-Wesley.

if, in fact, $\mu > 0.5$ ppm and the results of the sampling lead to that conclusion.

(d) If, in fact, the mean cadmium level in *Boletus pinicola* mushrooms is equal to 0.5 ppm, and we do not reject the null hypothesis that $\mu = 0.5$ ppm, we made a correct decision.

(e) If, in fact, the mean cadmium level in *Boletus pinicola* mushrooms is greater than to 0.5 ppm, and we do not reject the null hypothesis that $\mu = 0.5$ ppm, we made a Type II error.

9.23 (a) A Type I error would occur if, in fact, $\mu = 18$ mg, but the results of the sampling lead to the conclusion that $\mu < 18$ mg.

(b) A Type II error would occur if, in fact, $\mu < 18$ mg, but the results of the sampling fail to lead to that conclusion.

(c) A correct decision would occur if, in fact, $\mu = 18$ mg and the results of the sampling do not lead to the rejection of that fact; or if, in fact, $\mu < 18$ mg and the results of the sampling lead to that conclusion.

(d) If the mean iron intake equals the RDA of 18 mg, and we reject the null hypothesis that $\mu = 18$ mg, we made a Type I error.

(e) If, in fact, the mean iron intake is less than the RDA of 18 mg, and we reject the null hypothesis that $\mu = 18$ mg, we made a correct decision.

9.25 (a) A Type I error would occur if, in fact, $\mu = 16.7$ months, but the results of the sampling lead to the conclusion that µ $\neq 16.7$ months.

(b) A Type II error would occur if, in fact, $\mu \neq 16.7$ months, but the results of the sampling fail to lead to that conclusion.

(c) A correct decision would occur if, in fact, $\mu = 16.7$ months and the results of the sampling do not lead to the rejection of that fact; or if, in fact, $\mu \neq 16.7$ months and the results of the sampling lead to that conclusion.

(d) If, in fact, the mean length of imprisonment equals 16.7 months, and we do not reject the null hypothesis that $\mu = 16.7$ months, we made a correct decision.

(e) If, in fact, the mean length of imprisonment does not equal 16.7 months, and we do not reject the null hypothesis that $\mu = 16.7$ months, we made a Type II error.

9.27 (a) A Type I error would occur if, in fact, $\mu = 98.6°$ F, but the results of the sampling lead to the conclusion that $\mu \neq 98.6°$ F.

(b) A Type II error would occur if, in fact, $\mu \neq 98.6°$ F, but the results of the sampling fail to lead to that conclusion.

(c) A correct decision would occur if, in fact, $\mu = 98.6°$ F and the results of the sampling do not lead to the rejection of that fact; or if, in fact, $\mu \neq 98.6°$ F and the results of the sampling lead to that conclusion.

(d) If the mean temperature of all healthy humans equals 98.6° F, and we reject the null hypothesis that $\mu = 98.6°$ F, we made a Type I error.

(e) If, in fact, the temperature of all healthy humans is not equal to 98.6° F, and we reject the null hypothesis that $\mu = 98.6°$ F, we made a

Copyright © 2012 Pearson Education, Inc. Publishing as Addison-Wesley.

correct decision.

9.29 (a) A Type I error would occur if, in fact, $\mu = \$49.94$, but the results of the sampling lead to the conclusion that $\mu < \$49.94$.

(b) A Type II error would occur if, in fact, $\mu < \$49.94$, but the results of the sampling fail to lead to that conclusion.

(c) A correct decision would occur if, in fact, $\mu = \$49.94$ and the results of the sampling do not lead to the rejection of that fact; or if, in fact, $\mu < \$49.94$ and the results of the sampling lead to that conclusion.

(d) If the mean phone bill equals the 2007 mean of $\$49.94$, and we do not reject the null hypothesis that $\mu = \$49.94$, we made a correct decision.

(e) If, in fact, the mean cell phone bill is less than the 2007 mean of $\$49.94$, and we do not reject the null hypothesis that $\mu = \$49.94$, we made a Type II error.

9.31 (a) A Type I error would occur if, in fact, the defendant is innocent, but the jury concludes that the defendant is guilty.

(b) A Type II error would occur if, in fact, the defendant is guilty, but the jury fails to conclude that the defendant is guilty.

(c) If I were a defendant, I would want α to be small. Given that I am innocent, I certainly want there to be a small probability of the jury rejecting my innocence (i.e., finding me guilty).

(d) If I were a prosecutor, I would want β to be small. Given that the defendant is guilty, I want there to be a small probability that the jury would declare the defendant not guilty.

(e) If $\alpha = 0$, then an innocent person would never be declared guilty. If $\beta = 0$, then a guilty person would always be found guilty.

Exercises 9.2

9.33 (a) Rejection region: $z \geq 1.645$

(b) Nonrejection region: $z < 1.645$

(c) Critical value: $z = 1.645$

(d) Significance level: $\alpha = 0.05$

(e)

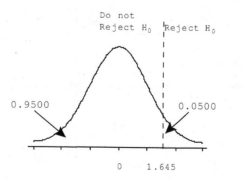

Copyright © 2012 Pearson Education, Inc. Publishing as Addison-Wesley.

   (f)   Right-tailed test

9.35 (a) Rejection region: $z \leq -2.33$

   (b) Nonrejection region: $z > -2.33$

   (c) Critical value: $z = -2.33$

   (d) Significance level: $\alpha = 0.01$

   (e)

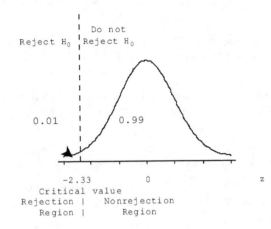

   (f) Left-tailed test

9.37 (a) Rejection region: $z \leq -1.645$ or $z \geq -1.645$

   (b) Nonrejection region: $-1.645 < z < -1.645$

   (c) Critical values: $z = -1.645$ and $z = 1.645$

   (d) Significance level: $\alpha = 0.10$

   (e)

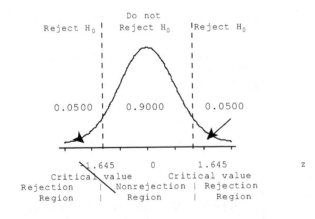

   (f)   Two-tailed test

Copyright © 2012 Pearson Education, Inc. Publishing as Addison-Wesley.

9.39 Critical values: $\pm z_{0.05} = \pm1.645$

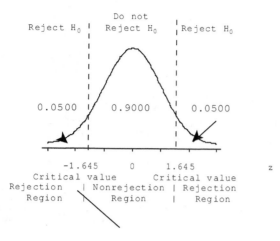

9.41 Critical value: $-z_{0.01} = -2.33$

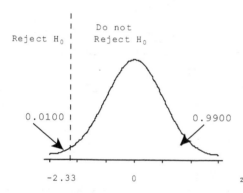

9.43 Critical value: $z_{0.01} = 2.33$

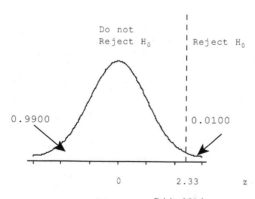

Copyright © 2012 Pearson Education, Inc. Publishing as Addison-Wesley.

Exercises 9.3

9.45 (1) It allows the reader to assess significance at any desired level, and (2) it permits the reader to evaluate the strength of the evidence against the null hypothesis.

9.47 The P-value for a one-sample z-test is obtained as:

(a) $P(z \leq$ observed z value) for a left-tailed test

(b) $P(z \geq$ observed z value) for a right-tailed test

(c) $2P(z \geq$ absolute value of the observed z value) for a two-tailed test.

9.49 (a) Do not reject the null hypothesis.

(b) Reject the null hypothesis.

(c) Reject the null hypothesis.

9.51 A P-value of 0.02 provides stronger evidence against the null hypothesis than does a value of 0.03. It says that if the null hypothesis is true, the data are less likely than they are when the P-value is 0.03.

9.53 (a) Strength of the evidence against the null hypothesis is moderate.

(b) There is weak or no evidence against the null hypothesis.

(c) Strength of the evidence against the null hypothesis is strong.

(d) Strength of the evidence against the null hypothesis is very strong.

9.55 (a) z = 2.03, P-value = 1.000 - 0.9788 = 0.0212. At a 5% significance level, we would reject the null hypothesis in favor of the alternative hypothesis.

(b) z = -0.31, P-value = 1.000 - 0.3783 = 0.6217. At a 5% significance level, we would not reject the null hypothesis in favor of the alternative hypothesis.

9.57 (a) z = -0.74, P-value = 0.2296. At a 5% significance level, we would not reject the null hypothesis in favor of the alternative hypothesis.

(b) z = 1.16, P-value = 0.8770. At a 5% significance level, we would not reject the null hypothesis in favor of the alternative hypothesis.

9.59 (a) z = -1.66, Left-tail probability = 0.0485

P-value = 0.0485 x 2 = 0.0970

At a 5% significance level, we would not reject the null hypothesis in favor of the alternative hypothesis.

(b) z = 0.52, Right-tail probability = 1.0000 - 0.6985 = 0.3015

P-value = 0.3015 x 2 = 0.6030

At a 5% significance level, we would not reject the null hypothesis in favor of the alternative hypothesis.

9.61 (a) The P-value is expressed as $P(z \leq z_0)$ if the hypothesis test is left-tailed.

(b) The P-value is expressed as $2 \cdot P(|z| \geq |z_0|)$ if the test is two-tailed.

9.63 Given that x can be transformed to z (and x_0 to z_0), we have:

1. $P(x \geq x_0) = P(z \geq z_0)$, for a right-tailed test

2. $P(x \leq x_0) = P(z \leq z_0)$, for a left-tailed test

3. $2 \cdot \min \{P(x \leq x_0), P(x \geq x_0)\} = 2 \cdot \min \{P(z \leq z_0), P(z \geq z_0)\}$

Copyright © 2012 Pearson Education, Inc. Publishing as Addison-Wesley.

$$= \begin{cases} 2 \cdot P(z \leq z_0) & \text{if } z_0 < 0 \\ 2 \cdot P(z \geq z_0) & \text{if } z_0 \geq 0 \end{cases}$$

$$\text{By symmetry} = \begin{cases} 2 \cdot P(z \geq -z_0) & \text{if } z_0 < 0 \\ 2 \cdot P(z \leq z_0) & \text{if } z_0 \geq 0 \end{cases} = 2 \cdot P(z \geq |z_0|)$$

By symmetry $= P(z \leq -|z_0|) + P(z \geq |z_0|) = P(|z| \geq |z_0|)$

Exercises 9.4

9.65 (a) The z-test in not an appropriate method for highly skewed data when the sample size is less than 30.

(b) The z-test is appropriate for large samples with no outliers even if the data are mildly skewed.

9.67 For the critical value approach, reject H_0 if z < -1.645;

$z = (20-22)/(4/\sqrt{32}) = -2.83$; therefore, reject H_0 and conclude that μ < 22. For the P-value approach, the P-value is 0.0023. Since the P-value is less than the significance level, reject H_0 and conclude that μ < 22.

9.69 For the critical value approach, reject H_0 if z > 1.645;

$z = (24-22)/(4/\sqrt{15}) = 1.94$; therefore, reject H_0 and conclude that μ > 22. For the P-value approach, the P-value is 1.000 − 0.9738 = 0.0262. Since the P-value is less than the significance level, reject H_0 and conclude that μ > 22.

9.71 For the critical value approach, reject H_0 if z < -1.96 or z > 1.96;

$z = (23-22)/(4/\sqrt{24}) = 1.22$; therefore, do not reject H_0. The data do not provide sufficient evidence to support H_a: $\mu \neq 22$. For the P-value approach, the right tailed probability is 0.1112 and the P-value would be 0.2224. Since the P-value is greater than the significance level, do not reject H_0. The data do not provide sufficient evidence to support H_a: $\mu \neq 22$.

9.73 n = 12, σ = 0.37 ppm, $\bar{x}$ = 6.31/12 = 0.526 ppm

Step 1: H_0: μ = 0.5 ppm, H_a: μ > 0.5 ppm

Step 2: α = 0.05

Step 3: $z = (0.526-0.5)/(0.37/\sqrt{12}) = 0.24$

Step 4: Critical-Value Approach: Critical value = $z_\alpha = z_{0.05} = 1.645$.

P-Value Approach: P-value is $P(Z > 0.24) = 1 - 0.5948 = 0.4052$.

Step 5: Critical-Value Approach: Since 0.24 < 1.645, do not reject H_0.

P-value Approach: Since 0.4052 > 0.05, do not reject H_0.

Step 6: At the 5% significance level, the data do not provide sufficient evidence to conclude that the mean cadmium level μ of *Boletus pinicola* mushrooms is greater than the safety limit of 0.5 ppm.

9.75 n = 45, $\bar{x}$ = 14.68, σ = 4.2

Step 1: H_0: μ = 18 mg, H_a: μ < 18 mg

Step 2: α = 0.01

Copyright © 2012 Pearson Education, Inc. Publishing as Addison-Wesley.

Step 3: $z = (14.68 - 18)/(4.2/\sqrt{45}) = -5.30$

Step 4:   Critical-Value Approach: Critical value $= -z_\alpha = -z_{0.01} = -2.33$.

P-Value Approach: P-value is $P(Z < -5.30) = 0.0000$.

Step 5:   Critical-Value Approach: Since -5.30 < -2.33, reject H_0.

P-value Approach: Since 0.0000 < 0.01, reject H_0.

Step 6:   At the 1% significance level, the data provide sufficient evidence to conclude that adult females under the age of 51 are, on the average, getting less than the RDA of 18 mg of iron. Considering that iron deficiency causes anemia and that iron is required for transporting oxygen in the blood, this result could have practical significance as well.

9.77   n = 100, $\bar{x}$ = 17.8 months, σ = 6.0 months

Step 1: H_0: μ = 16.7 months, H_a: $\mu \neq$ 16.7 months

Step 2: α = 0.05

Step 3: $z = (17.8 - 16.7)/(6.0/\sqrt{100}) = 1.83$

Step 4:   Critical-Value Approach: Critical value $= \pm z_{\alpha/2} = \pm z_{0.025} = \pm 1.96$.

P-Value Approach: P-value is $2 \cdot P(Z > 1.83) = 2 \cdot (1 - 0.9664) = 0.0672$.

Step 5:   Critical-Value Approach: Since 1.83 < 1.96, do not reject H_0.

P-value Approach: Since 0.0672 > 0.05, do not reject H_0.

Step 6:   At the 5% significance level, the data do not provide sufficient evidence to conclude that the mean length of imprisonment μ of motor-vehicle theft offenders in Sydney differs from the national mean in Australia.

9.79   (a)   Using Minitab, with the data in a column named GAIN, we choose **Stat ▶**

Basic Statistics ▶ 1-Sample z..., click in the **Samples in columns** text box and specify GAIN, click in the **Standard deviation** text box and type 0.42, and click in the **Test mean** text box and type 0.2. Click the **Options...** button, enter 95 in the **Confidence level** text box, click the arrow button at the right of the **Alternative** drop-down list box and select **greater than** and click **OK**. Click on the **Graphs** button and check the boxes for **Histogram of Data** and **Boxplot of data**. Then click **OK** twice. The result of the test is

```
Test of mu = 0.2 vs > 0.2
The assumed standard deviation = 0.42
```

Variable	N	Mean	StDev	SE Mean	95% Lower Bound	Z	P
GAIN	20	0.295000	0.499974	0.093915	0.140524	1.01	0.156

(b)   The histogram and boxplot were produced by the procedure in part (a).

Now choose **Stat ▶ Basic Statistics ▶ Normality test** and enter GAIN in

the **Variable** text box. Click **OK**. Then choose **Graph ▶ Stem-and-Leaf** ,

Copyright © 2012 Pearson Education, Inc. Publishing as Addison-Wesley.

enter <u>GAIN</u> in the **Graph variables** text box and click **OK**. The results are

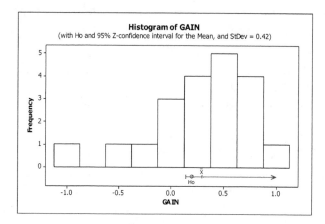

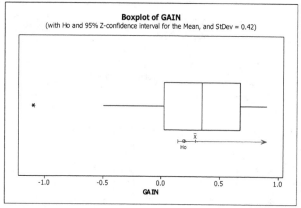

Stem-and-leaf of GAIN   N = 20
Leaf Unit = 0.10

```
LO -11

  2   -0   5
  3   -0   2
  4   -0   1
  6    0   01
 10    0   2233
 10    0   45
  8    0   6667
  4    0   8889
```

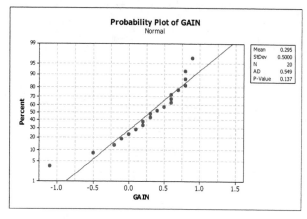

(c) Repeating the procedure of part (a), we obtain
     Test of mu = 0.2 vs > 0.2
     The assumed standard deviation = 0.42

Variable	N	Mean	StDev	SE Mean	95% Lower Bound	Z	P
GAIN	19	0.368421	0.387374	0.096355	0.209932	1.75	0.040

(d) The original sample size is only 20. The plots in part (b) indicate
     that the value -1.1 is a potential outlier. The z-test should not be
     used with the original data. This is further confirmed by the fact
     that using all of the data leads to z = 1.01, whereas, deleting the
     outlier leads to z = 1.75. This is enough of a change to alter our
     conclusion from not rejecting the null hypothesis to rejecting it. If
     there is no good reason for deleting the outlier, then the z-test is
     inappropriate for these data.

9.81 (a) Using Minitab, with the data in a column named TEMP, we choose **Stat ▶**

          Basic Statistics ▶ 1-Sample z..., click in the **Samples in columns** text
          box and specify <u>TEMP</u>, click in the **Standard deviation** text box and
          enter <u>0.63</u>, and click in the **Hypothesized** text box and enter <u>98.6</u>.
          Click the **Options...** button, enter <u>95</u> in the **Confidence level** text box,
          click the arrow button at the right of the **Alternative** drop-down list
          box and select **not equal** and click **OK**. Click on the **Graphs** button and
          check the boxes for **Histogram of Data** and **Boxplot of data**. Then click

Copyright © 2012 Pearson Education, Inc. Publishing as Addison-Wesley.

OK twice. Now choose **Stat ▶ Basic Statistics ▶ Normality test** and enter <u>CHARGE</u> in the **Variable** text box. Click **OK**. Then choose **Graph ▶ Stem-and-Leaf**, enter <u>CHARGE</u> in the **Graph variables** text box and click **OK**. The graphs are

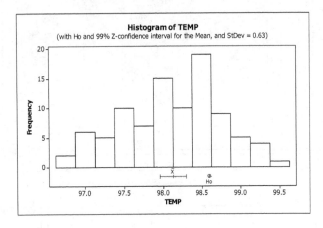

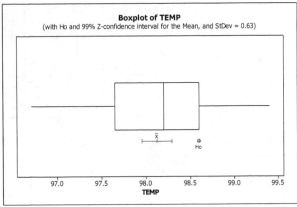

```
Stem-and-leaf of TEMP   N  = 93
Leaf Unit = 0.10

     1    96  7
     3    96  89
     8    97  00001
    13    97  22233
    19    97  444444
    26    97  6666777
    31    97  88889
    45    98  00000000000111
   (10)   98  2222222233
    38    98  4444445555
    28    98  66666666677
    17    98  8888888
    10    99  00001
     5    99  2233
     1    99  4
```

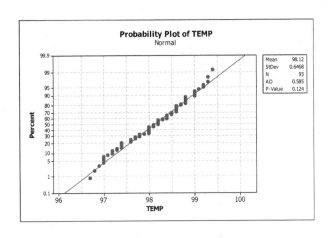

(b) Yes. The sample size 93 is large and the distribution of the data is quite symmetric.

(c) Yes. The procedure in part (a) also produced the results of the test which are

Test of mu = 98.6 vs not = 98.6
The assumed standard deviation = 0.63

Variable	N	Mean	StDev	SE Mean	99% CI	Z	P
TEMP	93	98.1237	0.6468	0.0653	(97.9554, 98.2919)	-7.29	0.000

The P-value is 0.000. Since the P-value is less than 0.01, we reject the null hypothesis and conclude that the mean body temperature of healthy humans is different from the generally accepted value of 98.6°F.

Copyright © 2012 Pearson Education, Inc. Publishing as Addison-Wesley.

9.83 (a) Using Minitab, with the data in a column named BILL, we choose **Stat ▶**

Basic Statistics ▶ 1-Sample z..., click in the **Samples in columns** text box and specify BILL, click in the **Standard deviation** text box and enter 25, and click in the **Hypothesized mean** text box and enter 49.94. Click the **Options...** button, enter 95 in the **Confidence level** text box, click the arrow button at the right of the **Alternative** drop-down list box and select **less than** and click **OK**. Click on the **Graphs** button and check the boxes for **Histogram of Data** and **Boxplot of data**. Then click **OK** twice. Now choose **Stat ▶ Basic Statistics ▶ Normality test** and enter BILL in the **Variable** text box. Click **OK**. Then choose **Graph ▶ Stem-and-Leaf**, enter BILL in the **Graph variables** text box and click **OK**. The graphs are

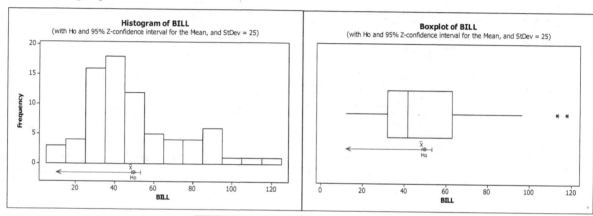

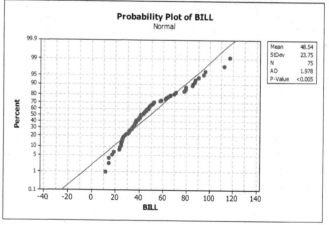

```
Stem-and-leaf of BILL   N  = 75
Leaf Unit = 1.0
   5    1   14478
  16    2   34555566689
  33    3   0112334556677 7799
 (17)   4   00011334556777899
  25    5   11278
  20    6   224669
  14    7   18999
   9    8   5577
   5    9   046
   2   10
   2   11   27
```

Copyright © 2012 Pearson Education, Inc. Publishing as Addison-Wesley.

(b) The results of the test carried out by the procedure in part (a) are

```
Test of mu = 49.94 vs < 49.94
The assumed standard deviation = 25
                                        95% Upper
Variable     N    Mean  StDev  SE Mean     Bound      Z      P
BILL        75   48.54  23.75    2.89      53.29   -0.48  0.314
```

The P-value for the test is 0.314. Since the P-value is greater than 0.05, we do not reject the null hypothesis. The data do not provide sufficient evidence that the mean local monthly cell phone bill has increased from the 2007 mean of $49.94.

(c) After deleting the two outliers 112.88 and 117.51, the graphs and test results are

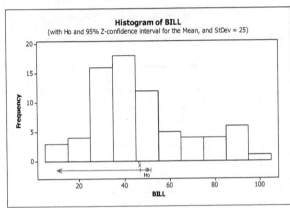

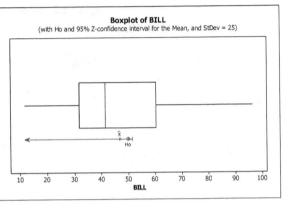

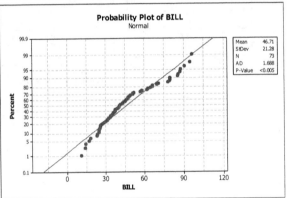

```
Stem-and-leaf of BILL   N  = 73
Leaf Unit = 1.0
 N* = 2
   3    1  144
   5    1  78
   7    2  34
  16    2  555566689
  23    3  0112334
  33    3  5566777799
  (8)   4  00011334
  32    4  556777899
  23    5  112
  20    5  78
  18    6  224
  15    6  669
  12    7  1
  11    7  8999
   7    8
   7    8  5577
   3    9  04
   1    9  6
```

Copyright © 2012 Pearson Education, Inc. Publishing as Addison-Wesley.

```
Test of mu = 49.94 vs < 49.94
The assumed standard deviation = 25


                                           95% Upper
Variable    N    Mean  StDev  SE Mean      Bound     Z      P
BILL        73  46.71  21.28     2.93      51.53  -1.10  0.135
```

(d) Although the value of z has been changed from -0.48 to -1.10 by deleting the two outliers, the conclusion remains the same. We do not reject the null hypothesis.

9.85 (a) $n = 28$, $\sigma = \$8.45$, $\overline{x} = \$1788.62/28 = \$63.88$

The 90% confidence interval is $63.88 \pm 1.645(8.45)/\sqrt{28} = (61.25, 66.51)$.

The hypothesized mean ($66.52) lies outside of the confidence interval, so we should reject the null hypothesis. Since the test statistic is $z = (63.88 - 66.52)/(8.45/\sqrt{28}) = -1.653$, which is less than the lower critical value of -1.645, the hypothesis test also leads to the conclusion that we should reject the null hypothesis.

(b) $n = 100$, $\overline{x} = 17.8$ months, $\sigma = 6.0$ months

The 90% confidence interval is $17.8 \pm 1.96(6.0)/\sqrt{100} = (16.6, 19.0)$.

The hypothesized mean (16.7) lies inside of the confidence interval, so we should not reject the null hypothesis. Since the test statistic is $z = (17.8 - 16.7)/(6.0/\sqrt{100}) = 1.83$, which is less than the upper critical value of 1.96, the hypothesis test also leads to the conclusion that we should not reject the null hypothesis.

9.87 (a) $n = 12$, $\sigma = 0.37$ ppm, $\overline{x} = 6.31/12 = 0.526$ ppm

The 95% lower level confidence bound is $0.526 - 1.645(0.37)/\sqrt{12} = 0.350$.

The hypothesized mean (0.5) lies above the lower confidence bound, so we should not reject the null hypothesis. Since the test statistic is $z = (0.526 - 0.5)/(0.37/\sqrt{12}) = 0.24$, which is less than the critical value of 1.645, the hypothesis test also leads to the conclusion that we should not reject the null hypothesis.

(b) $n = 30$, $\overline{x} = 78.3$, $\sigma = 11.2$

The 95% lower level confidence bound is $78.3 - 1.645(11.2)/\sqrt{30} = 74.94$.

The hypothesized mean (72) lies below the lower confidence bound, so we should reject the null hypothesis. Since the test statistic is $z = (78.3 - 72)/(11.2/\sqrt{30}) = 3.08$, which is greater than the critical value of 1.645, the hypothesis test also leads to the conclusion that we should reject the null hypothesis.

Exercises 9.5

9.89 (a) $0.01 < P < 0.025$

(b) Reject H_0 for $\alpha \geq 0.025$; Do not reject H_0 for $\alpha \leq 0.01$; Undecided for $0.01 < \alpha < 0.025$.

Copyright © 2012 Pearson Education, Inc. Publishing as Addison-Wesley.

9.91 (a) P < 0.005

(b) Reject H_0 for $\alpha \geq 0.005$; Undecided for $\alpha < 0.005$.

9.93 (a) 0.01 < P < 0.02

(b) Reject H_0 for $\alpha \geq 0.02$; Do not reject H_0 for $\alpha \leq 0.01$; Undecided for 0.01 < α < 0.02.

9.95 (a) df = 31; t = -2.828

(b) P < 0.005; Reject H_0; Evidence against H_0 is very strong.

9.97 (a) df = 14; t = 1.936

(b) 0.025 < P < 0.05; Reject H_0; Evidence against H_0 is strong.

9.99 (a) df = 23; t = 1.225

(b) P > 0.20; Do not reject H_0; Evidence against H_0 is weak or none.

9.101 H_0: μ = 4.55, H_a: $\mu \neq$ 4.55, α = 0.10; Critical values: $\pm$1.729

$$t = (\bar{x} - \mu)/(s/\sqrt{n}) = (4.760 - 4.55)/(2.297/\sqrt{20}) = 0.409 \;.$$

Since -1.729 < 0.409 < 1.729, we do not reject H_0. The P-value is P > 0.20. The data does not provide sufficient evidence to conclude that the amount of television watched per day last year by the average person differed from that in 2005.

9.103 n = 10, df = 9, $\bar{x}$ = 2.5, s = 0.149

Step 1: H_0: μ = 2.3, H_a: μ > 2.3

Step 2: α = 0.01

Step 3: $t = \dfrac{2.5 - 2.3}{0.149/\sqrt{10}} = 4.251$

Step 4: Critical Value Approach: Critical value = 2.821

P-Value Approach: P-value < 0.01

Step 5: Since 4.251 > 2.821, reject H_0. Since the P-value < α, reject H_0.

Step 6: At the 1% significance level, the data do provide sufficient evidence to conclude that the mean available limestone in soil treated with 100% MMBL effluent is greater than 2.30%. The practical significance of this result probably depends on what crop is to be grown in the soil.

9.105 n = 187, df = 186, $\bar{x}$ = 0.64, s = 0.15

Step 1: H_0: μ = 0.9, H_a: μ < 0.9

Step 2: α = 0.05

Step 3: $t = \dfrac{0.64 - 0.90}{0.15/\sqrt{187}} = -23.703$

Step 4: Critical Value Approach: Critical value = -1.653

P-Value Approach: P < 0.005

Step 5: Since -23.703 < -1.653, reject H_0. Since the p-value < α, reject H_0.

Step 6: At the 5% significance level, the data provide sufficient evidence to conclude that the mean ABI μ for women with peripheral arterial disease is less than the healthy ABI of 0.9.

Copyright © 2012 Pearson Education, Inc. Publishing as Addison-Wesley.

Thus, we conclude that such women do have an unhealthy ABI. The practical significance of this result is that the ABI may be a good tool for determining the possibility of peripheral arterial disease in women. There could also be other causes of a low ABI, so this test by itself may not be able to determine the precise ailment.

9.107 We used Minitab to produce the following histogram.

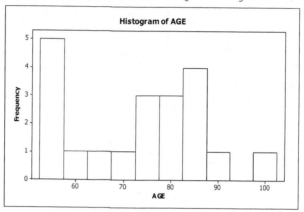

The sample size is only 20. Although there are no outliers, the distribution is not very close to being normally distributed. It does not appear to be reasonable to use a t-test with these data.

9.109 We used Minitab to produce the following normal probability plot.

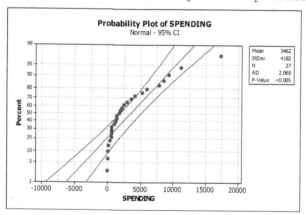

9.111 (a) Using Minitab, with the data in a column named PRESSURE, we choose **Stat**
▶ **Basic Statistics** ▶ **1-Sample t...**, click in the **Samples in columns** text box and specify PRESSURE, click in the **Test mean** text box and enter 80. Click the **Options...** button, enter 90 in the **Confidence level** text box, click the arrow button at the right of the **Alternative** drop-down list box and select **greater than** and click **OK**. Click on the **Graphs** button and check the boxes for **Histogram of Data** and **Boxplot of data**. Then click **OK** twice. Now choose **Stat** ▶ **Basic Statistics** ▶ **Normality test** and enter PRESSURE in the **Variable** text box. Click **OK**.

Then choose **Graph** ▶ **Stem-and-Leaf** , enter PRESSURE in the **Graph variables** text box and click **OK**. The results are

Copyright © 2012 Pearson Education, Inc. Publishing as Addison-Wesley.

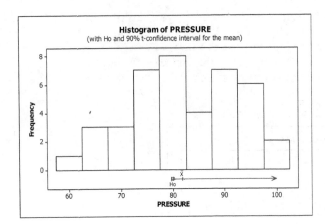

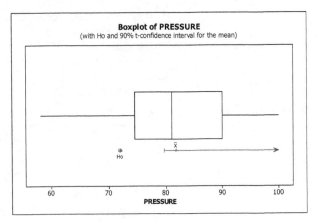

```
Stem-and-leaf of PRESSURE   N  = 41
Leaf Unit = 1.0


   1    5   8
   2    6   3
   5    6   569
  10    7   00334
  17    7   5677999
  (8)   8   00111334
  16    8   5899
  12    9   0001334
   5    9   5559
   1   10   0
```

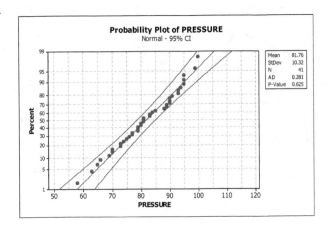

(b) The large sample size, lack of potential outliers, and the nearly linear probability plot all indicate that a t-test is reasonable for these data.

(c) The first procedure in part (a) also yielded the following test results:

Test of mu = 80 vs > 80

				90% Lower		
Variable	N	Mean	StDev	SE Mean	Bound	T	P
PRESSURE	41	81.7561	10.3242	1.6124	79.6551	1.09	0.141

We see that t = 1.09 and the P-value is 0.141. Since the P-value is greater than the significance level 0.10, we do not reject the null hypothesis and conclude that the data do not provide evidence that the mean diastolic blood pressure of bus drivers in Stockholm exceeds the normal pressure of 88 mm Hg.

9.113 (a) Using Minitab, with the data in a column named RENT, we choose **Stat ▶**

Basic Statistics ▶ 1-Sample t..., click in the **Samples in columns** text box and specify RENT, click in the **Test mean** text box and enter 779. Click the **Options...** button, enter 95 in the **Confidence level** text box, click the arrow button at the right of the **Alternative** drop-down list

Copyright © 2012 Pearson Education, Inc. Publishing as Addison-Wesley.

box and select **greater than** and click **OK**.

Test of mu = 779 vs > 779

Variable	N	Mean	StDev	SE Mean	95% Lower Bound	T	P
RENT	100	790.96	89.85	8.98	776.04	1.33	0.093

The P-value for the test is 0.093, larger than the significance lever 0.05, so we do not reject the null hypothesis. The data do not provide sufficient evidence to conclude that the mean rent for a two-bedroom unit in Maine is greater than the FMR of $779.

(b) After removing the outlier 492 and following the procedure in part (a), the results are

Test of mu = 779 vs > 779

Variable	N	Mean	StDev	SE Mean	95% Lower Bound	T	P
RENT	99	793.98	85.05	8.55	779.79	1.75	0.041

(c) Now the P-value is 0.041, leading to rejection of the null hypothesis. The sample mean increased by $3.02 and the standard deviation decreased by about $4.8.

(d) The sample size is large in both cases, yet the effect of the outlier is considerable. Caution should be used and perhaps the data should be analyzed using a method that is not influenced by outliers.

9.115 (a) Yes. The t-test is designed for use when the population is normally distributed and the population standard deviation is unknown.

(b) Yes. The Wilcoxon signed-rank test can be used when the population distribution is symmetric. Since a normal distribution is symmetric, a Wilcoxon signed-rank test is permissible.

(c) The t-test is better in this situation since it will give more accurate results when the population is normally distributed.

9.117 (a) n = 20, df = 19, $t_{\alpha/2}$ = 1.729, s = 2.291, $\bar{x}$ = 4.835

The 90% confidence interval is $4.835 \pm 1.729(2.291)/\sqrt{20} = (3.949, 5.721)$.

The hypothesized mean (4.66) lies within the confidence interval, so we should not reject the null hypothesis. Since the test statistic is

$t = (4.835 - 4.66)/(2.291/\sqrt{20}) = 0.342$, which is less than the critical value of 1.729, the hypothesis test also leads to the conclusion that we should not reject the null hypothesis.

(b) n = 25, df = 24, $t_{\alpha/2}$ = 2.064, s = $350.90, $\bar{x}$ = $1935.76
The 95% confidence interval is

$$1935.76 \pm 2.064(350.90)/\sqrt{25} = (1790.91, 2080.61).$$

The hypothesized mean ($1749) lies outside the confidence interval, so we should reject the null hypothesis. Since the test statistic is

$t = (1935.76 - 1749)/(350.90/\sqrt{25}) = 2.661$, which is greater than the critical value of 2.064, the hypothesis test also leads to the conclusion that we should reject the null hypothesis.

Copyright © 2012 Pearson Education, Inc. Publishing as Addison-Wesley.

9.119 (a) n = 6, df = 5, t_α = 2.015, s = 2.7, $\bar{x}$ = 182.7

The 95% lower confidence bound is $182.7 - 2.015(2.7)/\sqrt{6} = 180.479$.

The hypothesized mean (180) lies below the lower confidence bound, so we should reject the null hypothesis. Since the test statistic is $t = (182.7 - 180)/(2.7/\sqrt{6}) = 2.449$, which is greater than the critical value of 2.015, the hypothesis test also leads to the conclusion that we should reject the null hypothesis.

The 99% lower confidence bound is $182.7 - 3.365(2.7)/\sqrt{6} = 178.991$.

The hypothesized mean (180) lies above the lower confidence bound, so we should not reject the null hypothesis. Since the test statistic is $t = (182.7 - 180)/(2.7/\sqrt{6}) = 2.449$, which is less than the critical value of 3.365, the hypothesis test also leads to the conclusion that we should not reject the null hypothesis.

(b) n = 10, df = 9, t_α = 2.821, s = 0.149, $\bar{x}$ = 2.5

The 99% lower confidence bound is $2.5 - 2.821(0.149)/\sqrt{10} = 2.367$.

The hypothesized mean (2.30) lies below the lower confidence bound, so we should reject the null hypothesis. Since the test statistic is $t = (2.5 - 2.3)/(0.149/\sqrt{10}) = 4.245$, which is greater than the critical value of 2.821, the hypothesis test also leads to the conclusion that we should reject the null hypothesis.

Review Problems for Chapter 9

1.   (a) A null hypothesis always specifies a single value for the parameter of a population which is of interest.

(b) The alternative hypothesis reflects the purpose of the hypothesis test, which can be to determine that the parameter of interest is greater than, less than, or different from the single value specified in the null hypothesis.

(c) The test statistic is a quantity calculated from the sample, under the assumption that the null hypothesis is true, which is used as a basis for deciding whether or not to reject the null hypothesis.

(d) The significance level is the probability of making a Type I error. That is, it is the probability of rejecting a true null hypothesis.

2.   (a) The statement is expressing the fact that there is variability in the net weights of the boxes' content and some boxes may actually contain less than the printed weight on the box. However, the net weights for each day's production will average a bit more than the printed weight.

(b) To test the truth of this statement, we would use a null hypothesis that stated that the population mean net weight of the boxes was equal to the printed weight and an alternative hypothesis that stated that the population mean net weight of the boxes was greater than the printed weight.

(c) Null hypothesis: Population mean net weight = 76 oz

Alternative hypothesis: Population mean net weight > 76 oz

or

H_0: μ = 76 H_a: μ > 76

Copyright © 2012 Pearson Education, Inc. Publishing as Addison-Wesley.

3. (a) Roughly speaking, there is a range of values of the test statistic which one could reasonably expect to occur if the null hypothesis were true. If the value of the test statistic is one that would not be expected to occur when the null hypothesis is true, then we reject the null hypothesis.

    (b) To make this procedure objective and precise, we specify the probability with which we are willing to reject the null hypothesis when it is actually true. This is called the significance level of the test and is usually some small number like 0.05 or 0.01. Specifying the significance level allows us to determine the range of values of the test statistic that will lead to rejection of the null hypothesis. If the computed value of the test statistic falls in this "rejection region," then the null hypothesis is rejected. If it does not fall in the rejection region, then the null hypothesis is not rejected.

4. We would use the alternative hypothesis $\mu \neq \mu_0$ if we wanted to determine whether the population mean were <u>different from</u> the value μ_0 specified in the null hypothesis. We would use the alternative hypothesis $\mu > \mu_0$ if we wanted to determine whether the population mean were <u>greater than</u> from the value μ_0 specified in the null hypothesis. We would use the alternative hypothesis $\mu < \mu_0$ if we wanted to determine whether the population mean were <u>less than</u> the value μ_0 specified in the null hypothesis.

5. (a) A Type I error is made whenever the null hypothesis is true, but the value of the test statistic leads us to reject the null hypothesis. A Type II error is made whenever the null hypothesis is false, but the value of the test statistic leads us to not reject the null hypothesis.

    (b) The probability of a Type I error is represented by α and that of a Type II error by β.

    (c) If the null hypothesis is true, the test statistic can lead us to either reject or not reject the null hypothesis. The first is the correct decision, while the latter constitutes a Type I error. Thus a Type I error is the only type of error possible when the null hypothesis is true.

    (d) If the null hypothesis is not rejected, a correct decision has been made if the null hypothesis is, in fact, true. But if the null hypothesis is false, we have made a Type II error. Thus a Type II error is the only type of error possible when the null hypothesis is not rejected.

6. The probability of a Type II error is increased when the significance level is decreased for a fixed sample size.

7. (a) The rejection region is a set of values of the test statistic that lead to rejection of the null hypothesis.

    (b) The nonrejection region is a set of values of the test statistic that lead to not rejecting the null hypothesis.

    (c) The critical values are values of the test statistic that separate the rejection region from the nonrejection region.

8. True. You would reject the null hypothesis if your test statistic is within the rejection region or equal to the critical value.

9. Assuming that the null hypothesis is true, find the value of the test statistic for which the probability of obtaining a value less than this value (the critical value) is 0.05.

Copyright © 2012 Pearson Education, Inc. Publishing as Addison-Wesley.

10. (a) For a right tailed one-mean z-test at 1% significance, the critical value is $z_\alpha = z_{0.01} = 2.33$.

    (b) For a left tailed one-mean z-test at 1% significance, the critical value is $-z_\alpha = -z_{0.01} = -2.33$.

    (c) For a two tailed one-mean z-test at 1% significance, the critical value is $\pm z_{\alpha/2} = \pm z_{0.005} = \pm 2.575$.

11. (a) Rejection region: $z \geq 1.28$

    (b) Nonrejection region: $z < 1.28$

    (c) Critical value: $z = 1.28$

    (d) Significance level: $\alpha = 0.10$

    (e)

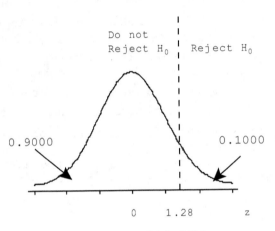

Critical Value
Nonrejection region | Rejection region

    (f) Right-tailed test

12. Step 1: State the null and alternative hypothesis.

    Step 2: Choose α .

    Step 3: Calculate the test statistic.

    Step 4: Calculate the critical value(s).

    Step 5: If the test statistic falls within the rejection region, reject the null hypothesis. Otherwise, do not reject the null hypothesis.

    Step 6: Interpret your conclusion in context of the problem.

13. The P-value of a hypothesis test is the probability, assuming that the null hypothesis is true, of getting a value of the test statistic that is as extreme or more extreme than the one actually obtained.

14. True. If the null hypothesis were true, a value of the test statistic with a P-value of 0.02 would be more extreme than one with a P-value of 0.03.

15. If the P-value is less than or equal to the significance level, reject the null hypothesis. Otherwise, do not reject the null hypothesis.

16. The P-value of a hypothesis test is also called the observed significance level since it represents the smallest possible significance level at which the null hypothesis could have been rejected.

Copyright © 2012 Pearson Education, Inc. Publishing as Addison-Wesley.

17. The P-value is the probability, assuming that the null hypothesis is true, of seeing a test statistic at least as extreme than that observed.

18. (a) The P-value for a left-tailed test with a test statistic of z = -1.25 is $P(Z < -1.25) = 0.1056$. This P-value of 0.1056 is greater than the significance level of 0.05. We would not reject the null hypothesis.

    (b) The P-value for a right-tailed test with a test statistic of z = 2.36 is $P(Z > 2.36) = 0.0091$. This P-value of 0.0091 is less than the significance level of 0.05. We would reject the null hypothesis.

    (c) The P-value for a two-tailed test with a test statistic of z = 1.83 is $2 \cdot P(Z > 1.83) = 2 \cdot (0.0336) = 0.0672$. This P-value of 0.0672 is greater than the significance level of 0.05. We would not reject the null hypothesis.

19. Step 1: State the null and alternative hypothesis.

    Step 2: Choose α.

    Step 3: Calculate the test statistic.

    Step 4: Calculate the P-value.

    Step 5: If the P-value is less than or equal to the significance level, reject the null hypothesis. Otherwise, do not reject the null hypothesis.

    Step 6: Interpret your conclusion in context of the problem.

20. The P-value of 0.062 is within the range of 0.05 < P-value < 0.10. Therefore, there is moderate evidence against the null hypothesis.

21. (a) A hypothesis test is exact if the actual significance level is the same as the one that is stated.

    (b) A hypothesis test is approximately correct if the actual significance level only approximately equals α.

22. A statistically significant result occurs when the value of the test statistic falls in the rejection region. A result has practical significance when it is statistically significant and the result also is different enough from results expected under the null hypothesis to be important to the consumer of the results. By taking large enough sample sizes, almost any result can be made statistically significant due to the increased ability of the test to detect a false null hypothesis, but small differences from the conditions expressed by the null hypothesis may not be important, that is, they may not have practical significance.

23. (a) If the population standard deviation is unknown, and the population is normal or the sample size is large, we can use the one-mean t-statistic, $t = (\bar{x} - \mu_0)/(s/\sqrt{n})$.

    (b) If the population standard deviation is known, and the population is normal or the sample size is large, we can use the one-mean z-statistic, $z = (\bar{x} - \mu_0)/(\sigma/\sqrt{n})$.

24. Let μ denote last year's mean cheese consumption by Americans.

    (a) H_0: μ = 30.0 lb

    (b) H_a: μ > 30.0 lb

    (c) This is a right-tailed test.

25. (a) A Type I error would occur if, in fact, μ = 30.0 lb, but the results

Copyright © 2012 Pearson Education, Inc. Publishing as Addison-Wesley.

of the sampling lead to the conclusion that $\mu > 30.0$ lb.

(b) A Type II error would occur if, in fact, $\mu > 30.0$ lb, but the results of the sampling fail to lead to that conclusion.

(c) A correct decision would occur if, in fact, $\mu = 30.0$ lb and the results of the sampling do not lead to the rejection of that fact; or if, in fact, $\mu > 30.0$ lb and the results of the sampling lead to that conclusion.

(d) If, in fact, last year's mean consumption of cheese for all Americans has not increased over the 2001 mean of 30.0 lb, and we reject the null hypothesis that $\mu = 30.0$ lb, we made a Type I error.

(e) If, in fact, last year's mean consumption of cheese for all Americans has increased over the 2001 mean of 30.0 lb, and we reject the null hypothesis that $\mu = 30.0$ lb, we made a correct decision.

26.   (a) $n = 35$, $\bar{x} = 1183/35 = 33.8$, $\sigma = 6.9$

Step 1: H_0: $\mu = 30.0$ lb, H_a: $\mu > 30.0$ lb

Step 2: $\alpha = 0.10$

Step 3: $z = (33.8 - 30.0)/(6.9/\sqrt{35}) = 3.26$

Step 4:   Critical Value Approach: Critical value = 1.28

P-value Approach: P-value is $P(Z > 3.26) = 0.0006$

Step 5:   Since 3.26 > 1.28, reject H_0. Since the P-value of 0.0006 is less than the significance level, reject H_0.

Step 6:   At the 10% significance level, the data provide sufficient evidence to conclude that last year's mean cheese consumption μ for all Americans has increased over the 2001 mean of 30.0 lb.

(b) Given the conclusion in part (a), if an error has been made, it must be a Type I error. This is because, given that the null hypothesis was rejected, the only error that could be made is the error of rejecting a true null hypothesis.

27. $n = 12$, $\bar{x} = \$404.0$, $s = \$86.8$

Step 1: H_0: $\mu = \$417$, H_a: $\mu < \$417$

Step 2: $\alpha = 0.05$

Step 3: $t = (404.0 - 417)/(86.8/\sqrt{12}) = -0.52$

Step 4:   Critical Value Approach: Critical value = -1.796

P-Value Approach: P > 0.10

Step 5:   Since -0.52 > -1.796, do not reject H_0. Since the P-value is greater than the significance level, do not reject H_0.

Step 6:   At the 5% significance level, the data do not provide sufficient evidence to conclude that the mean value lost because of purse snatching has decreased from the 2004 mean of $417.

28.   (a)   If the odds-makers are estimating correctly, the mean point-spread error is zero.

Copyright © 2012 Pearson Education, Inc. Publishing as Addison-Wesley.

(b) It seems reasonable to assume that the distribution of point spread errors is approximately normal. In any case, the sample size of 2109 is very large, so the t-test of H_0: $\mu = 0$ vs. H_a: $\mu \neq 0$ is appropriate. At the 5% significance level, the critical values are ± 1.960. Since $t = (-0.2 - 0.0)/(10.9/\sqrt{2109}) = 0.843$, we do not reject H_0.

(c) There is not sufficient evidence to conclude that the mean point-spread is different from zero.

29. Since the distribution is symmetric, use the Wilcoxon signed-rank test. The sample size is 50 and σ is known, so the z-test may be appropriate. Use caution, however, since it appears that there may be outliers.

30. The distribution is far from normal, being left-skewed. However, since the sample size is 37 and σ is unknown, it is probably reasonable to use the t-test.

31. Since we have normality and σ is known, use the z-test.

32. Since we have a large sample, use the z-test.

33. Since we have a large sample with no outliers and σ is unknown, use the t-test.

34. Neither. Note: Since we have a symmetric non-normal distribution, the Wilcoxon signed-rank test is appropriate.

35. Neither. **Note:** Since we have a symmetric non-normal distribution, the Wilcoxon signed-rank test is appropriate.

36. Neither. Note: Since we have a symmetric distribution, the Wilcoxon signed-rank test is appropriate.

37. The distribution looks skewed and the sample size is not large. Consult a statistician.

38. The distribution looks skewed and the sample size is not large. Consult a statistician.

39. (a,b) Using Minitab, choose **Stat ▶ Basic Statistics ▶ 1-Sample t**, enter CONSUMPTION in the **Samples in columns** text box and 64.5 in the **Test mean** text box. Click on the **Graphs** button and check the boxes for **Histogram of data** and **Boxplot of data**. Click **OK**. Click on the **Options** button, enter 95 in the **Confidence level** text box and select **Less than** from the **Alternative** drop down box. Click **OK** twice. Then choose **Stat ▶ Basic Statistics ▶ Normality test** and enter CONSUMPTION in the **Variable** text box and click **OK**. Finally, choose **Graph ▶ Stem-and-Leaf** and enter CONSUMPTION in the **Graph Variables** text box and click **OK**. The results are

Copyright © 2012 Pearson Education, Inc. Publishing as Addison-Wesley.

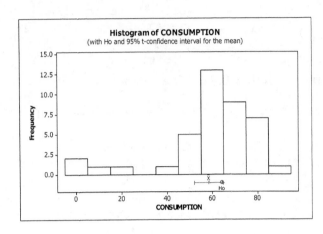

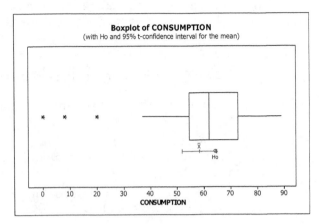

```
Stem-and-leaf of CONSUMPTION  N = 40
Leaf Unit = 1.0

LO 0, 0, 8, 20

     5   3   7
     5   4
     7   4   79
    10   5   014
    16   5   666667
   (7)   6   0112223
    17   6   57789
    12   7   1234
     8   7   5567789
     1   8
     1   8   9
```

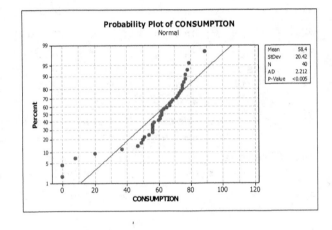

```
         Test of mu = 64.5 vs not = 64.5

         Variable       N    Mean    StDev   SE Mean      95% CI            T       P
         CONSUMPTION    40  58.4000  20.4172   3.2282  (51.8703, 64.9297)  -1.89  0.066
```

The P-value of the test is 0.066, which is greater than the 0.05 significance level. We do not reject H₀. The data do not provide sufficient evidence to conclude that the mean beef consumption this year is less than the 2002 mean of 64.5 lbs.

(c) After removing the four outliers, the test results are

```
         Test of mu = 64.5 vs not = 64.5

         Variable       N    Mean    StDev   SE Mean      95% CI            T       P
         CONSUMPTION    36  64.1111  11.0163   1.8360  (60.3837, 67.8385)  -0.21  0.833
```

(d) The outliers had a very large effect on the test results. Although the sample size was 40, four outliers is too many for the t-test to be appropriate. If the outliers are not recording errors, they represent legitimate observations from the population and should not be deleted. If they were deleted, the test results would not yield valid conclusions about the population. With so many outliers, it is just possible that the population itself is quite left skewed as is the sample.

(e) This is more appropriately done with a non-parametric test that is insensitive to outliers.

Copyright © 2012 Pearson Education, Inc. Publishing as Addison-Wesley.

40. (a) Using Minitab, choose **Stat ▶ Basic Statistics ▶ 1-Sample z**, enter <u>BMI</u> in the **Samples in columns** text box and <u>25</u> in the **Test mean** text box. Click on the **Graphs** button and check the boxes for **Histogram of data** and **Boxplot of data**. Click **OK**. Click on the **Options** button, enter <u>95</u> in the **Confidence level** text box and select **Greater than** from the **Alternative** drop down box. Click **OK** twice. Then choose **Stat ▶ Basic Statistics ▶ Normality test** and enter <u>BMI</u> in the **Variable** text box and click **OK**. Finally, choose **Graph ▶ Stem-and-Leaf** and enter <u>BMI</u> in the **Graph Variables** text box and click **OK**. The graphic results are

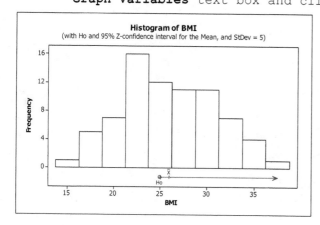

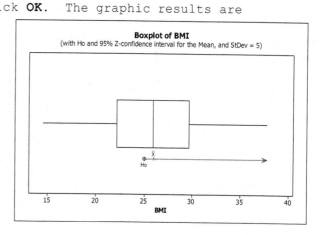

```
Stem-and-leaf of BMI   N  = 75
Leaf Unit = 1.0

    1    1  4
    4    1  677
    6    1  88
   16    2  0000111111
   29    2  2222222233333
   (9)   2  444444455
   37    2  666666777
   28    2  88888899999
   17    3  0000111
   10    3  22233
    5    3  4555
    1    3  7
```

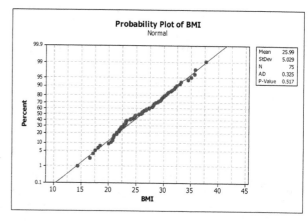

(b) Yes. There are no outliers, the sample size is large, and the data is reasonably normally distributed.

(c) The process in part (a) yielded the following test results:

```
Test of mu = 25 vs > 25
The assumed standard deviation = 5
```

Variable	N	Mean	StDev	SE Mean	95% Lower Bound	Z	P
BMI	75	25.9867	5.0293	0.5774	25.0370	1.71	0.044

Since the P-value of 0.044 is less than the significance level 0.05, we reject the null hypothesis. The data do provide sufficient evidence to

Copyright © 2012 Pearson Education, Inc. Publishing as Addison-Wesley.

conclude that the mean BMI of U.S. adults is greater than that for a healthy weight.

41.   (a)   Using Minitab, we choose **Graph ▶ Histogram**, choose the **Simple** version, enter <u>BEER</u> in the **Graph variables** text box, and click **OK**. The result is

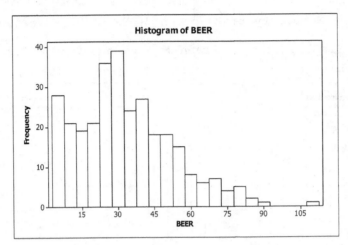

(b)   There may be an outlier at about 112.

(c)   Since the sample size is very large (300), we can use the z-test or the t-test. Since the standard deviation is unknown, we use the t-test. Using Minitab, we obtained the following result using all of the data.

Test of mu = 30.4 vs not = 30.4

Variable	N	Mean	StDev	SE Mean	99% CI	T	P
BEER	300	33.52	19.43	1.12	(30.61, 36.43)	2.78	0.006

After eliminating the potential outlier (112), we repeated the process and obtained

Test of mu = 30.4 vs not = 30.4

Variable	N	Mean	StDev	SE Mean	99% CI	T	P
BEER	299	33.26	18.92	1.09	(30.42, 36.09)	2.61	0.009

Clearly, elimination of the value 112 had little effect on the value of t or on the P-value. In either case, we reject the null hypothesis. The data do provide sufficient evidence to claim that the mean annual consumption of beer in Missouri is different from the national mean.

Copyright © 2012 Pearson Education, Inc. Publishing as Addison-Wesley.

Exercises 10.1

10.1 Answers will vary.

10.3 (a) μ_1, σ_1, μ_2, and σ_2 are parameters; $\bar{x}_1$, s_1, $\bar{x}_2$, and s_2 are statistics.

(b) μ_1, σ_1, μ_2, and σ_2 are fixed numbers; $\bar{x}_1$, s_1, $\bar{x}_2$, and s_2 are random variables.

10.5 It is the sampling distribution of the difference of the two sample means that allows us to determine whether the difference of the sample means can reasonably be attributed to sampling error or whether it is large enough for us to conclude that the population means are different.

10.7 H_0: $\mu_1 = \mu_2$, H_a: $\mu_1 > \mu_2$

10.9 (a) The variable is systolic blood pressure (SBP).

(b) The two populations are the adolescent offspring of diabetic mothers (ODM) and of nondiabetic mothers (ODN).

(c) H_0: $\mu_1 = \mu_2$, H_a: $\mu_1 > \mu_2$ where μ_1 is the mean SBP for ODM and μ_2 is the mean SBP for ODN.

(d) Right tailed

10.11 (a) The variable is household vehicle miles traveled (VMT).

(b) The two populations are households in the Midwest and in the South.

(c) H_0: $\mu_1 = \mu_2$, H_a: $\mu_1 \neq \mu_2$ where μ_1 is the mean VMT for households in the Midwest and μ_2 is the mean VMT for households in the South.

(d) two-tailed

10.13 (a) The variable is operative times in minutes.

(b) The two populations are operations performed using a dynamic system (Z-plate) and operations performed using a static system (ALPS plate).

(c) H_0: $\mu_1 = \mu_2$, H_a: $\mu_1 < \mu_2$ where μ_1 is the mean time of operations performed using the dynamic system and μ_2 is the mean time of operations performed using the static system.

(d) left-tailed

10.15 We can be 95% confident that the mean for population one is between 15 to 20 units greater than the mean for population two.

10.17 We can be 90% confident that the mean for population one is between 5 to 10 units less than the mean for population two.

10.19 We can be 99% confident that the mean for population one is between 20 units less to 15 units greater than the mean for population two.

10.21 (a) The mean of $\bar{x}_1 - \bar{x}_2$ is $\mu_1 - \mu_2 = 40 - 40 = 0$

The standard deviation of $\bar{x}_1 - \bar{x}_2$ is

$$\sigma_{\bar{x}_1 - \bar{x}_2} = \sqrt{\frac{\sigma_1^2}{n_1} + \frac{\sigma_2^2}{n_2}} = \sqrt{\frac{12^2}{9} + \frac{6^2}{4}} = 5$$

(b) No. The determination of the mean of $\bar{x}_1 - \bar{x}_2$ is the same for all populations. The formula for the standard deviation of $\bar{x}_1 - \bar{x}_2$ is dependent on the samples being independent, but not on the populations from which they came.

(c) No. It is not known that the two populations are normally distributed (which would lead to $\bar{x}_1 - \bar{x}_2$ being normally distributed), and the sample sizes of 9 and 4 are too small to claim that $\bar{x}_1$ and $\bar{x}_2$ are normally distributed (which would also lead to $\bar{x}_1 - \bar{x}_2$ being normally distributed).

Copyright © 2012 Pearson Education, Inc. Publishing as Addison-Wesley.

10.23 (a) The mean of $\bar{x}_1 - \bar{x}_2$ is $\mu_1 - \mu_2 = 40 - 40 = 0$

The standard deviation of $\bar{x}_1 - \bar{x}_2$ is $\sigma_{\bar{x}_1 - \bar{x}_2} = \sqrt{\dfrac{\sigma_1^2}{n_1} + \dfrac{\sigma_2^2}{n_2}} = \sqrt{\dfrac{12^2}{9} + \dfrac{6^2}{4}} = 5$

(b) Yes. Since x_1 and x_2 are each normally distributed, then so are $\bar{x}_1$ and $\bar{x}_2$. Since the difference of two normally distributed random variables is also normally distributed, so is $\bar{x}_1 - \bar{x}_2$. See also Key Fact 10.1.

(c) Since $\bar{x}_1 - \bar{x}_2$ is normally distributed with mean 0 and standard deviation 5, $P(-10 < \bar{x}_1 - \bar{x}_2 < 10) = P([-10 - 0]/5 < z < [10 - 0]/5)$
$= P(-2 < z < 2) = 0.9772 - 0.0228 = 0.9544$, so 95.44% of the differences in sample means from samples of size 9 and 4 from the two populations will lie between -10 and 10.

10.25 (a) We have Minitab take 2000 observations from a normally distributed population having mean 100 and standard deviation 16 by choosing **Calc ▶ Random Data ▶ Normal...**, typing 2000 in the **Generate rows of data** text box, typing X1 in the **Store in columns** text box, typing 100 in the **Mean** text box and 16 in the **Standard deviation** text box, and clicking **OK**.

(b) We repeat part (a) for a normally distributed population having mean 120 and standard deviation 12 by choosing **Calc ▶ Random Data ▶ Normal...**, typing 2000 in the **Generate rows of data** text box, typing X2 in the **Store in columns** text box, typing 120 in the **Mean** text box and 12 in the **Standard deviation** text box, and clicking **OK**.

(c) To determine the difference between each pair of observations in parts (a) and (b), we choose **Calc ▶ Calculator...**, type DIFF in the **Store results in variable** text box, type X1 - X2 in the **Expression** text box, and click **OK**. DIFF will be stored in C3.

(d) To obtain a histogram of the 2000 differences found in part (c), we choose **Graph ▶ Histogram...**, select the **Simple** version and click **OK**. Enter DIFF in the **Graph variables** text box and click **OK**.

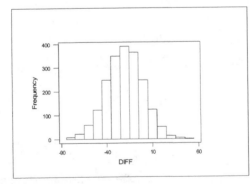

The histogram is bell-shaped because the random variable named DIFF is the difference of two random variables which themselves are normally distributed. A random variable which is the difference of two normally distributed random variables is itself normally distributed. Thus, its histogram is bell-shaped.

Copyright © 2012 Pearson Education, Inc. Publishing as Addison-Wesley.

Section 10.2

10.27 (a) The four conditions required for using the pooled t-procedure are (i) both samples are simple random samples, (ii) independence of the two samples, (iii) the variable under consideration is normally distributed in each of the two populations or both samples are large, and (iv) the population standard deviations are equal.

(b) Independence and randomness are absolutely essential. Moderate violations of the normality assumption are not serious problems even for small or moderate size samples. Moderate violations of the equal standard deviation requirement are not serious if the two sample sizes are roughly equal.

10.29 Not Reasonable. One sample standard deviation is more than twice the other one. This is a good indication that the population standard deviations are not equal. Since the two sample sizes are not roughly equal (6 and 14), nor are both of them large, there appears to be a serious violation of the equal standard deviations requirement of the pooled t-test.

10.31 Reasonable. The standard deviations are roughly equal and both sample sizes are large.

10.33 (a) $s_p = \sqrt{\dfrac{14(2.1)^2 + 14(2.3)^2}{15 + 15 - 2}} = \sqrt{4.85} = 2.2023$

Step 1: $H_0: \mu_1 = \mu_2$, $H_a: \mu_1 \neq \mu_2$

Step 2: $\alpha = 0.05$

Step 3: $t = \dfrac{10 - 12}{2.2023\sqrt{(1/15) + (1/15)}} = -2.487$

Step 4: df = 28, Critical values = ± 2.048
0.01 < P-value < 0.02.

Step 5: Since -2.487 < -2.048, reject H_0.

(b) 95% CI = $(10 - 12) \pm 2.048(2.2023)\sqrt{(1/15 + 1/15)} = (-3.647, -0.353)$

10.35 (a) $s_p = \sqrt{\dfrac{9(4)^2 + 14(5)^2}{10 + 15 - 2}} = \sqrt{21.47826} = 4.6345$

Step 1: $H_0: \mu_1 = \mu_2$, $H_a: \mu_1 > \mu_2$

Step 2: $\alpha = 0.05$

Step 3: $t = \dfrac{20 - 18}{4.6345\sqrt{(1/10) + (1/15)}} = 1.057$

Step 4: df = 23, Critical value = 1.714
P-value > 0.10.

Step 5: Since 1.057 < 1.714, do not reject H_0.

(b) 90% CI = $(20 - 18) \pm 1.714(4.6345)\sqrt{(1/10 + 1/15)} = (-1.243, 5.243)$

10.37 (a) $s_p = \sqrt{\dfrac{19(4)^2 + 14(5)^2}{20 + 15 - 2}} = \sqrt{19.81818} = 4.4518$

Step 1: $H_0: \mu_1 = \mu_2$, $H_a: \mu_1 < \mu_2$

Step 2: $\alpha = 0.05$

Step 3: $t = \dfrac{20 - 24}{4.4518\sqrt{(1/20) + (1/15)}} = -2.631$

Step 4: df = 33, Critical value = -1.692
0.005 < P-value < 0.010.

Step 5: Since -2.631 < -1.697, reject H_0.

Copyright © 2012 Pearson Education, Inc. Publishing as Addison-Wesley.

(b) 90% CI $= (20-24) \pm 1.697(4.4518)\sqrt{(1/20+1/15)} = (-6.580, -1.420)$

10.39 Population 1: Fraud, $n_1 = 10$, $\bar{x}_1 = 10.12$, $s_1 = 4.90$

Population 2: Firearms, $n_2 = 10$, $\bar{x}_2 = 18.78$, $s_2 = 4.64$

$$s_p = \sqrt{\frac{9(4.90)^2 + 9(4.64)^2}{10+10-2}} = \sqrt{22.7698} = 4.7718$$

Step 1: H_0: $\mu_1 = \mu_2$, H_a: $\mu_1 < \mu_2$

Step 2: $\alpha = 0.05$

Step 3: $t = \dfrac{10.12-18.78}{4.7718\sqrt{(1/10)+(1/10)}} = -4.058$

Step 4:   df = 18, Critical value = -1.734
          P-value < 0.005.

Step 5:   Since -4.06 < -1.734, reject H_0.

Step 6:   At the 5% significance level, the data do provide sufficient evidence to conclude that, on the average, the mean time served for fraud is less than that served for firearms offenses.

10.41 Population 1: Fortified, $n_1 = 14$, $\bar{x}_1 = 9.0$, $s_1 = 37.4$

Population 2: Unfortified, $n_2 = 12$, $\bar{x}_2 = 1.6$, $s_2 = 34.6$

$$s_p = \sqrt{\frac{13(37.4)^2 + 11(34.6)^2}{24}} = \sqrt{1306.36} = 36.1436$$

Step 1: H_0: $\mu_1 = \mu_2$, H_a: $\mu_1 > \mu_2$

Step 2: $\alpha = 0.05$

Step 3: $t = \dfrac{9.0-1.6}{36.1436\sqrt{1/14+1/12}} = 0.520$

Step 4:   Critical value = 1.711
          P-value > 0.10

Step 5:   Since 0.520 < 1.711, do not reject H_0.

Step 6:   At the 5% significance level, the data do not provide sufficient evidence that drinking fortified orange juice reduces PTH level more than drinking unfortified orange juice.

10.43 Population 1: Cropland, $n_1 = 126$, $\bar{x}_1 = 14.06$, $s_1 = 4.83$

Population 2: Wetland, $n_2 = 98$, $\bar{x}_2 = 15.36$, $s_2 = 4.95$

$$s_p = \sqrt{\frac{125(4.83)^2 + 97(4.95)^2}{222}} = \sqrt{23.84169} = 4.8828$$

Step 1: H_0: $\mu_1 = \mu_2$, H_a: $\mu_1 \neq \mu_2$

Step 2: $\alpha = 0.05$

Step 3: $t = \dfrac{14.06-15.36}{4.8828\sqrt{1/126+1/98}} = -1.977$

Step 4:   df = 222, Critical values = ±1.971 (using technology)
          For the p-value approach, $0.02 < 2\{P(t < -1.977)\} < 0.05$.

Step 5:   Since -1.977 < -1.971, reject H_0. Because the P-value is smaller than the significance level, reject H_0.

Copyright © 2012 Pearson Education, Inc. Publishing as Addison-Wesley.

Step 6: At the 5% significance level, the data provide sufficient evidence that there is a difference in the mean number of native species in the two regions.

In each of Exercises 10.45-10.50, the value of s_p has been obtained in Exercise 10.39-10.44, respectively.

10.45 $t_{\alpha/2} = 1.734$, 90%CI $= (10.12\text{-}18.78) \pm 1.734(4.7718)\sqrt{1/10+1/10} = (-12.36, -4.96)$

10.47 $t_{\alpha/2} = 1.711$, 90%CI $= (9.0\text{-}1.6) \pm 1.711(36.1436)\sqrt{1/14+1/12} = (-16.93, 31.73)$

10.49 $t_{\alpha/2} = 1.971$, 95%CI $= (14.06\text{-}15.36) \pm 1.971(4.8828)\sqrt{1/126+1/98} = (-2.596, -0.004)$

10.51 (a) Using Minitab, we choose **Graph ▶ Probability Plot**, select **Single** and click **OK**. Enter VEGETARIAN and OMNIVORE in the **Graph Variables** text box. Click on the **Multiple Graphs** button and select **On separate graphs**. Click **OK** twice. Note that the standard deviations are shown on the plots. Then choose **Graph ▶ Boxplot**, click on the **Simple** plot in the **Multiple Y's** row and click **OK**..Enter VEGETARIAN and OMNIVORE in the **Graph Variables** text box and click **OK**. The results are

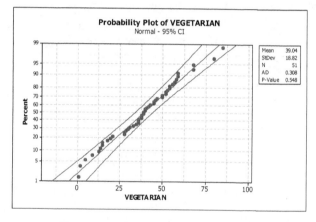

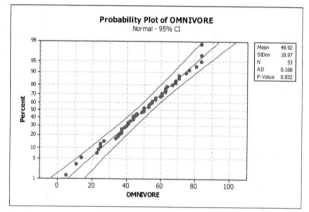

The standard deviations are 18.82 for Vegetarians and 18.97 for Omnivores.

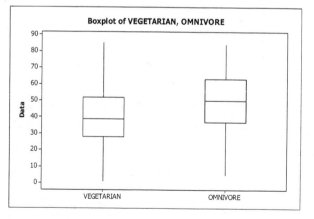

(b) Choose **Stat ▶ Basic Statistics ▶ 2-Sample t**, click on **Samples in different columns** and enter VEGETARIANS in the **First** textbox and OMNIVORES in the **Second** text box. Check the box for **Assume equal variances** and click on the **Options** button. Enter 99 in the **Confidence level** textbox, 0 in the **Test difference** text box and select **not equal** in the **Alternative drop down box**. Click **OK** twice. The results are

Copyright © 2012 Pearson Education, Inc. Publishing as Addison-Wesley.

```
Two-sample T for VEGETARIAN vs OMNIVORE

            N   Mean  StDev  SE Mean
VEGETARIAN  51  39.0  18.8   2.6
OMNIVORE    53  49.9  19.0   2.6


Difference = mu (VEGETARIAN) - mu (OMNIVORE)
Estimate for difference:  -10.8853
99% CI for difference:  (-20.6147, -1.1559)
T-Test of difference = 0 (vs not =): T-Value = -2.94  P-Value = 0.004
DF = 102
Both use Pooled StDev = 18.8964
```

Since the P-value of 0.004 is less than the significance level of 0.01, the data do provide sufficient evidence that there is a difference between the mean daily protein intakes of female vegetarians and female omnivores.

(c) The 99% confidence interval is (-20.6147, -1.1559).

(d) Yes. The probability plots indicate that normality is reasonable, the sample sizes are large, and there are no outliers. Furthermore, the two sample standard deviations are almost equal, making the pooled t procedure a good choice.

10.53 (a) Using Minitab, we choose **Graph ▶ Probability Plot**, select **Single** and click **OK**. Enter STINGER and REGULAR in the **Graph Variables** text box. Click on the **Multiple Graphs** button and select **On separate graphs**. Click **OK** twice. Note that the standard deviations are shown on the

plots. Then choose **Graph ▶ Boxplot**, click on the **Simple** plot in the **Multiple Y's** row and click **OK**...Enter STINGER and REGULAR in the **Graph Variables** text box and click **OK**. The results are

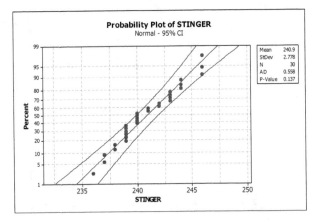
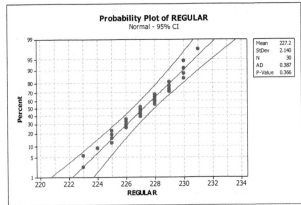

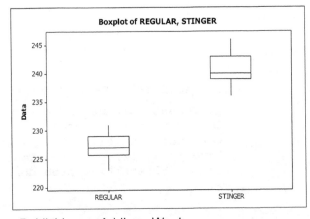

The standard deviations are 2.778 for STINGER and 2.140 for REGULAR.

Copyright © 2012 Pearson Education, Inc. Publishing as Addison-Wesley.

(b) Choose **Stat ▶ Basic Statistics ▶ 2-Sample t**, click on **Samples in different columns** and enter <u>STINGER</u> in the **First** text box and <u>REGULAR</u> in the **Second** text box. Check the box for **Assume equal variances** and click on the **Options** button. Enter <u>99</u> in the **Confidence level** textbox, <u>0</u> in the **Test difference** text box and select **greater than** in the **Alternative drop down** box. Click **OK** twice. The results are

```
Two-sample T for STINGER vs REGULAR

          N    Mean   StDev  SE Mean
STINGER  30  240.93   2.78    0.51
REGULAR  30  227.20   2.14    0.39


Difference = mu (STINGER) - mu (REGULAR)
Estimate for difference:  13.7333
99% lower bound for difference:  12.2015
T-Test of difference = 0 (vs >): T-Value = 21.45  P-Value = 0.000  DF = 58
Both use Pooled StDev = 2.4798
```
Since the P-value of 0.000 is less than the significance level of 0.01, the data do provide sufficient evidence that the Stinger tee improves total distance traveled.

(c) Repeat the process in part (b), but select **not equal** in the **Alternative drop down** box. The confidence interval is
    99% CI for difference: (12.0281, 15.4386)

(d) Yes. The probability plots indicate that normality is reasonable, the sample sizes are large, and there are no outliers. Furthermore, the two sample standard deviations are almost equal, making the pooled t procedure a good choice.

10.55 If $n_1 = n_2$, say both equal n, then

$$s_p^2 = \frac{(n_1-1)s_1^2 + (n_2-1)s_2^2}{n_1+n_2-2} = \frac{(n-1)s_1^2 + (n-1)s_2^2}{n+n-2}$$

$$= \frac{(n-1)(s_1^2+s_2^2)}{2n-2} = \frac{(n-1)(s_1^2+s_2^2)}{2(n-1)} = \frac{(s_1^2+s_2^2)}{2}$$

Thus, if $n_1 = n_2$, then s_p^2 is just the mean of s_1^2 and s_2^2.

10.57 (a) The test value of t = -0.927 fell between the critical values ±2.052, leading to nonrejection of the null hypothesis. The 95% confidence interval was (-4.69, 1.77). This includes zero, so the null hypothesis is not rejected.

(b) The test value of t = -1.977 fell between the critical values ±1.982, leading to nonrejection of the null hypothesis. The 95% confidence interval was (-2.603, 0.003). This includes zero, so the null hypothesis is not rejected. Note: These results are so close to rejection that if intermediate rounding is done during the calculations, it is possible to obtain results rejecting the null hypothesis and getting a confidence interval that does not include zero.

10.59 (a) The test value of t = 0.520 fell to the left of the critical value 1.711, leading to nonrejection of the null hypothesis. The 95% lower confidence bound is -16.928. This is not positive, so the null hypothesis is not rejected.

(b) The test value of t = = 2.707 fell to the right of the critical value 2.362, leading to rejection of the null hypothesis. The 99% lower confidence bound is 0.790. This is positive, so the null hypothesis is rejected.

Copyright © 2012 Pearson Education, Inc. Publishing as Addison-Wesley.

10.61 (a) Yes. The pooled-t test may be used when the sample sizes are large and the standard deviations are equal, even though the populations are not normally distributed.

(b) Yes. The Mann-Whitney test is appropriate when the two distributions of the variable have the same shape.

(c) The Mann-Whitney procedure is preferred when the populations are not normal.

Exercises 10.3

10.63 (a) Pooled t-test. When the standard deviations are equal, the pooled t-test is more powerful than the non-pooled t-test.

(b) Nonpooled t-test. Use of the pooled t-test would result in a larger Type I error probability than the one specified.

(c) Pooled t-test. The closeness of the sample standard deviations indicates that the population standard deviations are probably close to being equal. Using the pooled t-test will not cause serious problems under these conditions when the sample sizes are equal.

(d) Nonpooled t-test. The sample standard deviations are enough different to raise doubt about the equality of the population standard deviations.

10.65 (a) Step 1: $H_0: \mu_1 = \mu_2, \quad H_a: \mu_1 \neq \mu_2$

Step 2: $\alpha = 0.05$

Step 3: $t = \dfrac{10-12}{\sqrt{(2^2/15)+(5^2/15)}} = -1.438$

Step 4: $\Delta = \dfrac{\left[\left(\dfrac{s_1^2}{n_1}\right)+\left(\dfrac{s_2^2}{n_2}\right)\right]^2}{\dfrac{\left(\dfrac{s_1^2}{n_1}\right)^2}{n_1-1}+\dfrac{\left(\dfrac{s_2^2}{n_2}\right)^2}{n_2-1}} = \dfrac{\left[\left(\dfrac{2^2}{15}\right)+\left(\dfrac{5^2}{15}\right)\right]^2}{\dfrac{\left(\dfrac{2^2}{15}\right)^2}{15-1}+\dfrac{\left(\dfrac{5^2}{15}\right)^2}{15-1}} = 18.37; df = 18$

Critical values = ± 2.101
$0.100 < $ P-value < 0.200.

Step 5: Since $-2.101 < -1.438 < 2.101$, do not reject H_0.

(b) 95% CI = $(10-12)\pm 2.101\sqrt{(2^2/15+5^2/15)} = (-4.921, 0.921)$

10.67 (a) Step 1: $H_0: \mu_1 = \mu_2, \quad H_a: \mu_1 > \mu_2$

Step 2: $\alpha = 0.05$

Step 3: $t = \dfrac{20-18}{\sqrt{(4^2/10)+(5^2/15)}} = 1.107$

Step 4: $\Delta = \dfrac{\left[\left(\dfrac{s_1^2}{n_1}\right)+\left(\dfrac{s_2^2}{n_2}\right)\right]^2}{\dfrac{\left(\dfrac{s_1^2}{n_1}\right)^2}{n_1-1}+\dfrac{\left(\dfrac{s_2^2}{n_2}\right)^2}{n_2-1}} = \dfrac{\left[\left(\dfrac{4^2}{10}\right)+\left(\dfrac{5^2}{15}\right)\right]^2}{\dfrac{\left(\dfrac{4^2}{10}\right)^2}{10-1}+\dfrac{\left(\dfrac{5^2}{15}\right)^2}{15-1}} = 22.10; df = 22$

Critical value = 1.717
P-value > 0.100.

Step 5: Since $1.107 < 1.717$, do not reject H_0.

Copyright © 2012 Pearson Education, Inc. Publishing as Addison-Wesley.

(b) $90\% \text{ CI } = (20-18) \pm 1.717\sqrt{(4^2/10+5^2/15)} = (-1.103, 5.103)$

10.69 (a)   Step 1: $H_0: \mu_1 = \mu_2, \quad H_a: \mu_1 < \mu_2$

Step 2: $\alpha = 0.05$

Step 3: $t = \dfrac{20-24}{\sqrt{(6^2/20)+(2^2/15)}} = -2.782$

Step 4: $\Delta = \dfrac{\left[\left(\dfrac{s_1^2}{n_1}\right)+\left(\dfrac{s_2^2}{n_2}\right)\right]^2}{\dfrac{\left(\dfrac{s_1^2}{n_1}\right)^2}{n_1-1}+\dfrac{\left(\dfrac{s_2^2}{n_2}\right)^2}{n_2-1}} = \dfrac{\left[\left(\dfrac{6^2}{20}\right)+\left(\dfrac{2^2}{15}\right)\right]^2}{\dfrac{\left(\dfrac{6^2}{20}\right)^2}{20-1}+\dfrac{\left(\dfrac{2^2}{15}\right)^2}{15-1}} = 24.322; \; df = 24$

Critical value = -1.711
$0.005 < \text{P-value} < 0.010.$

Step 5:   Since -2.782 < -1.711, reject H_0.

(b) $90\% \text{ CI } = (20-24) \pm 1.711\sqrt{(6^2/20+2^2/15)} = (-6.460, -1.540)$

10.71 Population 1: Chronic, $n_1 = 32$, $\bar{x}_1 = 28.8$, $s_1 = 9.2$

Population 2: Remitted, $n_2 = 20$, $\bar{x}_2 = 22.1$, $s_2 = 5.7$

Step 1: $H_0: \mu_1 = \mu_2, \quad H_a: \mu_1 \neq \mu_2$

Step 2: $\alpha = 0.10$

Step 3: $t = \dfrac{25.8-22.1}{\sqrt{(9.2^2/32)+(5.7^2/20)}} = 1.791$

Step 4: $\Delta = \dfrac{\left[\left(\dfrac{s_1^2}{n_1}\right)+\left(\dfrac{s_2^2}{n_2}\right)\right]^2}{\dfrac{\left(\dfrac{s_1^2}{n_1}\right)^2}{n_1-1}+\dfrac{\left(\dfrac{s_2^2}{n_2}\right)^2}{n_2-1}} = \dfrac{\left[\left(\dfrac{9.2^2}{32}\right)+\left(\dfrac{5.7^2}{20}\right)\right]^2}{\dfrac{\left(\dfrac{9.2^2}{32}\right)^2}{32-1}+\dfrac{\left(\dfrac{5.7^2}{20}\right)^2}{20-1}} = 49.99995; \; df = 49$

Critical values = ± 1.677
$0.050 < \text{P-value} < 0.100.$

Step 5:   Since 1.791 > 1.677, reject H_0.
At the 10% significance level, the data do provide sufficient evidence to conclude that there is a difference between the mean age at arrest of East German prisoners with chronic PTSD and remitted PTSD. For the P-value approach, $0.05 < 2\{P(t > 1.791)\} < 0.10$. Therefore, because the P-value is smaller than the significance level, reject H_0.

10.73 Population 1: Dynamic, $n_1 = 14$, $\bar{x}_1 = 7.36$, $s_1 = 1.22$

Population 2: Static, $n_2 = 6$, $\bar{x}_2 = 10.50$, $s_2 = 4.59$

$H_0: \mu_1 = \mu_2, \quad H_a: \mu_1 < \mu_2$

Step 2: $\alpha = 0.05$

Step 3: $t = \dfrac{7.36-10.50}{\sqrt{(1.22^2/14)+(4.59^2/6)}} = -1.651$

Copyright © 2012 Pearson Education, Inc. Publishing as Addison-Wesley.

Step 4: $\Delta = \dfrac{\left[\left(\dfrac{s_1^2}{n_1}\right)+\left(\dfrac{s_2^2}{n_2}\right)\right]^2}{\dfrac{\left(\dfrac{s_1^2}{n_1}\right)^2}{n_1-1}+\dfrac{\left(\dfrac{s_2^2}{n_2}\right)^2}{n_2-1}} = \dfrac{\left[\left(\dfrac{1.22^2}{14}\right)+\left(\dfrac{4.59^2}{6}\right)\right]^2}{\dfrac{\left(\dfrac{1.22^2}{14}\right)^2}{14-1}+\dfrac{\left(\dfrac{4.59^2}{6}\right)^2}{6-1}} = 5.305;\ df\ =\ 5$

Critical value = -2.015
0.050 < P-value < 0.100.
Step 5: Since -1.651 > -2.015, do not reject H_0.
At the 5% significance level, the data do not provide sufficient
evidence to conclude that the mean number of acute postoperative days
in the hospital is smaller with the dynamic system than with the static
system.
For the P-value approach, 0.05 < P(t<-1.651) < 0.10. Therefore, since
the P-value is larger than the significance level, do not reject H_0.

10.75 Population 1: Psychotic patients, n_1 = 10, $\bar{x}_1$ = 0.02426, s_1 = 0.00514

Population 2: Non-psychotic patients, n_2 = 15, $\bar{x}_2$ = 0.01643, s_2 = 0.00470

(a) H_0: $\mu_1 = \mu_2$, H_a: $\mu_1 > \mu_2$
Step 2: α = 0.01
Step 3: $t = \dfrac{0.02426 - 0.01643}{\sqrt{(0.00514^2/10)+(0.00470^2/15)}} = 3.860$

Step 4: $\Delta = \dfrac{\left[\left(\dfrac{s_1^2}{n_1}\right)+\left(\dfrac{s_2^2}{n_2}\right)\right]^2}{\dfrac{\left(\dfrac{s_1^2}{n_1}\right)^2}{n_1-1}+\dfrac{\left(\dfrac{s_2^2}{n_2}\right)^2}{n_2-1}} = \dfrac{\left[\left(\dfrac{0.00514^2}{10}\right)+\left(\dfrac{0.01643^2}{15}\right)\right]^2}{\dfrac{\left(\dfrac{1.00514^2}{10}\right)^2}{10-1}+\dfrac{\left(\dfrac{0.01643^2}{15}\right)^2}{15-1}} = 18.20;\ df\ =\ 18$

Critical value = 2.552 (approximately)
P-value < 0.005.
Step 5: Since 3.860 > 2.552, reject H_0.
At the 1% significance level, the data do provide sufficient evidence
to conclude that dopamine activity is higher, on average, in psychotic
patients than in non-psychotic patients.
For the P-value approach, P(t > 3.860) < 0.005. Therefore, since the
P-value is smaller than the significance level, reject H_0.

10.77 $t_{\alpha/2} = 1.677$; 90% CI = $(25.8\text{-}22.1) \pm 1.677\sqrt{9.2^2/32 + 5.7^2/20} = (0.235, 7.165)$

10.79 $t_{\alpha/2} = 2.015$; 90% CI = $(7.36\text{-}10.50) \pm 2.015\sqrt{1.22^2/14 + 4.59^2/6} = (-6.973, 0.693)$

10.81 $t_{\alpha/2} = 2.552$; 98% CI = $(0.02426\text{-}0.01643) \pm 2.552\sqrt{0.00514^2/10 + 0.00470^2/15}$
$= (0.00265, 0.01301)$

10.83 (a) Choose nonpooled t-prodedures. There is clearly much more variability
in the data for males than in the data for females.
(b) No. The two sample sizes are small and moderate. The data for the
males is not normally distributed or has a large outlier at 39.2.
Neither t-procedure is reasonable under these circumstances.

10.85 Population 1: Dynamic, n_1 = 14, $\bar{x}_1$ = 7.36, s_1 = 1.22

Population 2: Static, n_2 = 6, $\bar{x}_2$ = 10.50, s_2 = 4.59

Copyright © 2012 Pearson Education, Inc. Publishing as Addison-Wesley.

(a) $s_p = \sqrt{\dfrac{13(1.22)^2 + 5(4.59)^2}{18}} = \sqrt{6.9272} = 2.6320$, df = 18

Step 1: H_0: $\mu_1 = \mu_2$, H_a: $\mu_1 < \mu_2$

Step 2: α = 0.05

Step 3: $t = \dfrac{7.36 - 10.50}{2.6320\sqrt{1/14 + 1/6}} = -2.445$

Step 4:   Critical value = -1.734
         0.01 < P(t<-2.445) < 0.025. The P-value is smaller than the
       significance level.

Step 5:   Since -2.445 < -1.734, reject H_0.

Step 6:   At the 5% significance level, the data do provide sufficient
         evidence to conclude that the mean number of acute
         postoperative days in the hospital is smaller with the
         dynamic system than with the static system.

(b)   The pooled t-test resulted in rejecting the null hypothesis while the
       nonpooled t-test resulted in not rejecting the null hypothesis.

(c)   The nonpooled t-test is more appropriate. One sample standard
       deviation is almost four times as large as the other, making it highly
       unlikely that the two population standard deviations are equal. The
       fact that the two sample sizes are also quite different makes it
       essential that the pooled t-test not be used.

10.87 (a) Using Minitab, we choose **Graph ▶ Probability Plot**, select **Single** and
          click **OK**. Enter INTEGRATED and STANDARD in the **Graph Variables** text
          box. Click on the **Multiple Graphs** button and select **On separate
          graphs**. Click **OK** twice. Note that the standard deviations are shown
          on the plots. Then choose **Graph ▶ Boxplot**, click on the **Simple** plot in
          the **Multiple Y's** row and click **OK**...Enter INTEGRATED and STANDARD in
          the **Graph Variables** text box and click **OK**. The results are

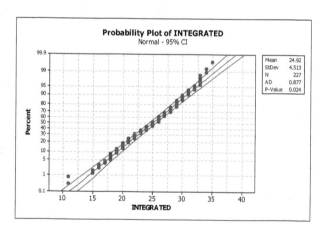
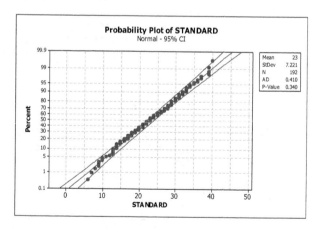

Copyright © 2012 Pearson Education, Inc. Publishing as Addison-Wesley.

The standard deviations are
4.513 for INTEGRATED and 7.221
for STANDARD.

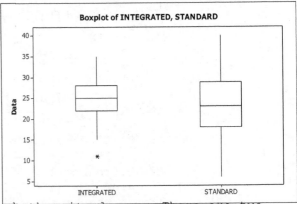

(b)   The sample sizes are 227 and 192, both quite large. There are two
potential outliers on the low end of the INTEGRATED data, but
otherwise, normality seems reasonable. The ratio of the larger
standard deviation to the smaller is about 1.6, less than the
guideline of 2 for using the pooled t-test. The pooled t-test is a
reasonable choice.

(c)   Choose **Stat ▶ Basic Statistics ▶ 2-Sample t**, click on **Samples in
different columns** and enter <u>INTEGRATED</u> in the **First** text box and
<u>STANDARD</u> in the **Second** text box. Check the box for **Assume equal
variances** and click on the **Options** button. Enter <u>99</u> in the **Confidence
level** textbox, <u>0</u> in the **Test difference** text box and select **greater
than** in the **Alternative** drop down box. Click **OK** twice. Then repeat
this procedure with the **Assume equal variances** box unchecked. The
results are

```
Two-sample T for INTEGRATED vs STANDARD

              N    Mean   StDev  SE Mean
INTEGRATED   227   24.92   4.51    0.30
STANDARD     192   23.00   7.22    0.52


Difference = mu (INTEGRATED) - mu (STANDARD)
Estimate for difference:  1.91630
99% lower bound for difference:  0.56312
T-Test of difference = 0 (vs >): T-Value = 3.31   P-Value = 0.001   DF = 417
Both use Pooled StDev = 5.9097


Two-sample T for INTEGRATED vs STANDARD

              N    Mean   StDev  SE Mean
INTEGRATED   227   24.92   4.51    0.30
STANDARD     192   23.00   7.22    0.52


Difference = mu (INTEGRATED) - mu (STANDARD)
Estimate for difference:  1.91630
99% lower bound for difference:  0.51063
T-Test of difference = 0 (vs >): T-Value = 3.19  P-Value = 0.001  DF = 309
```

Both tests yield a P-value of 0.001, less than the significance level
0.01. Reject the null hypothesis. The data do provide sufficient
evidence that clients preferred the integrated treatment.

(d)   To get the confidence intervals, repeat the process in part (b),
entering <u>98</u> in the **Confidence level** text box and selecting **not equal** in
the **Alternative** drop down box. The confidence interval portions of the
results are

            Pooled   98% CI for difference: (0.56312, 3.26947)

Copyright © 2012 Pearson Education, Inc. Publishing as Addison-Wesley.

Nonpooled 98% CI for difference: (0.51063, 3.32197)
The confidence intervals are similar with the Nonpooled interval being slightly longer. Neither interval includes zero.

10.89 (a) Using Minitab, we choose **Graph ▶ Probability Plot**, select **Single** and click **OK**. Enter ITALIANS and ETRUSCANS in the **Graph Variables** text box. Click on the **Multiple Graphs** button and select **On separate graphs**. Click **OK** twice. Note that the standard deviations are shown on the plots. Then choose **Graph ▶ Boxplot**, click on the **Simple** plot in the **Multiple Y's** row and click **OK**...Enter ITALIANS and ETRUSCANS in the **Graph Variables** text box and click **OK**. The results are

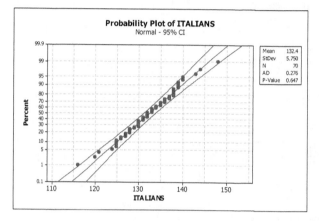

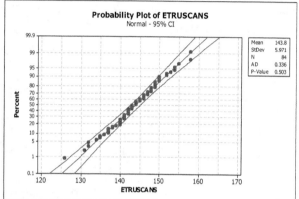

Standard deviation are 5.750 for ITALIANS and 5.971 for ETRUSCANS.

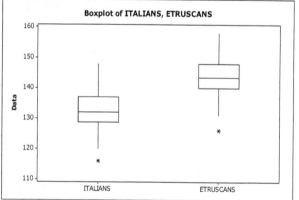

(b) Pooled t-procedures. Both sample sizes are large, the probability plots indicate that normality is reasonable, there is only one potential outlier in each sample, and the standard deviations are almost equal. It is likely that both procedures will yield similar results.

(c) Choose **Stat ▶ Basic Statistics ▶ 2-Sample t**, click on **Samples in different columns** and enter ITALIANS in the **First** text box and ETRUSCANS in the **Second** text box. Check the box for **Assume equal variances** and click on the **Options** button. Enter 95 in the **Confidence level** textbox, 0 in the **Test difference** text box and select **Not equal** in the **Alternative** drop down box. Click **OK** twice. Then repeat this procedure with the **Assume equal variances** box unchecked. The results are
Two-sample T for ITALIANS vs ETRUSCANS

Copyright © 2012 Pearson Education, Inc. Publishing as Addison-Wesley.

```
              N    Mean   StDev  SE Mean
ITALIANS     70   132.44   5.75    0.69
ETRUSCANS    84   143.77   5.97    0.65


Difference = mu (ITALIANS) - mu (ETRUSCANS)
Estimate for difference:  -11.3310
95% CI for difference:  (-13.2083, -9.4537)
T-Test of difference = 0 (vs not =): T-Value = -11.92   P-Value = 0.000
DF = 152
Both use Pooled StDev = 5.8714


Two-sample T for ITALIANS vs ETRUSCANS

              N    Mean   StDev  SE Mean
ITALIANS     70   132.44   5.75    0.69
ETRUSCANS    84   143.77   5.97    0.65


Difference = mu (ITALIANS) - mu (ETRUSCANS)
Estimate for difference:  -11.3310
95% CI for difference:  (-13.2022, -9.4597)
T-Test of difference = 0 (vs not =): T-Value = -11.97   P-Value = 0.000
DF = 148
```

Both procedures yield a P-value = 0.000, which is less than the significance level 0.05. The data do provide sufficient evidence that the mean maximum head breadths of modern Italian males and Etruscan males are different.

(d) The confidence intervals are shown in part (c) as
   Pooled   95% CI for difference: (-13.2083, -9.4537)
   Nonpooled   95% CI for difference: (-13.2022, -9.4597)
   The results of the two procedures are nearly identical.

10.91 (a) From Exercise 10.55, when the sample sizes are equal, in the pooled procedure, we have

$$s_p^2 = \frac{s_1^2 + s_2^2}{2}; \text{ if } n_1 = n_2, \text{ then } t = \frac{\overline{x}_1 - \overline{x}_2}{\sqrt{\frac{s_1^2 + s_2^2}{2}(\frac{1}{n} + \frac{1}{n})}} = \frac{\overline{x}_1 - \overline{x}_2}{\sqrt{\frac{s_1^2 + s_2^2}{2}(\frac{2}{n})}} = \frac{\overline{x}_1 - \overline{x}_2}{\sqrt{\frac{s_1^2 + s_2^2}{n}}}.$$

In the nonpooled procedure, we have

$$t = \frac{\overline{x}_1 - \overline{x}_2}{\sqrt{\frac{s_1^2}{n} + \frac{s_2^2}{n}}} = \frac{\overline{x}_1 - \overline{x}_2}{\sqrt{\frac{s_1^2 + s_2^2}{n}}}, \text{ which is identical to the pooled t-statistic.}$$

(b) If the standard deviations are not equal, the computed degrees of freedom in the nonpooled t-procedure could be quite different from the degrees of freedom in the pooled t-procedure. This will lead to different critical values and/or P-values and possibly to different conclusions.

10.93 (a) The confidence interval in Exercise 10.77 consists of all positive numbers and does not contain zero. Therefore, the hypothesis test for H_0: $\mu_1 = \mu_2$, H_a: $\mu_1 \neq \mu_2$ should reject the null hypothesis. The results of Exercise 10.71 confirm that by leading to a rejection of the null hypothesis.

(b) The confidence interval in Exercise 10.82 consists of all negative numbers and does not contain zero. Therefore, the hypothesis test for H_0: $\mu_1 = \mu_2$ vs. H_a: $\mu_1 \neq \mu_2$ should reject the null hypothesis. The results of Exercise 10.76 confirm that by leading to a rejection of the null hypothesis.

Copyright © 2012 Pearson Education, Inc. Publishing as Addison-Wesley.

10.95 (a)   Population 1: CCB, $n_1 = 51$, $\bar{x}_1 = 115.1$, $s_1 = 79.4$
Population 2: LLM, $n_2 = 19$, $\bar{x}_2 = 24.3$, $s_2 = 10.5$
From Exercise 10.72, df = 54 and $\alpha = 0.01$

$$t_\alpha = 2.397; \ (115.1-24.3)-2.397\sqrt{79.4^2/51+10.5^2/19} = 63.53$$

The 99% lower confidence bound for $\mu_1 - \mu_2$ is positive. Therefore, the right-tailed hypothesis test for H_0: $\mu_1 = \mu_2$ vs. H_a: $\mu_1 > \mu_2$ will be rejected at 1% significance. This compares to the result of the hypothesis test in Exercise 10.72.

(b)   Population 1: Psychotic patients, $n_1 = 10$, $\bar{x}_1 = 0.02426$, $s_1 = 0.00514$
Population 2: Non-psychotic patients, $n_2 = 15$, $\bar{x}_2 = 0.01643$,
$s_2 = 0.00470$
From Exercise 10.75, df = 18 and $\alpha = 0.01$

$$t_\alpha = 2.552; \ (0.02426-0.01643)-2.552\sqrt{0.00514^2/10+0.00470^2/15} = 0.0026$$

The 99% lower confidence bound for $\mu_1 - \mu_2$ is positive. Therefore, the right-tailed hypothesis test for H_0: $\mu_1 = \mu_2$ vs. H_a: $\mu_1 > \mu_2$ will be rejected at 1% significance. This compares to the result of the hypothesis test in Exercise 10.75.

Exercises 10.4

10.97 The samples must be paired (this is essential), and the population of all paired differences must be normally distributed or the sample must be large (the procedure works reasonably well even for small or moderate size samples if the paired-difference variable is not normally distributed provided that the deviation from normality is small).

10.99 (a) TV viewing time   (b) Married men and married women
(c) Married couples
(d) Paired difference variable = TV viewing time of Man - TV viewing time of Woman

(e) H_0: $\mu_M = \mu_W$; H_a: $\mu_M < \mu_W$ (f) Left-tailed

10.101 (a) Price of a home
(b) Homes near a sports stadium
(c) Prices of homes before (1) and after (2) construction of a sports stadium
(d) Paired difference variable = Price before construction - Price after construction

(e) H_0: $\mu_1 = \mu_2$; H_a: $\mu_1 \neq \mu_2$ (f) Two-tailed

10.103 (a)   Step 1: H_0: $\mu_1 = \mu_2$, H_a: $\mu_1 \neq \mu_2$
Step 2: $\alpha = 0.10$
Step 3:   The paired differences, d = x_1 - x_2, are

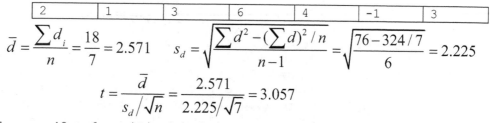

| 2 | 1 | 3 | 6 | 4 | -1 | 3 |

$$\bar{d} = \frac{\sum d_i}{n} = \frac{18}{7} = 2.571 \quad s_d = \sqrt{\frac{\sum d^2 - (\sum d)^2/n}{n-1}} = \sqrt{\frac{76-324/7}{6}} = 2.225$$

$$t = \frac{\bar{d}}{s_d/\sqrt{n}} = \frac{2.571}{2.225/\sqrt{7}} = 3.057$$

Step 4:   df = 6; critical values = ± 1.943
0.02 < P-value < 0.05

Copyright © 2012 Pearson Education, Inc. Publishing as Addison-Wesley.

Step 5: Since 3.057 > 1.943 and the P-value is less than 10%, reject H_0.

10.105 (a) Step 1: H_0: $\mu_1 = \mu_2$, H_a: $\mu_1 > \mu_2$

Step 2: $\alpha = 0.10$

Step 3: The paired differences, $d = x_1 - x_2$, are

4	-1	1	5	3	0	-5	-6

$$\bar{d} = \frac{\sum d_i}{n} = \frac{1}{8} = 0.125 \qquad s_d = \sqrt{\frac{\sum d^2 - (\sum d)^2 / n}{n-1}} = \sqrt{\frac{113 - 1/8}{7}} = 4.016$$

$$t = \frac{\bar{d}}{s_d / \sqrt{n}} = \frac{0.125}{4.016 / \sqrt{8}} = 0.088$$

Step 4: df = 7; critical values = 1.415
P-value > 0.10

Step 5: Since 0.088 < 1.415 and the P-value is greater than 10%, do not reject H_0.

10.107 (a) Step 1: H_0: $\mu_1 = \mu_2$, H_a: $\mu_1 < \mu_2$

Step 2: $\alpha = 0.10$

Step 3: The paired differences, $d = x_1 - x_2$, are

-3	-3	-2	3	-6	0	-2	-7	-1

$$\bar{d} = \frac{\sum d_i}{n} = \frac{-21}{9} = -2.333 \qquad s_d = \sqrt{\frac{\sum d^2 - (\sum d)^2 / n}{n-1}} = \sqrt{\frac{121 - 441/9}{8}} = 3.000$$

$$t = \frac{\bar{d}}{s_d / \sqrt{n}} = \frac{-2.333}{3.000 / \sqrt{9}} = -2.333$$

Step 4: df = 8; critical values = -1.397
0.01 < P-value < 0.025

Step 5: Since -2.333 < -1.397 and the P-value is less than 10%, reject H_0.

10.109 (a) The variable is the height of the plant.

(b) The two populations are the cross-fertilized plants and the self-fertilized plants.

(c) The paired-difference variable is the difference in heights of the cross-fertilized plants and the self-fertilized plants grown in the same pot.

(d) Yes. They represent the difference in heights of two plants grown under the same conditions (same pot), one from each category of plants.

(e) Step 1: H_0: $\mu_1 = \mu_2$, H_a: $\mu_1 \neq \mu_2$

Step 2: $\alpha = 0.05$

Step 3: The paired differences are given.

$$t = \frac{\bar{d}}{s_d / \sqrt{n}} = \frac{20.93}{37.74 / \sqrt{15}} = 2.148$$

Step 4: df = 14; critical values = ± 2.145
0.02 < P-Value < 0.05

Step 5: Since 2.148 > 2.145, reject H_0. The data do provide sufficient evidence at the 5% significance level that there is a difference between the mean heights of cross-fertilized and self-fertilized plants.

Copyright © 2012 Pearson Education, Inc. Publishing as Addison-Wesley.

(f) For the 0.01 significance level, the critical values are ±2.977 and the P-value is the same as in part (e). Since t = 2.148 is not in the rejection region, do not reject the null hypothesis at the 0.01 significance level.

10.111 Population 1: Weights after treatment for anorexia nervosa
Population 2: Weights before treatment for anorexia nervosa

Step 1: H_0: $\mu_1 = \mu_2$, H_a: $\mu_1 > \mu_2$

Step 2: $\alpha = 0.05$

Step 3: The paired differences 'Weight after - Weight before' are
11.0, 5.5, 9.4, 13.6, -2.9, 10.7, -0.1, 7.4, 21.5, -5.3, -3.8, 11.4, 13.4, 13.1, 9.0, 3.9, 5.7

For these differences, $\sum d = 123.5$ and $\sum d^2 = 1716.9$

$$\bar{d} = \sum d / n = 123.5 / 17 = 7.26$$

$$s_d = \sqrt{\frac{\sum d^2 - \left(\sum d\right)^2 / n}{n-1}} = \sqrt{\frac{1716.9 - 123.5^2 / 17}{17-1}} = 7.16$$

$$t = \frac{\bar{d}}{s_d / \sqrt{n}} = \frac{7.26}{7.16 / \sqrt{17}} = 4.181$$

Step 4: df = 16; critical values = 1.746
P-Value < 0.005

Step 5: Since 4.181 > 1.746, reject H_0. The data do provide sufficient evidence at the 5% significance level that family therapy is effective in helping anorexic young women gain weight.
For the P-value approach, since the P-value is smaller than the significance level, reject H_0.

10.113 Population 1: Corneal thickness in normal eyes
Population 2: Corneal thickness in glaucoma eyes

Step 1: H_0: $\mu_1 = \mu_2$, H_a: $\mu_1 > \mu_2$

Step 2: $\alpha = 0.10$

Step 3: The paired differences, d = $x_1 - x_2$, are
-4, 0 12, 18, -4, -12, 6, 16

For these differences, $\sum d = 32$ and $\sum d^2 = 936$

$$\bar{d} = \sum d / n = 32 / 8 = 4.0$$

$$s_d = \sqrt{\frac{\sum d^2 - \left(\sum d\right)^2 / n}{n-1}} = \sqrt{\frac{936 - 32^2 / 8}{8-1}} = 10.7438$$

$$t = \frac{\bar{d}}{s_d / \sqrt{n}} = \frac{4.0}{10.7438 / \sqrt{8}} = 1.053$$

Copyright © 2012 Pearson Education, Inc. Publishing as Addison-Wesley.

Step 4: df = 7, Critical value = 1.415
P-Value > 0.10

Step 5: Since 1.053 < 1415, do not reject H_0. The data do not provide sufficient evidence at the 10% significance level that the mean corneal thickness is greater in normal eyes than in eyes with glaucoma.
For the P-value approach, since the P-value is larger than the significance level, do not reject H_0.

10.115 From Exercise 10.109, df = 14.

(a)

$$20.93 \pm 2.145 \cdot \frac{37.74}{\sqrt{15}} = 20.93 \pm 20.90 = 0.03 \text{ to } 41.83 \text{ eighths of an inch}$$

We can be 95% confident that the difference, $\mu_1 - \mu_2$, between the mean heights of cross-fertilized and self-fertilized Zea mays is somewhere between 0.03 and 41.83 eighths of an inch.

(b)

$$20.93 \pm 2.977 \cdot \frac{37.74}{\sqrt{15}} = 20.93 \pm 29.01 = -8.08 \text{ to } 49.94 \text{ eighths of an inch}$$

We can be 99% confident that the difference, $\mu_1 - \mu_2$, between the mean heights of cross-fertilized and self-fertilized Zea mays is somewhere between -8.08 and 49.94 eighths of an inch.

10.117 From Exercise 10.111, df = 16.

$$7.26 \pm 1.746 \cdot \frac{7.16}{\sqrt{17}} = 7.26 \pm 3.03 = 4.23 \text{ to } 10.29 \text{ pounds}$$

We can be 90% confident that the mean weight gain, $\mu_1 - \mu_2$, resulting from family-therapy treatment by anorexic young women is somewhere between 4.23 and 10.29 pounds.

10.119 From Exercise 10.113, , df = 7.

$$4.0 \pm 1.415 \cdot \frac{10.7434}{\sqrt{8}} = 4.0 \pm 5.37 = (-1.37, 9.37) \text{ microns}$$

We can be 80% confident that the difference, $\mu_1 - \mu_2$, in corneal thickness of patients who had glaucoma in one eye but not the other is between -1.37 and 9.37 microns.

10.121 Using Minitab, with the differences in a column named DIFF, choose **Graph ▶ Probability plot...**, select the **Single** version and click **OK**. Enter DIFF in the **Graph variables text box** and click **OK**. Then choose **Graph ▶ Boxplot**, select the **One Y Simple** version and click **OK**. Enter DIFF in **Graph variables** text box and click **OK**.

Copyright © 2012 Pearson Education, Inc. Publishing as Addison-Wesley.

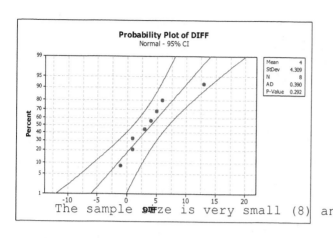

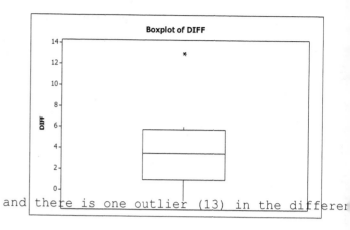

The sample size is very small (8) and there is one outlier (13) in the differen

10.123 Using Minitab, enter DIFF at the top of column C3, then choose **Calc ▶ Calculator**, enter DIFF in the **Store result in variable** box, enter 'RESOLUTION'-'ONSET' in the Expression box, and click **OK**. Choose **Graph ▶ Probability plot...**, select the **Single** version and click **OK**. Enter ONSET RESOLUTION DIFF in the **Graph variables** text box and click **OK**. Then choose **Graph ▶ Boxplot**, select the **Multiple Y's Simple** version and click **OK**. Enter ONSET and RESOLUTION in the **Graph Variables** box and click **OK**. Finally, choose **Graph ▶ Boxplot**, select the **One Y Simple** version and click **OK**. Enter DIFF in **Graph variables** text box and click **OK**. The results are

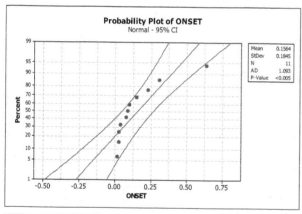

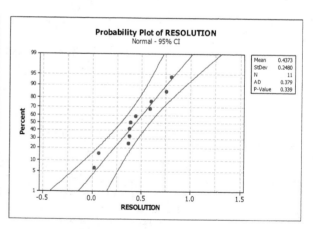

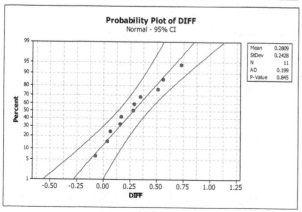

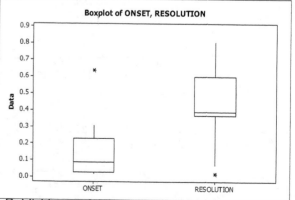

Copyright © 2012 Pearson Education, Inc. Publishing as Addison-Wesley.

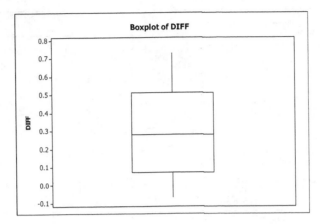

(b) No. The sample size is small, there is an outlier, and the normal probability plot is not linear.

(c) No. The sample size is small and there is an outlier. The probability plot is closer to linear than for the ONSET data, but overall, t-procedures are not advisable.

(d) Yes. There are no outliers and the probability plot is reasonably linear.

(e) They imply that the assumptions for using a paired t-procedure may be met even though the assumptions for using a t-procedure may not be met for the original variables.

10.125 (a) Using Minitab, choose **Stat ▶ Basic Statistics ▶ Paired t...,** enter MEN in the **First sample** box and WOMEN in the **Second Sample** box. Click on the **Options** button and enter 99 in the **Confidence** level box, 0.0 in the **Test mean** box, and **greater than** in the **Alternative** drop down box. Click **OK** twice. The results are

Paired T for MEN - WOMEN

	N	Mean	StDev	SE Mean
MEN	75	35.4800	9.7695	1.1281
WOMEN	75	32.0800	7.0265	0.8114
Difference	75	3.40000	7.23542	0.83547

99% lower bound for mean difference: 1.41341
T-Test of mean difference = 0 (vs > 0): T-Value = 4.07   P-Value = 0.000

The P-value is 0.000, which is less than the significance level 0.01. Therefore, the data do provide sufficient evidence that the mean age of Norwegian men at the time of marriage exceeds that of Norwegian women.

(b) The 99% confidence interval is found by repeating the procedure of part (a) using **not equal** in the **Alternative** droop down box. The result is
99% CI for mean difference: (1.19108, 5.60892)
We can be 99% confident that the mean difference in ages of Norwegian men and women at the time of marriage lies somewhere between 1.19 and 5.61 years.

(c) We remove the pairs (18,32) and (53,29) from the data and repeat parts (a) and (b). The results are
Paired T for MEN - WOMEN

	N	Mean	StDev	SE Mean
MEN	73	35.4795	9.4650	1.1078
WOMEN	73	32.1233	7.1140	0.8326
Difference	73	3.35616	6.61095	0.77375

99% lower bound for mean difference: 1.51520
T-Test of mean difference = 0 (vs > 0): T-Value = 4.34   P-Value = 0.000

Copyright © 2012 Pearson Education, Inc. Publishing as Addison-Wesley.

99% CI for mean difference: (1.30893, 5.40340)
By removing the outliers, the t-value increases from 4.07 to 4.34, the P-value remains at 0.000, and the conclusion remains the same. The 99% confidence interval is shortened by about 0.3 years and the mean difference changes from 3.40 to 3.36.

10.127 Data is obtained from two populations whose members can be naturally paired. By letting d represent the difference between the values in each pair, we can reduce the data set from pairs of numbers to a single number representing each pair. Since the mean of the paired differences is the same as the difference of the two population means, we can test the equality of the two population means by testing to see if the mean of the paired differences is zero (or some other value if appropriate). If the paired differences are normally distributed, then the one-sample t-test can be used to carry out the test.

10.129 The paired Wilcoxon signed-rank test. The sample size is small. The outlier will have a greater effect on the paired t-test than on the Wilcoxon test.

10.131 (a) Wilcoxon signed-rank test. A uniform distribution is symmetric, but not normal. Therefore the Wilcoxon test is appropriate.
(b) Paired t-test. The sample size is very large. The t-test may be used when we have a simple random paired sample and a large sample.
(c) If the sample size is large, the t-test is appropriate. If the sample size is small, the assumptions are not met for either test, but the Wilcoxon signed-rank test might be appropriate if the degree of skewness is small. It would be less affected by the moderate skewness than the paired t-test would be.

Review Problems for Chapter 10

1. Randomly sample independently from both populations, compute the means of both samples and reject the null hypothesis if the sample means differ by too much. Otherwise, do not reject the null hypothesis.

2. Sample independent pairs of observations from the two populations, compute the difference of the two observations in each pair, compute the mean of the differences, and reject the null hypothesis if the sample mean of the differences differs from zero by too much. Otherwise, do not reject the null hypothesis.

3. (a) The pooled t-test requires that the population standard deviations be equal whereas the nonpooled t-test does not.
(b) It is absolutely essential that the assumption of independence be satisfied.
(c) The normality assumption is especially important for both t-tests for small samples. With large samples, the Central Limit Theorem applies and the normality assumption is less important.
(d) Unless we are quite sure that the population standard deviations are equal, the nonpooled t-procedures should be used instead of the pooled t-procedures.

4. A paired sample may reduce the estimate of the standard error of the mean of the differences, making it more likely that a difference of a given size will be judged significant.

5. (a) Population 1: Male; $n_1 = 13$, $\bar{x}_1 = 2127$, $s_1 = 513$

   Population 2: Female; $n_2 = 14$, $\bar{x}_2 = 1843$, $s_2 = 446$

   Step 1: H_0: $\mu_1 = \mu_2$, H_a: $\mu_1 > \mu_2$
   Step 2: $\alpha = 0.05$,
   Step 3:

Copyright © 2012 Pearson Education, Inc. Publishing as Addison-Wesley.

$$s_p = \sqrt{\frac{12(513)^2 + 13(446)^2}{25}} = 479.33$$

$$t = \frac{2127 - 1843}{479.33\sqrt{\frac{1}{13} + \frac{1}{14}}} = 1.538$$

Step 4: df = 25, Critical value = 1.708
0.05 < P(t > 1.538) < 0.10

Step 5: Since 1.538 < 1.708 and since the P-value is larger than the significance level, do not reject H_0.

Step 6: At the 5% significance level, the data do not provide enough evidence to conclude that the mean right-leg strength of males exceeds that of females.

6.

$$2127 - 1843 \pm 1.708(479.33)\sqrt{\frac{1}{13} + \frac{1}{14}}$$

$$284 \pm 315.33$$

$$-31.33 \text{ to } 599.33$$

We can be 90% confident that the difference, $\mu_1 - \mu_2$, between the mean right-leg strengths of males and females is between -31.3 and 599.3 newtons.

7. Population 1: Florida, n_1 = 24, $\bar{x}_1$ = 5.46, s_1 = 1.59
Population 2: Virginia, n_2 = 44, $\bar{x}_2$ = 7.59, s_2 = 2.68

$$\Delta = \frac{\left[\left(\frac{s_1^2}{n_1}\right) + \left(\frac{s_2^2}{n_2}\right)\right]^2}{\frac{\left(\frac{s_1^2}{n_1}\right)^2}{n_1 - 1} + \frac{\left(\frac{s_2^2}{n_2}\right)^2}{n_2 - 1}} = \frac{\left[\left(\frac{5.46^2}{24}\right) + \left(\frac{7.59^2}{44}\right)\right]^2}{\frac{\left(\frac{5.46^2}{24}\right)^2}{24 - 1} + \frac{\left(\frac{7.59^2}{44}\right)^2}{44 - 1}} = 65.45; \; df = 65$$

Step 1: $H_0: \mu_1 = \mu_2$, $H_a: \mu_1 < \mu_2$
Step 2: α = 0.01
Step 3:

$$t = \frac{5.4583 - 7.5909}{\sqrt{\frac{1.5874^2}{24} + \frac{2.6791^2}{44}}} = -4.119$$

Step 4: Critical value = -2.385
P(t < -4.119) < 0.005

Step 5: Since -4.119 < -2.385 and also the P-value is smaller than the significance level, reject H_0.

Step 6: At the 1% significance level, the data provide sufficient evidence to conclude that the average litter size of cottonmouths in Florida is less than that in Virginia.

Copyright © 2012 Pearson Education, Inc. Publishing as Addison-Wesley.

8.

$$(5.46-7.49)\pm2.385\sqrt{\frac{1.59^2}{24}+\frac{2.68^2}{44}}$$

$$-2.13\pm1.24$$

$$-3.37 \text{ to } -0.89$$

We can be 98% confident that the average difference, $\mu_1 - \mu_2$, between cottonmouth litter sizes in Florida and Virginia is somewhere between -3.37 and -0.89.

9. (a) We used Minitab to produce the following normal probability plot and boxplot of the differences.

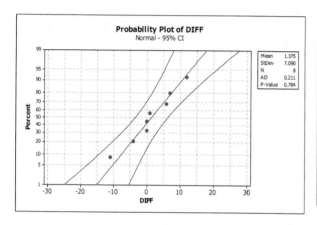
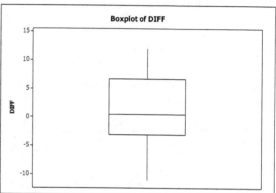

(b) Yes. The normal probability plot is reasonably linear and the boxplot shows no outliers.

(c) Step 1: H_0: $\mu_1 = \mu_2$, H_a: $\mu_1 \neq \mu_2$

Step 2: $\alpha = 0.10$

Step 3: Paired differences, d = x_1 - x_2:

12	7	1	0
-11	0	6	-4

From the probability plot, we see that the sample mean of the differences is 1.375 and the standard deviation is 7.090. Therefore,

$$t = \frac{1.375}{7.090/\sqrt{8}} = 0.549$$

Step 4: Critical values = ±1.895
2{P(t > 0.549)} > 0.20

Step 5: Since -1.895 < 0.549 < 1.895 and also since the P-value is greater than the significance level, do not reject H_0.

Step 6: At the 10% significance level, the data do not provide sufficient evidence to conclude that there is a difference in the mean length of time that ice stays on the two lakes.

10.

$$1.375\pm1.860\cdot7.090/\sqrt{8}=1.375\pm4.749=-3.374 \text{ to } 6.124 \text{ days}$$

We can be 90% confident that the difference, $\mu_1-\mu_2$, between the mean

Copyright © 2012 Pearson Education, Inc. Publishing as Addison-Wesley.

numbers of days ice stays on Lakes Mendota and Mononar is somewhere between −3.374 and 6.124 days.

11. To determine which procedure should be used to perform the hypothesis test, ask the following questions in sequence:

	Answer to question	
Question to ask	Yes	No
Are the samples paired?	x	
Are the differences normal?		x
Are the differences symmetric?		x
Is the sample large?	x	

It appears that the population of paired differences is neither normal nor symmetric, but the sample size is large. Thus, use the paired t-test.

12. To determine which procedure should be used to perform the hypothesis test, ask the following questions in sequence:

	Answer to question	
Question to ask	Yes	No
Are the samples paired?		x
Are the populations normal?	x	
Are the populations the same shape?		x
Is the sample large?		x

Since the two independent populations are normal, use the nonpooled t-test.

13. To determine which procedure should be used to perform the hypothesis test, ask the following questions in sequence:

	Answer to question	
Question to ask	Yes	No
Are the samples paired?		x
Are the populations normal?		x
Are the populations the same shape?	x	

Since the independent samples have the same nonnormal shape, use the Mann-Whitney test.

14. To determine which procedure should be used to find the required confidence interval, ask the following questions in sequence:

	Answer to question	
Question to ask	Yes	No
Are the samples paired?		x
Are the populations normal?		x
Are the samples large?	x	

Copyright © 2012 Pearson Education, Inc. Publishing as Addison-Wesley.

It appears that one of the populations is neither normal nor symmetric, but the sample sizes are large. Thus, use the nonpooled t-test.

15. To determine which procedure should be used to perform the hypothesis test, ask the following questions in sequence:

Question to ask	Answer to question	
	Yes	No
Are the samples paired?		x
Are the populations normal?		x
Are the populations the same shape?		x
Is the sample large?		x

Because the two samples are small, nonnormal, and do not have the same shape, none of the procedures studied are appropriate. A statistician must be consulted.

16. To determine which procedure should be used to perform the hypothesis test, ask the following questions in sequence:

Question to ask	Answer to question	
	Yes	No
Are the samples paired?	x	
Are the differences normal?		x
Are the differences symmetric?		x
Is the sample large?		x

It appears the paired differences are neither normal or symmetric and the sample size is small. None of the procedures studied is appropriate. Thus, a statistician must be consulted.

17. (a) Using Minitab, choose **Graph ▶ Probability plot...**, select the **Single** version and click **OK**. Enter MALE FEMALE in the **Graph Variables** text box, and click **OK**. Then choose **Graph ▶ Boxplot**, select the **Multiple Y's Simple** version and click **OK**. Enter MALE FEMALE **in the Graph variables** text box, and click **OK**. The standard deviations will be produced on the probability plot legend. The results are

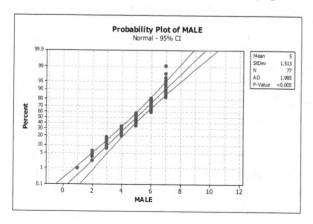
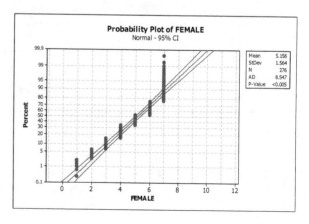

Copyright © 2012 Pearson Education, Inc. Publishing as Addison-Wesley.

The standard deviations are
1.513 for MALE and 1.564 for
FEMALE.

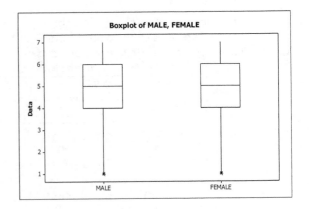

(b) Pooled t-procedures. The sample sizes are both large, there are no
potential outliers in either sample, and the sample standard
deviations are nearly equal.

(c) To carry out the pooled t-test, choose **Stat ▶ Basic Statistics ▶ 2-
Sample t**, click on **Samples in different columns**, enter <u>MALE</u> in the
First text box and <u>FEMALE</u> in the **Second** text box, select **not equal** in
the **Alternative** box, enter <u>90</u> in the **Confidence level** text box, click
on **Assume equal variances** to make certain that there is an **X** in the
box, and click **OK**. The results are

Two-sample T for MALE vs FEMALE

```
            N   Mean   StDev   SE Mean
MALE       77   5.00   1.51     0.17
FEMALE    276   5.16   1.56     0.094
```

```
Difference = mu (MALE) - mu (FEMALE)
Estimate for difference:  -0.155797
95% CI for difference:  (-0.549387, 0.237793)
T-Test of difference = 0 (vs not =): T-Value = -0.78   P-Value = 0.437   DF =
351
Both use Pooled StDev = 1.5528
```

The P-value is 0.437, which is greater than the significance level
0.05. Do not reject the null hypothesis. The data do not provide
sufficient evidence to conclude that there is a difference in mean
dating satisfaction of male and female college students.

(d) A 95% confidence interval for the difference between mean dating
satisfaction of male and female college students was produced as part
of the process in part (b). The interval is (-0.55, 0.24).

(e) Yes. Although the data for both genders are left-skewed and discrete
(possible values include only the whole numbers from 1 to 7), the large
sample sizes justify the use of the pooled t-procedures.

18. (a) Using Minitab, choose **Graph ▶ Probability plot...**, select the **Single**
version and click **OK**. Enter <u>OVER 65</u> and <u>CONTROL</u> in the **Graph Variables**
text box, and click **OK**. Then choose **Graph ▶ Boxplot**, select the
Multiple Y's Simple version and click **OK**. Enter <u>OVER 65</u> and <u>CONTROL</u> **in
the Graph variables** text box, and click **OK**. The standard deviations
will be produced on the probability plot legend. The results are

Copyright © 2012 Pearson Education, Inc. Publishing as Addison-Wesley.

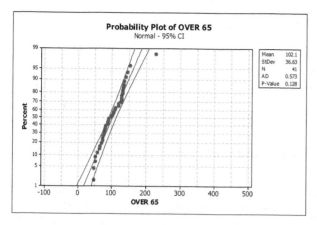

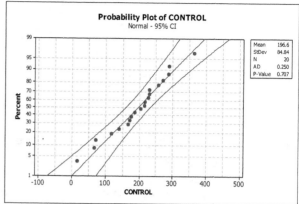

The standard deviations are 36.63 for OVER 65 and 84.84 for CONTROL,

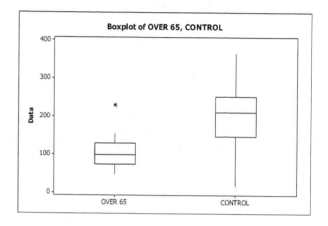

(b) The nonpooled t-procedure is preferable since the ratio of the standard deviations is greater than 2.

(c) To carry out the nonpooled t-test, choose **Stat ▶ Basic Statistics ▶ 2-Sample t**, click on **Samples in different columns**, enter 'OVER 65' in the **First** text box and CONTROL in the **Second** text box, and leave the **Assume equal variances** box unchecked. Click on the **Options** button, enter 99 in the **Confidence level** text box, select **less than** in the **Alternative** box, and click **OK** twice. The results are

```
Two-sample T for OVER 65 vs CONTROL

            N    Mean   StDev   SE Mean
OVER 65    41   102.1    36.6       5.7
CONTROL    20   196.6    84.8        19


Difference = mu (OVER 65) - mu (CONTROL)
Estimate for difference:  -94.4768
99% upper bound for difference:  -44.7729
T-Test of difference = 0 (vs <): T-Value = -4.77  P-Value = 0.000  DF = 22
```

Since the P-value = 0.000, which is less than the significance level 0.01, reject the mull hypothesis. The data do provide sufficient evidence to conclude that, on average, men over 65 have a lower IGF-1 level than younger men.

(d) To obtain the 99% confidence interval, follow the procedure in part (c), but select not equal in the Alternative drop down box. The confidence interval is given by
    99% CI for difference: (-150.3322, -38.6215)

Copyright © 2012 Pearson Education, Inc. Publishing as Addison-Wesley.

We can be 99% confident that the difference in IGF-1 levels, on average, between men over 65 and younger men is somewhere between -150.3 and -38.6.

(e) Yes. The CONTROL sample of size 20 is reasonably normally distributed with no outliers. The OVER 65 sample of size 41 is also close to normal with one outlier. The effect of the outlier is not great enough to alter the conclusion (Deleting it changes the value of t from -4.77 to -4.99).

19. (a) Using Minitab, choose **Stat ▶ Basic Statistics ▶ Paired t**, click on **Samples in columns**, enter MEN in the **First** text box and WOMEN in the **Second** text box. Click on the **Options** button, enter 95 in the **Confidence level** text box, select **greater than** in the **Alternative** box, and click **OK**. Click on the **Graphs** button, check the box for **Boxplot of differences**, and click **OK** twice. The results are
Paired T for MEN - WOMEN

	N	Mean	StDev	SE Mean
MEN	50	974	820	116
WOMEN	50	844	818	116
Difference	50	130.1	93.8	13.3

95% lower bound for mean difference: 107.9
T-Test of mean difference = 0 (vs > 0): T-Value = 9.81 P-Value = 0.000
Since the P-value = 0.000, which is less than the significance level 0.05, reject the null hypothesis. The data do provide sufficient evidence to conclude that, on average, the weekly earnings of male full-time and salary workers exceed that of women.

(b) Follow the procedure in part (a), but enter 90 in the **Confidence level** text box and select **not equal** from the **Alternative** drop down box. The resulting interval is
90% CI for mean difference: (107.9, 152.3)
We can be 90% confident that the mean difference in weekly earnings of male and female full-time wage and salary workers is somewhere between $107.9 and $152.3.

(c) To create a column of differences, click in the **Sessions Window** and then click on the **Editor** tab and on **Enable Commands**. An MTB> prompt will appear at the bottom of the Sessions Window. After the prompt, enter Let c3=c1-c2 and press ENTER. Click at the top of Column 3 and name the column DIFF. Choose **Graph ▶ Probability plot...**, select the **Single** version and click **OK**. Enter DIFF in the **Graph Variables** text box, and click **OK**. The standard deviations will be produced on the probability plot legend. A boxplot was produced by the procedure in part (a). Now choose **Graph ▶ Stem-and-Leaf...**, enter DIFF in the **Graph Variables** text box, and click **OK**. The results are

Copyright © 2012 Pearson Education, Inc. Publishing as Addison-Wesley.

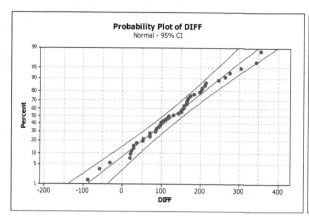

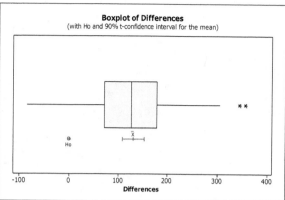

```
Stem-and-leaf of DIFF   N  = 50
Leaf Unit = 10

   2    -0   85
   3    -0   3
   9     0   222233
  20     0   55777888999
  (7)    1   0011234
  23     1   5555566667789
  10     2   00114
   5     2   67
   3     3   04
   1     3   5
```

(d)   Yes. The sample size is large (50), there are only two mild outliers
on the right, and the normal probability plot is close to linear. The
outliers may cause some concern since they are in direction that helps
to reject the null hypothesis. If so, a nonparametric method may be
used.

Copyright © 2012 Pearson Education, Inc. Publishing as Addison-Wesley.

Exercises 11.1

11.1 Answers will vary.

11.3 A population proportion p is a parameter since it is a descriptive measure for a population. A sample proportion $\hat{p}$ is a statistic since it is a descriptive measure for a sample.

11.5 (a) $p = 2/5 = 0.4$

(b)

Sample	Number of females x	Sample proportion $\hat{p}$
J,G	1	0.5
J,P	0	0.0
J,C	0	0.0
J,F	1	0.5
G,P	1	0.5
G,C	1	0.5
G,F	2	1.0
P,C	0	0.0
P,F	1	0.5
C,F	1	0.5

(c) The population proportion is marked by the vertical line below.

(d) $\mu_{\hat{p}} = \sum \hat{p}/10 = 4.0/10 = 0.4$

(e) The answers to (a) and (d) are the same. $\hat{p}$ is a sample mean. The mean of the sampling distribution of $\hat{p}$ is the same as the mean of the population, which is p.

11.7 (b)

Sample	Number of females x	Sample proportion $\hat{p}$
J,P,C	0	0
J,P,G	1	1/3
J,P,F	1	1/3
J,C,G	1	1/3
J,C,F	1	1/3
J,G,F	2	2/3
P,C,G	1	1/3

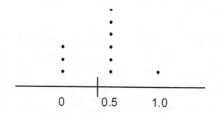

Copyright © 2012 Pearson Education, Inc. Publishing as Addison-Wesley.

P,C,F	1	1/3
P,G,F	2	2/3
C,G,F	2	2/3

(c) The population proportion is marked by the vertical line below.

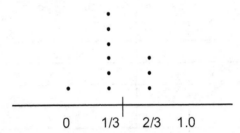

(d) $\mu_{\hat{p}} = \sum \hat{p}/10 = (12/3)/10 = 0.4$

(e) The answers to (a) and (d) are the same. $\hat{p}$ is a sample mean. The mean of the sampling distribution of $\hat{p}$ is the same as the mean of the population, which is p.

11.9 (b)

Sample	Number of females X	Sample proportion $\hat{p}$
J,P,C,G,F	2	0.4

(c) The population proportion is marked by the vertical line below.

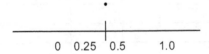

(d) $\mu_{\hat{p}} = \sum \hat{p}/1 = 0.4/1 = 0.4$

(e) The answers to (a) and (d) are the same. $\hat{p}$ is a sample mean. The mean of the sampling distribution of $\hat{p}$ is the same as the mean of the population, which is p.

11.11 (a) The population consists of all #1 NBA draft picks since 1947.
(b) The specified attribute is being a other than a U.S. national.
(c) The 11.3% is a population proportion since all of the #1 picks are known. There is no need to sample this group.

11.13 (a) $\hat{p} = 0.79$; $z_{0.005} = 2.575$; the margin of error is

Copyright © 2012 Pearson Education, Inc. Publishing as Addison-Wesley.

$$E = z_{0.005} \cdot \sqrt{\hat{p}(1-\hat{p})/n} = 2.575\sqrt{0.79(0.21)/21355} = 0.00718$$

(b) The margin of error will be smaller for a 90% confidence interval. Specifically, 2.575 will be replaced by 1.645 in the formula for E and everything else stays the same. More generally speaking, in order to have a higher level of confidence in an interval, one needs to have a wider interval.

11.15 (a) 0.4 (b) 0.2 (c) 0.5

(d) For (a), $0.4 < \hat{p} < 0.6$; For (b), $0.2 < \hat{p} < 0.8$; For (c), none

11.17 (a) $\hat{p} = 8/40 = 0.2$

(b) x = 8 and n - x = 32. Both are at least 5, so the one-proportion z-interval procedure is appropriate.

(c) The 95% confidence interval is

$$\hat{p} \pm z_{\alpha/2}\sqrt{\hat{p}(1-\hat{p})/n} = 0.2 \pm 1.96\sqrt{0.2(1-0.2))/40} = (0.076, 0.324)$$

11.19 (a) $\hat{p} = 35/50 = 0.70$

(b) x = 35 and n - x = 15. Both are at least 5, so the one-proportion z-interval procedure is appropriate.

(c) The 99% confidence interval is

$$\hat{p} \pm z_{\alpha/2}\sqrt{\hat{p}(1-\hat{p})/n} = 0.70 \pm 2.575\sqrt{0.70(1-0.70))/50} = (0.533, 0.867)$$

11.21 (a) $\hat{p} = 16/20 = 0.8$

(b) x = 16 and n - x = 4. n - x is less than 5, so the one-proportion z-interval procedure is not appropriate.

11.23 $\hat{p} = 410/603 = 0.68$; The 95% confidence interval is

$$\hat{p} \pm z_{\alpha/2}\sqrt{\hat{p}(1-\hat{p})/n} = 0.68 \pm 1.96\sqrt{0.68(1-0.68))/603} = (0.643, 0.717)$$

11.25 n = 500, x = 38, and n - x = 462. Both x and n - x are at least 5.

$\hat{p}$ = x/n = 38/500 = 0.076; $z_{\alpha/2} = z_{0.025} = 1.96$

(a) $$0.076 - 1.96\sqrt{0.076(1-0.076)/500} \text{ to } 0.076 + 1.96\sqrt{0.076(1-0.076)/500}$$

$$0.0528 \text{ to } 0.0992$$

(b) We can be 95% confident that the proportion of U.S. asthmatics who are allergic to sulfites is somewhere between 0.0528 and 0.0992.

11.27 n = 1000, x = 800, and n - x = 200. Both x and n - x are at least 5.

$\hat{p}$ = x/n = 800/1000 = 0.80; $z_{\alpha/2} = z_{0.005} = 2.575$

(a) $$0.80 - 2.575\sqrt{0.80(1-0.80)/1000} \text{ to } 0.80 + 2.575\sqrt{0.80(1-0.80)/1000}$$

$$0.767 \text{ to } 0.833$$

(b) We can be 99% confident that the percentage of all registered voters who favor the creation of standards on CAFO pollution is somewhere between 76.7% and 83.3%.

11.29 Here, n = 100, x = 96, and n - x = 4. To use Procedure 11.1, both x and

n - x must be at least 5. Since n - x < 5, Procedure 11.1 should not have been used.

11.31 $\hat{p}$ = 647/1043 = 0.620, E = 0.029

$\hat{p}$ - E to $\hat{p}$ + E

0.620 - 0.029 to 0.620 + 0.029   In percentage terms, this is

59.1%   to   64.9%

Copyright © 2012 Pearson Education, Inc. Publishing as Addison-Wesley.

11.33 From Exercise 11.25, $\hat{p} = 0.076$; $z_{\alpha/2} = 1.96$; $n = 500$

(a) $E = z_{\alpha/2}\sqrt{\hat{p}(1-\hat{p})/n} = 1.96\sqrt{0.076(1-0.076)/500} = 0.0232$

(b) $E = 0.01$; $z_{\alpha/2} = z_{0.025} = 1.96$;

$$n = \hat{p}(1-\hat{p})\frac{z_{\alpha/2}^2}{E^2} = 0.5(1-0.5)\frac{1.96^2}{0.01^2} = 9604$$

(c) $n = 9604$; $\hat{p} = 0.071$; $z_{\alpha/2} = z_{0.025} = 1.96$

$$0.071 - 1.96\sqrt{0.071(1-0.071)/9604} \text{ to } 0.3192 + 1.96\sqrt{0.071(1-0.071)/9604}$$

$$0.0659 \text{ to } 0.0761$$

(d) The margin of error for the estimate is 0.0051. As expected, this is less than the required margin of error of 0.01 specified in part (b).

(e) $\hat{p}_g = 0.10$; $z_{\alpha/2} = z_{0.025} = 1.96$

part (b) $E = 0.01$, $z_{0.025} = 1.96$; sample size is:

$$n = \hat{p}(1-\hat{p})\frac{z_{\alpha/2}^2}{E^2} = 0.10(1-0.10)\frac{1.96^2}{0.01^2} = 3457.44 \rightarrow 3458$$

Thus the required sample size is n = 3458.
part (c)
$$0.071 - 1.96\sqrt{0.071(1-0.071)/3458} \text{ to } 0.3192 + 1.96\sqrt{0.071(1-0.071)/3458}$$

$$0.0624 \text{ to } 0.0796$$

part (d) The margin of error is 0.0086. As expected, this is less than what was specified in part (b).

(f) By employing the guess for $\hat{p}$ in part (d) we can reduce the required sample size by more than 6000 (from 9604 to 3458), saving considerable time and money. Moreover, the margin of error only rises from 0.0051 to 0.0086. The risk of using the guess 0.10 for $\hat{p}$ is that if the actual value of $\hat{p}$ turns out to be between 0.10 and 0.90, then the achieved margin of error will exceed the specified 0.01.

11.35 (a) The confidence interval from Exercise 11.27 was (76.7%, 83.3%). To find the error, take the width of the confidence interval and divide by 2.   E = (83.3% - 76.7%)/2 = 3.3%.

(b) $E = 0.015$; $z_{\alpha/2} = z_{0.005} = 2.575$;

$$n = \hat{p}(1-\hat{p})\frac{z_{\alpha/2}^2}{E^2} = 0.5(1-0.5)\frac{2.575^2}{0.015^2} = 7367.36 \rightarrow 7368$$

(c) $n = 7368$; $= 0.822$; $z_{\alpha/2} = z_{0.005} = 2.575$

$$0.822 - 2.575\sqrt{0.822(1-0.822)/7368} \text{ to } 0.822 + 2.575\sqrt{0.822(1-0.822)/7368}$$

$$0.811 \text{ to } 0.833$$

(d) The margin of error for the estimate is 0.011, which is less than the 0.015 required in part (b).

(e) $\hat{p}_g$ is between 0.75 and .85; $z_{\alpha/2} = z_{0.005} = 2.575$

part (b) E = 0.015, $z_{0.005} = 2.575$; sample size is:

Copyright © 2012 Pearson Education, Inc. Publishing as Addison-Wesley.

$$n = \hat{p}(1-\hat{p})\frac{z_{\alpha/2}^2}{E^2} = 0.75(1-0.75)\frac{2.575^2}{0.015^2} = 5525.52 \rightarrow 5526$$

Thus the required sample size is n = 5526.
part (c)

$$0.822 - 2.575\sqrt{0.822(1-0.822)/5526} \; to \; 0.822 + 2.575\sqrt{0.822(1-0.822)/5526}$$

$$0.809 \; to \; 0.835$$

part (d)
> The margin of error is 0.013, which is less than the 0.015 specified in part (b).

(f) By employing the guess for $\hat{p}$ in part (d) we can reduce the required sample size by 1842, from 7368 to 5526, saving considerable time and money. Moreover, the margin of error only rises from 0.011 to 0.013. The risk of using the guess 0.75 for $\hat{p}$ is that if the actual value of $\hat{p}$ turns out to be between 0.25 and 0.75, then the achieved margin of error will exceed the specified 0.015.

11.37 (a) E = 0.01; $z_{\alpha/2}$ = $z_{0.025}$ = 1.96;

$$n = \hat{p}(1-\hat{p})\frac{z_{\alpha/2}^2}{E^2} = 0.5(1-0.5)\frac{1.96^2}{0.01^2} = 9604$$

(b) $\hat{p}_g$ is between 0.5% and 4.9%; $z_{\alpha/2}$ = $z_{0.025}$ = 1.96; E = 0.01

$$n = \hat{p}(1-\hat{p})\frac{z_{\alpha/2}^2}{E^2} = 0.049(1-0.049)\frac{1.96^2}{0.01^2} = 1790.15 \rightarrow 1791$$

Thus the required sample size is n = 1,791.
(c) The sample size has decreased from 9604 to 1791.
(d) If the actual value of $\hat{p}$ turns out to be between 0.049 and 0.951, then the achieved margin of error will exceed the specified error of 0.01.

11.39 $\hat{p}$= 444/1010 = 0.440; $z_{0.025}$ = 1.96; the 95% confidence interval is

$$0.440 - 1.96\sqrt{0.440(1-0.440)/1010} \; to \; 0.440 + 1.96\sqrt{0.440(1-0.440)/1010}$$

$$0.409 \; to \; 0.470$$

We can be 95% confident that the proportion of U.S. adults who approved of the way that President George W. Bush was doing his job was somewhere between 0.409 and 0.470.

11.41 $\hat{p}$= 276/1063 = 0.260; $z_{0.05}$ = 1.645; the 90% confidence interval is

$$0.260 - 1.645\sqrt{0.260(1-0.260)/1063} \; to \; 0.260 + 1.645\sqrt{0.260(1-0.260)/1063}$$

$$0.238 \; to \; 0.282$$

We can be 90% confident that the proportion of U.S. adults who would purchase or lease a new car from a manufacturer that had declared bankruptcy is somewhere between 0.238 and 0.282.

11.43 The Central Limit Theorem

11.45 The sample size is directly proportional to the quantity p(1 - p), but the final margin of error is directly proportional to $\hat{p}$ (1 - $\hat{p}$). The quantity p(1 - p) takes on its maximum value when p = 0.5. If $\hat{p}$ is closer to .5 than is p, then $\hat{p}$(1 - $\hat{p}$) will be greater than p(1 - p) and the confidence

Copyright © 2012 Pearson Education, Inc. Publishing as Addison-Wesley.

interval using $\hat{p}(1 - \hat{p})$ will be larger than the desired one associated with $p(1 - p)$.

11.47 (a) $\tilde{p} = (8+2)/(40+4) = 0.227$

The 95% one-proportion plus-four z-interval is

$$\tilde{p} \pm z_{\alpha/2}\sqrt{\tilde{p}(1-\tilde{p})/(n+4)} = 0.227 \pm 1.96\sqrt{0.227(1-0.227))/44} = (0.103, 0.351)$$

(b) This interval has the same level of precision as the interval found in Exercise 11.17. However, it is centered at 0.227 rather than 0.20.

11.49 (a) $\tilde{p} = (35+2)/(50+4) = 0.685$

The 99% one-proportion plus-four z-interval is

$$\tilde{p} \pm z_{\alpha/2}\sqrt{\tilde{p}(1-\tilde{p})/(n+4)} = 0.685 \pm 2.575\sqrt{0.685(1-0.685))/54} = (0.522, 0.848)$$

(b) This interval is slightly more precise than the interval found in Exercise 11.19. Also, it is centered at 0.685 rather than 0.70.

11.51 (a) $\tilde{p} = (16+2)/(20+4) = 0.75$

The 90% one-proportion plus-four z-interval is

$$\tilde{p} \pm z_{\alpha/2}\sqrt{\tilde{p}(1-\tilde{p})/(n+4)} = 0.75 \pm 1.645\sqrt{0.75(1-0.75))/24} = (0.605, 0.895)$$

(b) It was not appropriate to find the confidence interval in Exercise 11.21.

11.53 $\tilde{p} = (758+2)/(1245+4) = 0.608$

The 95% one-proportion plus-four z-interval is

$$\tilde{p} \pm z_{\alpha/2}\sqrt{\tilde{p}(1-\tilde{p})/(n+4)} = 0.608 \pm 1.96\sqrt{0.608(1-0.608))/1249} = (0.581, 0.635)$$

We can be 95% confident that the proportion of all U.S. adults who oppose, at the time, providing more government money in the financial bailout of banks is between 0.581 and 0.635.

11.55 77% of 434 is 334. Thus, $x = 334$ and $n = 434$.

$$\tilde{p} = (334+2)/(434+4) = 0.767$$

The 90% one-proportion plus-four z-interval is

$$\tilde{p} \pm z_{\alpha/2}\sqrt{\tilde{p}(1-\tilde{p})/(n+4)} = 0.767 \pm 1.645\sqrt{0.767(1-0.767))/438} = (0.734, 0.800)$$

We can be 90% confident that 73.4% to 80.0% of all new mothers breast-feed their infants at least briefly.

Exercises 11.2

11.57 Procedure 11.2 is a special case of the one-mean z-test for a population mean. If y is a variable taking on the values 0 for a failure and 1 for a success, then $\bar{y} = (\sum y)/n = x/n = \hat{p}$. Applying Procedure 9.1 (z-test for a population mean) is the same as Procedure 11.2.

11.59 (a) The sample proportion is $\hat{p}$ = x/n = 8/40 = 0.2.

(b) np_o = 40(0.3) = 12; n(1 - p_0) = 40(1-0.30) = 28
Since both are at least 5, we can employ Procedure 11.2.

(c) $z = \dfrac{0.2-0.3}{\sqrt{.03(1-0.3)/40}} = -1.38$; For α = 0.10, the critical value is

$-z_{\alpha}$ = -1.28. Since -1.38 < -1.28, reject H_0.

11.61 (a) The sample proportion is $\hat{p}$ = x/n = 35/50 = 0.70.

(b) np_o = 50(0.6) = 30; n(1 - p_0) = 50(1-0.60) = 20
Since both are at least 5, we can employ Procedure 11.2.

(c) $z = \dfrac{0.70-0.60}{\sqrt{0.6(1-0.6)/50}} = 1.44$; For α = 0.05, the critical value is

Copyright © 2012 Pearson Education, Inc. Publishing as Addison-Wesley.

z_α = 1.645. Since 1.44 < 1.645, do not reject H_0.

11.63 (a) The sample proportion is $\hat{p}$ = x/n = 16/20 = 0.80.

(b) np_o = 20(0.7) = 14; $n(1 - p_0)$ = 20(1-0.70) = 6
Since both are at least 5, we can employ Procedure 11.2.

(c) $z = \dfrac{0.80 - 0.70}{\sqrt{0.7(1-0.7)/20}} = 0.98$; For α = 0.05, the critical values are

$\pm z_{\alpha/2}$ = ±1.96. Since -1.96 < 0.98 < 1.96, do not reject H_0

11.65 (a) The sample proportion is $\hat{p}$ = x/n = 459/850 = 0.540.

(b) α = 0.05, p_0 = 0.50
np_o = 850(0.50) = 425; $n(1 - p_0)$ = 850(1-0.50) = 425
Since both are at least 5, we can employ Procedure 11.2.

Step 1: H_0: p = 0.50, H_a: p > 0.50
Step 2: α = 0.05

Step 3: $z = \dfrac{0.540 - 0.500}{\sqrt{0.5(1-0.5)/850}} = 2.33$

Step 4: Since α = 0.05, the critical value is z_α = 1.645

For the p-value approach, P(z > 2.33) = 0.0099
Step 5: Since 2.33 > 1.645, reject H_0. P-value < α. Thus, we reject H_0.
Step 6: The test results are statistically significant at the 5% level; that is, at the 5% significance level, the data do provide sufficient evidence to conclude that a majority of Generation Y Web users use the Internet to download music.

11.67 The sample proportion is $\hat{p}$ = x/n = 205/1283 = 0.1598.

α = 0.10, p_0 = 0.136
np_o = 1283(0.136) = 174.5; $n(1 - p_0)$ = 1283(1 - 0.136) = 1108.5
Since both are at least 5, we can employ Procedure 11.2.
Step 1: H_0: p = 0.136, H_a: p ≠ 0.136
Step 2: α = 0.10

Step 3: $z = \dfrac{0.1598 - 0.1360}{\sqrt{0.136(1-0.136)/1283}} = 2.49$

Step 4: Since α = 0.10, the critical values are $\pm z_{\alpha/2}$ = ±1.645

For the p-value approach, 2P(Z > 2.49) = 2(0.0064) = 0.0128
Step 5: Since 2.49 > 1.645, reject H_0. P-value < α. Thus, we reject H_0.
Step 6: The test results are statistically significant at the 10% level; that is, at the 10% significance level, the data provide sufficient evidence to conclude that the percentage of 18-25 year-olds who currently use marijuana or hashish has changed from the 2000 percentage of 13.6%.

11.69 (a) The sample proportion is $\hat{p}$ = 0.650.

α = 0.05, p_0 = 0.72
np_o = 1003(0.72) = 722.2; $n(1 - p_0)$ = 1003(1 - 0.72) = 280.8
Since both are at least 5, we can employ Procedure 11.2.
Step 1: H_0: p = 0.72, H_a: p < 0.72
Step 2: α = 0.05

Copyright © 2012 Pearson Education, Inc. Publishing as Addison-Wesley.

Step 3: $z = \dfrac{0.65 - 0.72}{\sqrt{0.72(1-0.72)/1003}} = -4.94$

Step 4: Since $\alpha = 0.05$, the critical value is $-z_\alpha = -1.645$

For the p-value approach, $P(Z < -4.94) = 0.0000$

Step 5: Since $-4.94 < -1.645$, reject H_0.   P-value $< \alpha$.   Thus, we reject H_0.

Step 6: The test results are statistically significant at the 5% level; that is, at the 5% significance level, the data do provide sufficient evidence to conclude that the percentage of Americans who approve of labor unions has decreased since 1936.

(b) The sample proportion is $\hat{p} = 0.650$.

$\alpha = 0.05$, $p_0 = 2/3 = 0.667$
$np_0 = 1003(0.67) = 672$; $n(1 - p_0) = 1003(1 - 0.67) = 331$
Since both are at least 5, we can employ Procedure 11.2.
Step 1: H_0: $p = 0.67$, H_a: $p < 0.67$
Step 2: $\alpha = 0.05$
Step 3: $z = \dfrac{0.65 - 0.67}{\sqrt{0.67(1-0.67)/1003}} = -1.35$

Step 4: Since $\alpha = 0.05$, the critical value is $-z_\alpha = -1.645$

For the p-value approach, $P(Z < -1.35) = 0.0885$
Step 5: Since $-1.35 > -1.645$, do not reject H_0.   P-value $> \alpha$.   Thus, we do not reject H_0.
Step 6: The test results are not statistically significant at the 5% level; that is, at the 5% significance level, the data do not provide sufficient evidence to conclude that the percentage of Americans who approve of labor unions has decreased since 1963.

11.71 $n = 609$, $p_0 = 0.5$, $x = 341$, $\hat{p} = 341/609 = 0.560$

$np_0 = 304.5$, $n(1 - p_0) = 304.5$
Since both are at least 5, we can employ Procedure 11.2.
Step 1: H_0: $p = 0.5$, H_a: $p > 0.5$
Step 2: $\alpha = 0.01$
Step 3: $z = \dfrac{0.560 - 0.500}{\sqrt{0.500(1-0.500)/609}} = 2.96$

Step 4: Since $\alpha = 0.01$, the critical value is $z_\alpha = 2.33$

For the p-value approach, $P(z > 2.96) = 0.0015$
Step 5: Since $2.96 > 2.33$, reject H_0. P-value $< \alpha$.   Thus, reject H_0.
Step 6: The test results are statistically significant at the 1% level; that is, at the 1% significance level, the data do provide sufficient evidence to conclude that the headline is justified.

Using Minitab, choose **Stat, Basic Statistics**, then **1 Proportion**. Choose **Summarized data.** Enter 341 in the **Number of Events** textbox. Enter 609 in the **Number of trials** textbox. Check **Perform Hypothesis test**. Enter 0.50 in the **Hypothesized proportion** textbox. Click **Options**. Enter 99.0 in the **Confidence level** textbox. Choose **greater than** in the **Alternative** pull-down

Copyright © 2012 Pearson Education, Inc. Publishing as Addison-Wesley.

menu. Check **Use test and interval based on normal distribution**. Click **OK** twice. The results are

Test of p = 0.5 vs p > 0.5

Sample	X	N	Sample p	99% Lower Bound	Z-Value	P-Value
1	341	609	0.559934	0.513140	2.96	0.002

Using the normal approximation.
The test statistic is 2.96 and the P-value is 0.002, which confirms our results stated above for Exercise 11.71.

11.73 $n = 11$, $p_0 = 0.5$, $x = 5$, $\hat{p} = 5/11 = 0.455$

$np_0 = 5.5$, $n(1 - p_0) = 5.5$
Since both are at least 5, we can employ Procedure 11.2.
Step 1: H_0: $p = 0.5$, H_a: $p < 0.5$
Step 2: $\alpha = 0.05$

Step 3: $z = \dfrac{0.455 - 0.5}{\sqrt{0.5(1 - 0.5)/11}} = -0.30$

Step 4: Since $\alpha = 0.05$, the critical value is $-z_\alpha = -1.645$

For the p-value approach, $P(z < -0.30) = 0.3821$
Step 5: Since $-0.30 > -1.645$, do not reject H_0. P-value = $0.3821 > \alpha$. Thus, do not reject H_0.
Step 6: The test results are not statistically significant at the 5% level; that is, at the 5% significance level, the data do not provide sufficient evidence to conclude that less than half of all young children drowning in Victorian dams located on farms are girls.

Using Minitab, choose **Stat**, **Basic Statistics**, then **1 Proportion**. Choose **Summarized data**. Enter 5 in the **Number of Events** textbox. Enter 11 in the **Number of trials** textbox. Check **Perform Hypothesis test**. Enter 0.5 in the **Hypothesized proportion** textbox. Click **Options**. Enter 95.0 in the **Confidence level** textbox. Choose **less than** in the **Alternative** pull-down menu. Check **Use test and interval based on normal distribution**. Click **OK** twice. The results are

Test of p = 0.5 vs p < 0.5

Sample	X	N	Sample p	95% Upper Bound	Z-Value	P-Value
1	5	11	0.454545	0.701490	-0.30	0.382

Using the normal approximation.
The test-statistic is -0.30 and the P-value is 0.382, which confirms our results stated above for Exercise 11.73.

11.75 (a) Using Minitab, choose **Stat**, **Basic Statistics**, then **1 Proportion**. Choose **Summarized data**. Enter 921 in the **Number of Events** textbox. Enter 1001 in the **Number of trials** textbox. Check **Perform Hypothesis test**. Enter 0.90 in the **Hypothesized proportion** textbox. Click **Options**. Enter 95.0 in the **Confidence level** textbox. Choose **greater than** in the **Alternative** pull-down menu. Check **Use test and interval based on normal distribution**. Click **OK** twice. The results are

Test of p = 0.9 vs p > 0.9

Sample	X	N	Sample p	95% Lower Bound	Z-Value	P-Value
1	921	1001	0.920080	0.905982	2.12	0.017

Copyright © 2012 Pearson Education, Inc. Publishing as Addison-Wesley.

Using the normal approximation.
The test-statistic is 2.12 and the P-value is 0.017. This P-value is less than a significance level of 5%. Therefore, we would reject the null hypothesis. At 5% significance, there is evidence that more than 9 out of 10 Americans always wash up after using the bathroom.

(b)   Using the results in part (a) and changing the significance level to 1%, the test statistic and the P-value would still be 2.12 and 0.017 respectively. The P-value is greater than a significance level of 1%. Therefore, we would not reject the null hypothesis. At 1% significance, there is not significant evidence that more than 9 out of 10 Americance always wash up after using the bathroom.

Exercises 11.3

11.77 We need to decide whether the difference between two sample proportions can reasonably be attributed to sampling error or are the population proportions really different.

11.79 (a) Using sunscreen before going out in the sun
(b) Teen-age girls and teen-age boys
(c) The two proportions are sample proportions. The reference is specifically to those teen-age girls and boys who were surveyed

11.81 (a) The parameters are p_1 and p_2. The rest are statistics.
(b) The fixed numbers are p_1 and p_2. The rest are variables.

11.83 (a) $\hat{p}_1 = x_1/n_1 = 18/40 = 0.45$; $\hat{p}_2 = x_2/n_2 = 30/40 = 0.75$;

$\hat{p}_p = (x_1 + x_2)/(n_1 + n_2) = 48/80 = 0.6$

(b) $x_1 = 18$, $n_1 - x_1 = 22$, $x_2 = 30$, $n_2 - x_2 = 10$.
All are at least 5, so z-procedures are appropriate.

(c) $z = \dfrac{0.45 - 0.75}{\sqrt{0.6(1-0.6)[(1/40)+(1/40)]}} = -2.74$

$\alpha = 0.10$, critical value is -1.28; since $-2.74 < -1.28$, reject H_0.

(d) The 80% confidence interval is

$$\hat{p}_1 - \hat{p}_2 \pm z_{\alpha/2}\sqrt{\hat{p}_1(1-\hat{p}_1)/n_1 + \hat{p}_2(1-\hat{p}_2)/n_2}$$

$$(0.45 - 0.75) \pm 1.28\sqrt{0.45(1-0.4)/40 + 0.75(1-0.75)/40} = (-0.43, \; -0.17)$$

11.85 (a) $\hat{p}_1 = x_1/n_1 = 15/20 = 0.75$; $\hat{p}_2 = x_2/n_2 = 18/30 = 0.60$;

$\hat{p}_p = (x_1 + x_2)/(n_1 + n_2) = 33/50 = 0.66$

(b) $x_1 = 15$, $n_1 - x_1 = 5$, $x_2 = 18$, $n_2 - x_2 = 12$.

All are at least 5, so z-procedures are appropriate.

(c) $z = \dfrac{0.75 - 0.60}{\sqrt{0.66(1-0.66)[(1/20)+(1/30)]}} = 1.10$

$\alpha = 0.05$, critical value is 1.645; since $1.10 < 1.645$, do not reject H_0.

(d) The 90% confidence interval is

$$\hat{p}_1 - \hat{p}_2 \pm z_{\alpha/2}\sqrt{\hat{p}_1(1-\hat{p}_1)/n_1 + \hat{p}_2(1-\hat{p}_2)/n_2}$$

$$(0.75 - 0.60) \pm 1.645\sqrt{0.75(1-0.75)/20 + 0.60(1-0.60)/30} = (-0.067, \; 0.367)$$

11.87 (a) $\hat{p}_1 = x_1/n_1 = 30/80 = 0.375$; $\hat{p}_2 = x_2/n_2 = 15/20 = 0.750$;

$\hat{p}_p = (x_1 + x_2)/(n_1 + n_2) = 45/100 = 0.45$

Copyright © 2012 Pearson Education, Inc. Publishing as Addison-Wesley.

(b) $x_1 = 30$, $n_1 - x_1 = 50$, $x_2 = 15$, $n_2 - x_2 = 5$.

All are at least 5, so z-procedures are appropriate.

(c) $z = \dfrac{0.375 - 0.750}{\sqrt{0.45(1-0.45)[(1/80)+(1/20)]}} = -3.02$

$\alpha = 0.05$, critical values are ± 1.96; since $-3.02 < -1.96$, reject H_0.

(d) The 95% confidence interval is

$$\hat{p}_1 - \hat{p}_2 \pm z_{\alpha/2}\sqrt{\hat{p}_1(1-\hat{p}_1)/n_1 + \hat{p}_2(1-\hat{p}_2)/n_2}$$

$$(0.375 - 0.750) \pm 1.96\sqrt{0.375(1-0.375)/80 + 0.750(1-0.750)/20} =$$

$$(-0.592, \; -0.158)$$

11.89 (a)   Population 1:   Women who took multivitamins containing folic acid

$\hat{p}_1 = 35/2701 = 0.01296$

Population 2:   Women who received only trace elements

$\hat{p}_2 = 47/2052 = 0.02290$

$\hat{p}_p = (35 + 47)/(2,701 + 2,052) = 0.01725$

Step 1: $H_0: p_1 = p_2$, $H_a: p_1 < p_2$

Step 2: $\alpha = 0.01$

Step 3: $z = \dfrac{0.01296 - 0.0220}{\sqrt{0.01725(1-0.01725)[(1/2701)+(1/2052)]}} = -2.61$

Step 4:   Since $\alpha = 0.01$, the critical value is $-z_\alpha = -2.33$

For the P-value approach, $P(z < -2.61) = 0.0045$

Step 5:   Since $-2.61 < -2.33$, reject H_0.   P-value $< \alpha$.   Thus, we reject H_0.

Step 6:   The test results are significant at the 1% level; that is, at the 1% significance level, the data do provide sufficient evidence to conclude that the women who take folic acid are at lesser risk of having children with major birth defects than those women who do not.

(b)   This is a designed experiment.   The researchers decided which women would take daily multivitamins.

(c)   Yes.   By using the basic principles of design (control, randomization, and replication) the doctors can conclude that the reduction in the rates of major birth defects in the folic acid group is likely caused by the folic acid.

11.91 Population 1: Drivers of age 25-34, $\hat{p}_1 = 270/1000 = 0.270$

Population 2: Drivers of age 45-64, $\hat{p}_2 = 330/1100 = 0.300$

$\hat{p}_p = (270 + 330)/(1000 + 1100) = 0.286$

Step 1: $H_0: p_1 = p_2$, $H_a: p_1 \neq p_2$

Step 2: $\alpha = 0.10$

Step 3: $z = \dfrac{0.270 - 0.300}{\sqrt{0.286(1-0.286)[(1/1000)+(1/1000)]}} = -1.52$

Copyright © 2012 Pearson Education, Inc. Publishing as Addison-Wesley.

Step 4: Since $\alpha = 0.10$, the critical values is $\pm z_{\alpha} = \pm 1.645$

For the P-value approach, $2P(z < -1.52) = 0.1286$

Step 5: Since $-1.645 < -1.52 < 1.645$, do not reject H_0. P-value $> \alpha$. Thus, we do not reject H_0.

Step 6: The test results are not significant at the 10% level; that is, at the 10% significance level, the data do not provide sufficient evidence to conclude that there is a difference in seat-belt usage between drivers 25-34 years old and those 45-64 years old.

11.93 Population 1: Bachelors degree, $\hat{p}_1 = 386/750 = 0.5147$

Population 2: Graduate degree, $\hat{p}_2 = 237/500 = 0.4740$

$$\hat{p}_p = (386 + 237)/(750 + 500) = 0.4984$$

(a) The assumptions for using the two-sample z-test are simple random samples, independent samples, and x_1, $n_1 - x_1$, x_2, and $n_2 - x_2$ must all be greater than or equal to 5.

(b) Step 1: H_0: $p_1 = p_2$, H_a: $p_1 > p_2$

Step 2: $\alpha = 0.05$

Step 3: $z = \dfrac{0.5147 - 0.4740}{\sqrt{0.4984(1-0.4984)[(1/750)+(1/500)]}} = 1.41$

Step 4: Since $\alpha = 0.05$, the critical value is $z_{\alpha} = 1.645$

For the P-value approach, $P(z > 1.41) = 0.0793$

Step 5: Since $1.41 < 1.645$, do not reject H_0. P-value $> \alpha$. Thus, we do not reject H_0.

Step 6: The test results are not significant at the 5% level; that is, at the 5% significance level, the data do not provide sufficient evidence to conclude that a higher percentage of adults with Bachelors degrees are overweight than of adults with graduate degrees.

(c) At the 10% significance level, the P-value of 0.0793 is less than the significance level, so we reject the null hypothesis. The data do provide sufficient evidence that a higher percentage of adults with Bachelors degrees are overweight than of adults with graduate degrees.

11.95 From Exercise 11.89, the 99% confidence interval is

$$(0.012958 - 0.022904) \pm 2.33 \sqrt{\frac{0.012958(1-0.012958)}{2701} + \frac{0.022904(1-0.022904)}{2052}}$$

$$-0.009946 \pm 0.009215 \ or \ -0.019161 \ to \ -0.000731$$

We can be 98% confident that the difference $p_1 - p_2$ between the rates of major birth defects for babies born to women who have taken folic acid and those born to women who have not taken folic acid is somewhere between -0.019161 and -0.000731.

11.97 From Exercise 11.91, the 90% confidence interval is

$$\hat{p}_1 - \hat{p}_2 \pm z_{\alpha/2} \sqrt{\hat{p}_1(1-\hat{p}_1)/n_1 + \hat{p}_2(1-\hat{p}_2)/n_2}$$

$$(0.270 - 0.300) \pm 1.645 \sqrt{0.270(1-0.270)/1000 + 0.300(1-0.300)/100}$$

$$= (-0.0624, \ 0.0024)$$

We can be 90% confident that the difference between proportions of seat belt users for drivers in the age groups 25-34 years and 45-64 years is somewhere

Copyright © 2012 Pearson Education, Inc. Publishing as Addison-Wesley.

between -0.0624- and 0.0024.

11.99 (a) From Exercise 11.93, the 90% confidence interval is

$$\hat{p}_1 - \hat{p}_2 \pm z_{\alpha/2} \sqrt{\hat{p}_1(1-\hat{p}_1)/n_1 + \hat{p}_2(1-\hat{p}_2)/n_2}$$

$$(0.515-0.474) \pm 1.645 \sqrt{0.515(1-0.515)/750 + 0.474(1-0.474)/500}$$

$$= (-0.007, \quad 0.088)$$

(b) The 80% confidence interval is

$$\hat{p}_1 - \hat{p}_2 \pm z_{\alpha/2} \sqrt{\hat{p}_1(1-\hat{p}_1)/n_1 + \hat{p}_2(1-\hat{p}_2)/n_2}$$

$$(0.515-0.474) \pm 1.28 \sqrt{0.515(1-0.515)/750 + 0.474(1-0.474)/500}$$

$$= (0.004, \quad 0.078)$$

11.101 (a) Using Minitab, choose **Stat ▶ Basic statistics ▶ 2 Proportions...**, select the **Summarized data** option button, click in the **Trials** text box for **First sample** and enter 300, click in the **Events** text box for **First sample** and type 215, click in the **Trials** text box for **Second sample** and type 250, click in the **Events** text box for **Second sample** and type 186, click the **Options...** button, click in the **Confidence level** text box and type 95, click in the **Test difference** text box and type 0, click the arrow button at the right of the **Alternative** drop-down list box and select **not equal**, select the **Use pooled estimate of p for test** check box, click **OK**, and click **OK**. The resulting output is

```
Sample   X     N    Sample p
1       215   300   0.716667
2       186   250   0.744000

Difference = p (1) - p (2)
Estimate for difference:  -0.0273333
95% CI for difference:  (-0.101675, 0.0470087)
Test for difference = 0 (vs not = 0):  Z = -0.72  P-Value = 0.473
```

Since the P-value = 0.473, which is greater than the significance level 0.05, do not reject the null hypothesis. The data do not provide sufficient evidence to conclude that there is a difference between the labor-force participation rates of U.S. and Canadian women.

(b) The confidence interval is part of the output for part (a). We can be 95% confident that the difference in labor-force participation rates between of U.S. and Canadian women is somewhere between -0.102 and 0.047.

11.103 (a) E = (-0.031 - (-0.129))/2 = 0.049; We can be 90% confident that when we estimate p_1 - p_2 with $\hat{p}_1$ - $\hat{p}_2$ that our estimate is not off by more than 0.049.

(b) $$E = 1.645 \sqrt{\frac{0.369(1-0.369)}{747} + \frac{0.449(1-0.449)}{434}} = 0.049$$

(c) $$n = 0.5 \frac{1.645^2}{0.01^2} = 13530.125 \rightarrow 13531$$

Thus, the sample size is 13,531.

Copyright © 2012 Pearson Education, Inc. Publishing as Addison-Wesley.

(d) $n = 13,531$, $\hat{p}_1 = 0.223$, $\hat{p}_2 = 0.272$

$$(0.383 - 0.437) \pm 1.645 \sqrt{\frac{0.383(1-0.383)}{13531} + \frac{0.437(1-0.437)}{13531}}$$

$$-0.054 \pm 0.010 \ or \ -0.064 \ to \ -0.044$$

(e) $E = (-0.044 - (-0.064))/2 = 0.010$ (if we carried more significant digits, we would find that $E = 0.0098$); this is just slightly smaller than the specified error in part (c). This is expected since 0.383 and 0.437 are both considerably smaller than the assumed 0.5 probability in part (c).

(f) $\hat{p}_1 = 0.41$, $\hat{p}_2 = 0.49$

part (c)

$$n = \{0.41(1-0.41) + 0.49(1-0.49)\} \cdot \frac{1.645^2}{0.01^2} = 13308.23 \rightarrow 13309$$

Thus, the sample size is 13309.

part (d)

$n = 13309$, $\hat{p}_{1g} = 0.383$, $\hat{p}_{2g} = 0.437$

$$(0.383 - 0.437) \pm 1.645 \sqrt{\frac{0.383(1-0.383)}{13309} + \frac{0.437(1-0.437)}{1339}}$$

$$-0.0540 \pm 0.0099 \ or \ -0.064 \ to \ -0.044$$

part (e)

$E = (-0.044 - (-0.064)/2 = 0.010$. As you can see from the calculation above, the actual margin of error is 0.0099, just slightly smaller than the required 0.01.

(g) By employing the guesses of $\hat{p}_1$ and $\hat{p}_2$, the sample size is reduced from 13,531 to 13,309. Moreover, the margin of error only rises from 0.0098 to 0.0099. The confidence interval is unchanged for all practical purposes.

11.105 (a) $\tilde{p}_1 = (18+1)/(40+2) = 0.452$ and $\tilde{p}_2 = (30+1)/(40+2) = 0.738$

The 80% two-proportions plus-four z-interval is

$$(\tilde{p}_1 - \tilde{p}_2) \pm z_{\alpha/2} \sqrt{\tilde{p}_1(1-\tilde{p}_1)/(n+2) + \tilde{p}_2(1-\tilde{p}_2)/(n+2)}$$

$$= (0.452 - 0.738) \pm 1.28 \sqrt{0.452(1-0.452))/42 + 0.738(1-0.738)/42} = (-0.417, -0.155)$$

(b) This interval is close to the same level of precision as the interval found in Exercise 11.83. However, it is centered at -0.286 rather than -0.300.

11.107 (a) $\tilde{p}_1 = (15+1)/(20+2) = 0.727$ and $\tilde{p}_2 = (18+1)/(30+2) = 0.594$

The 90% two-proportions plus-four z-interval is

$$(\tilde{p}_1 - \tilde{p}_2) \pm z_{\alpha/2} \sqrt{\tilde{p}_1(1-\tilde{p}_1)/(n+2) + \tilde{p}_2(1-\tilde{p}_2)/(n+2)}$$

$$= (0.727 - 0.594) \pm 1.645 \sqrt{0.727(1-0.727))/22 + 0.594(1-0.594)/32} = (-0.079, 0.345)$$

(b) This interval is slightly more precise than the interval found in Exercise 11.85. In addition, it is centered at 0.133 rather than 0.150.

11.109 (a) $\tilde{p}_1 = (30+1)/(80+2) = 0.378$ and $\tilde{p}_2 = (15+1)/(20+2) = 0.727$

The 95% two-proportions plus-four z-interval is

Copyright © 2012 Pearson Education, Inc. Publishing as Addison-Wesley.

$$(\tilde{p}_1 - \tilde{p}_2) \pm z_{\alpha/2}\sqrt{\tilde{p}_1(1-\tilde{p}_1)/(n+2) + \tilde{p}_2(1-\tilde{p}_2)/(n+2)}$$

$$= (0.378 - 0.727) \pm 1.96\sqrt{0.378(1-0.378))/82 + 0.727(1-0.727)/22} = (-0.563, -0.135)$$

(b) This interval is more precise than the interval found in Exercise 11.87. In addition, it is centered at -0.349 rather than -0.375.

11.111 Let population 1 be the civilian labor forces of Finland and population 2 be the civilian labor forces of Denmark.

$$\tilde{p}_1 = (7+1)/(100+2) = 0.078 \text{ and } \tilde{p}_2 = (3+1)/(75+2) = 0.052$$

The 95% two-proportions plus-four z-interval is

$$(\tilde{p}_1 - \tilde{p}_2) \pm z_{\alpha/2}\sqrt{\tilde{p}_1(1-\tilde{p}_1)/(n+2) + \tilde{p}_2(1-\tilde{p}_2)/(n+2)}$$

$$= (0.078 - 0.052) \pm 1.96\sqrt{0.078(1-0.078))/102 + 0.052(1-0.052)/77} = (-0.046, 0.098)$$

We can be 95% confident that the difference between the unemployment rates in Finland and Denmark is between -4.6% and 9.8%. In other words, we can be 95% confident that the unemployment rate in Finland is between 4.6% smaller to 9.8% larger than the unemployment rate in Denmark.

11.113 Let population 1 be elderly people taking calcium channel blockers and population 2 be elderly people taking beta blockers.

$$\tilde{p}_1 = (27+1)/(202+2) = 0.137 \text{ and } \tilde{p}_2 = (28+1)/(424+2) = 0.068$$

The 95% two-proportions plus-four z-interval is

$$(\tilde{p}_1 - \tilde{p}_2) \pm z_{\alpha/2}\sqrt{\tilde{p}_1(1-\tilde{p}_1)/(n+2) + \tilde{p}_2(1-\tilde{p}_2)/(n+2)}$$

$$= (0.137 - 0.068) \pm 1.645\sqrt{0.137(1-0.137))/204 + 0.068(1-0.068)/426}$$

$$= (0.025, 0.113)$$

We can be 90% confident that the difference between the cancer rates of elderly people taking calcium channel blockers and those taking beta blockers is between 2.5% and 11.3%. In other words, we can be 90% confident that the cancer rate of elderly people taking calcium channel blockers is between 2.5% to 11.3% larger than the cancer rate of elderly people taking beta blockers.

Chapter 11 Review Problems

1. (a) Feeling that marijuana should be legalized for medicinal use in patients with cancer and other painful and terminal diseases.
    (b) Americans
    (c) Proportion of all Americans who feel that marijuana should be legalized for medicinal use in patients with cancer and other painful and terminal diseases.
    (d) Proportion of Americans in the sample who feel that marijuana should be legalized for medicinal use in patients with cancer and other painful and terminal diseases. Clearly the population of all Americans is much larger than 83,957.

2. It is often impossible to take a census of an entire population. It is also expensive and time-consuming.

3. (a) "Number of successes" stands for the number of members of the sample that exhibit the specified attribute.
    (b) "Number of failures" stands for the number of members of the sample that do not exhibit the specified attribute.

4. (a) population proportion
    (b) normal
    (c) number of successes, number of failures, 5

5. The margin of error for the estimate of a population proportion tells us what the maximum difference between the sample proportion and the population proportion is <u>likely</u> to be. It is not an absolute maximum, and how likely

Copyright © 2012 Pearson Education, Inc. Publishing as Addison-Wesley.

it is depends on the confidence level used.

6. (a) Getting the "holiday blues"
    (b) Men and women
    (c) The proportion of men in the population who get the "holiday blues" and the proportion of women in the population who get the "holiday blues"
    (d) The proportion of men in the sample who get the "holiday blues" and the proportion of women in the sample who get the "holiday blues"
    (e) They are sample proportions since the information came from a poll, not a census. Also, it could not be a population proportion because I was not asked.

7. (a) The mean of all possible differences between the two sample proportions equals the <u>difference of the population proportions</u>.
    (b) For large samples, the possible differences between the two sample proportions have approximately a <u>normal</u> distribution.

8. The 95% confidence interval for the percentage of U.S. adults who would get a smallpox shot if it were available is 40% $\pm$ 3.0% or from 37% to 43%.

9. (a) $n = 0.5\dfrac{1.96^2}{0.01^2} = 19208$

    (b) $n = \{0.75(1-0.75) + 0.75(1-0.75)\}\dfrac{1.96^2}{0.01^2} = 14406$

10. $n = 1218$, $\hat{p} = 733/1218 = 0.60$, $z_{\alpha/2} = z_{0.025} = 1.96$

    $$0.602 - 1.96\sqrt{0.602(1-0.602)/1218} \text{ to } 0.602 + 1.96\sqrt{0.602(1-0.602)/1218}$$

    $$0.574 \text{ to } 0.629$$

    We can be 95% confident that the proportion, p, of students who expect difficulty finding a job is somewhere between 0.574 and 0.629.

11. (a) The error is found by taking the width of the confidence interval and dividing by 2. So E = (0.629 - 0.574)/2 = 0.028.

    (b) E = 0.02; $z_{\alpha/2} = z_{0.025} = 1.96$; $\hat{p}_g = 0.50$

    $$n = 0.5^2 \frac{z_{\alpha/2}^2}{E^2} = 0.25 \cdot \frac{1.96^2}{0.02^2} = 2401$$

    (c) $n = 2401$; $\hat{p} = 0.587$; $z_{\alpha/2} = z_{0.025} = 1.96$

    $$0.587 - 1.96\sqrt{0.587(1-0.587)/2401} \text{ to } 0.587 + 1.96\sqrt{0.587(1-0.587)/2401}$$

    $$0.567 \text{ to } 0.607$$

    (d) The margin of error for the estimate is 0.020, the same as what is required in part (b).

    (e) $\hat{p}_g = 0.56$; $z_{\alpha/2} = z_{0.025} = 1.96$

    part (b)   E = 0.02, $z_{0.025}$ = 1.96; sample size is:

    $$n = 0.56(1-0.56)\frac{z_{\alpha/2}^2}{E^2} = 0.2464 \cdot \frac{1.96^2}{0.02^2} = 2366.42 \to 2367$$

    Thus the required sample size is n = 2367.

    part (c)
    $$0.587 - 1.96\sqrt{0.587(1-0.587)/2367} \text{ to } 0.587 + 1.96\sqrt{0.587(1-0.587)/2367}$$

    $$0.567 \text{ to } 0.607$$

    part (d) The margin of error is 0.02 (actually it's 0.0198), which is the same as what is specified in part (b).

Copyright © 2012 Pearson Education, Inc. Publishing as Addison-Wesley.

(f) By employing the guess for $\hat{p}$ in part (d) we can reduce the required

sample size by 34, from 2401 to 2367, saving a little time and money. Moreover, the margin of error stays the same. The risk of using the guess 0.56 for $\hat{p}$ is that if the actual value of $\hat{p}$ turns out to be

between .44 and .56, then the achieved margin of error will exceed the specified 0.02.

12. n = 2512, x = 578, α = 0.05, $\hat{p}$ = 578/2512 = 0.2301

np_0 = 2512(0.25) = 628; $n(1 - p_0)$ = 2512(1-0.25) = 1884
Since both are at least 5, we can employ Procedure 11.2.
(a) Step 1: H_0: p = 0.25, H_1: p < 0.25
    Step 2: α = 0.05

$$\text{Step 3: } z = \frac{0.23 - 0.25}{\sqrt{0.25(1-0.25)/2512}} = -2.31$$

Step 4: Since α = 0.05, the critical value is $-z_\alpha$ = -1.645

For the P-value approach: P(z < -2.31) = 0.0104.
Step 5: Since -2.31 < -1.645, reject H_0. Since P-value < α, reject H_0.
Step 6: The test results are statistically significant at the 5% level; that is, at the 5% significance level, the data do provide sufficient evidence to conclude that less than one in four Americans believe that juries "almost always" convict the guilty and free the innocent

13. (a) Observational study. The researchers had no control over any of the factors of the study.
    (b) Height may not be the only factor to be considered. Although there does appear to be an association, we don't know if it is a direct association.

14. Population 1: first poll, $\hat{p}_1$ = 0.48,

Population 2: second poll, $\hat{p}_2$ = 0.60

$$\hat{p}_p = (0.48 + 0.60)/2 = 0.54$$

(a) Step 1: H_0: p_1 = p_2, H_a: p_1 < p_2
    Step 2: α = 0.01

$$\text{Step 3: } z = \frac{0.48 - 0.60}{\sqrt{0.54(1-0.54)}\sqrt{(1/600)+(1/600)}} = -4.17$$

P(z < -4.17) = 0.0000.
Step 4: Since α = 0.01, the critical value $-z_\alpha$ = -2.33
    P(z < -4.17) = 0.0000.
Step 5: Since -4.17 < -2.33, reject H_0. P-value < α. Thus, we reject H_0.
Step 6: The test is significant at the 1% level; that is the data do provide evidence to conclude that the percentage of Maricopa County residents who thought that the state's economy would improve over the next 2 years was less during the time of the first poll than during the time of the second poll.

15. (a) From Exercise 14

$$(0.48 - 0.60) \pm 2.326\sqrt{\frac{0.48(1-0.48)}{600} + \frac{0.60(1-0.60)}{600}}$$

$$-0.120 \pm 0.066 \ or \ -0.186 \ to \ -0.054$$

Copyright © 2012 Pearson Education, Inc. Publishing as Addison-Wesley.

(b) We can be 95% confident that the difference $p_1 - p_2$ between the proportions of Maricopa County residents who thought that the state's economy would improve over the next 2 years during the time of the 1992 poll and during the time of the poll is somewhere between -0.186 and -0.054.

16. (a) E = (-0.054 - (-0.186))/2 = 0.066
   We can be 98% confident that the error in estimating the difference between the two population proportions, $p_1 - p_2$, by the difference between the two sample proportions, -0.12, is at most 0.066.

   (b) $$E = 2.326\sqrt{\frac{0.48(1-0.48)}{600} + \frac{0.60(1-0.60)}{600}} = 0.066$$

   (c) E = 0.03, α = 0.02
   $$n = 0.5 \cdot \frac{2.326^2}{0.03^2} = 3005.715 \rightarrow 3006$$

   (d) n = 3006, $\hat{p}_1$ = 0.475, $\hat{p}_2$ = 0.603

   $$(0.475 - 0.603) \pm 2.326\sqrt{\frac{0.475(1-0.475)}{3006} + \frac{0.603(1-0.603)}{3006}}$$
   $$-0.128 \pm 0.030 \ or \ -0.158 \ to \ -0.098$$

   (e) E = 0.030 which is the same as that required in part (c).

17. (a) $\hat{p} = 4/103 = 0.039$
   The 95% one-proportion z-interval is
   $$\hat{p} \pm z_{\alpha/2}\sqrt{\hat{p}(1-\hat{p})/n} = 0.039 \pm 1.96\sqrt{0.039(1-0.039))/103} = (0.002, 0.076)$$

   (b) $\tilde{p} = (4+2)/(103+4) = 0.056$
   The 95% one-proportion plus-four z-interval is
   $$\tilde{p} \pm z_{\alpha/2}\sqrt{\tilde{p}(1-\tilde{p})/(n+4)} = 0.056 \pm 1.96\sqrt{0.056(1-0.056))/107} = (0.012, 0.100)$$

   (c) The discrepancy between the two methods is primarily due to using a z-interval when the conditions for using such an interval are not met. To get a good approximation to an exact interval, both x and n - x should be at least 5. In this case, x is only 4.

   (d) I would use the one-proportion plus-four z-interval because the conditions to use the one-proportion z-interval are not met.

18. Using Minitab, choose **Stat ▶ Basic statistics ▶ 1 Proportion...**, select the **Summarized data** option button, click in the **Number of trials** text box and enter 2435, click in the **Number of events** text box and enter 317, click the **Options...** button, click in the **Confidence level** text box and enter 95, ensure that the **Alternative** drop down box contains **not equal**, check the box for **Use test and interval based on normal distribution**, click **OK**, and click **OK**. The resulting output is

   Test and CI for One Proportion

   Test of p = 0.5 vs p not = 0.5

Sample	X	N	Sample p	95% CI	Z-Value	P-Value
1	317	2435	0.130185	(0.116819, 0.143551)	-36.50	0.000

   We can be 95% confident that the **percentage** of U.S. adults who would participate in an office pool for March Madness is somewhere between 11.68% and 14.36%.

19. Using Minitab, choose **Stat ▶ Basic statistics ▶ 1 Proportion...**, select the **Summarized data** option button, click in the **Number of trials** text box and

Copyright © 2012 Pearson Education, Inc. Publishing as Addison-Wesley.

enter <u>1961</u>, click in the **Number of events** text box and enter <u>1137</u>, click the **Options...** button, click in the **Confidence level** text box and enter <u>95</u>, enter 0.5 in the Test proportion text box, select **greater than** from the **Alternative** drop down box, check the box for **Use test and interval based on normal distribution**, click **OK**, and click **OK**. The resulting output is

Test of p = 0.5 vs p > 0.5

```
                               95%
                             Lower
   Sample      X      N   Sample p   Bound   Z-Value  P-Value
     1       1137   1961  0.579806  0.561472   7.07    0.000
```

Since the P-value = 0.000, which is less than the significance level 0.05, reject the null hypothesis. The data provide sufficient evidence to conclude that a majority of U.S. adults do not believe that abstinence programs are effective in reducing or preventing AIDS.

20. Using Minitab, choose **Calc ▶ Basic statistics ▶ 2 Proportions...**, select the **Summarized data** option button, click in the **Trials** text box for **First sample** and enter <u>56</u>, click in the **Events** text box for **First sample** and enter <u>32</u>, click in the **Trials** text box for **Second sample** and enter <u>70</u>, click in the **Events** text box for **Second sample** and enter <u>9</u>, click the **Options...** button, click in the **Confidence level** text box and enter <u>95</u>, click in the **Test difference** text box and type <u>0</u>, click the arrow button at the right of the **Alternative** drop-down list box and select **not equal**, select the **Use pooled estimate of p for test** check box, click **OK**, and click **OK**. The resulting output is

Test and CI for Two Proportions

```
   Sample    X    N   Sample p
     1      32   56   0.571429
     2       9   70   0.128571
```

```
Difference = p (1) - p (2)
Estimate for difference:  0.442857
95% CI for difference:  (0.291371, 0.594343)
Test for difference = 0 (vs not = 0):  Z = 5.27  P-Value = 0.000
```

(a) Since the P-value = 0.000, which is less than the significance level 0.05, reject the null hypothesis. The data provide sufficient evidence that there is a difference in the cure rates of the two types of treatment.

(b) The 95% confidence interval for the difference in cure rates is found in the output for part (a) as (0.291, 0.594).

21. Using Minitab, choose **Calc ▶ Basic statistics ▶ 2 Proportions...**, select the **Summarized data** option button, click in the **Trials** text box for **First sample** and enter <u>4368</u>, click in the **Events** text box for **First sample** and enter <u>803</u>, click in the **Trials** text box for **Second sample** and enter <u>4692</u>, click in the **Events** text box for **Second sample** and enter <u>1147</u>, click the **Options...** button, click in the **Confidence level** text box and enter <u>99</u>, click in the **Test difference** text box and type <u>0</u>, click the arrow button at the right of the **Alternative** drop-down list box and select **less than**, select the **Use pooled estimate of p for test** check box, click **OK**, and click **OK**. The resulting output is

```
   Sample    X     N    Sample p
     1      803   4368  0.183837
     2     1147   4692  0.244459
```

```
Difference = p (1) - p (2)
Estimate for difference:  -0.0606217
```

Copyright © 2012 Pearson Education, Inc. Publishing as Addison-Wesley.

```
        99% upper bound for difference:  -0.0406483
        Test for difference = 0 (vs < 0):  Z = -7.06   P-Value = 0.000
```
The test statistic is -7.06 and the P-value is 0.000, which is less than the significance level of 1%. Therefore, we reject the null hypothesis. At 1% significance there is evidence that finasteride reduces the risk of prostate cancer.

Exercises 12.1

12.1 A variable has a chi-square distribution if its distribution has the shape of a special type of right-skewed curve, called a chi-square curve.

12.3 The χ^2-curve with 20 degrees of freedom more closely resembles a normal curve. This follows from Property 4 of Key Fact 12.1, "As the number of degrees of freedom becomes larger, χ^2-curves look increasingly like normal curves."

12.5 (a) $\chi^2_{0.025} = 32.852$ (b) $\chi^2_{0.95} = 10.117$

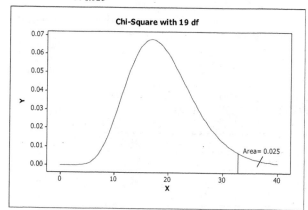
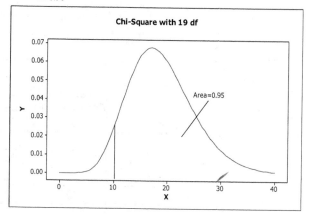

12.7 (a) $\chi^2_{0.05} \doteq 18.307$ (b) $\chi^2_{0.975} = 3.247$

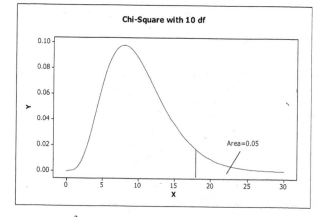
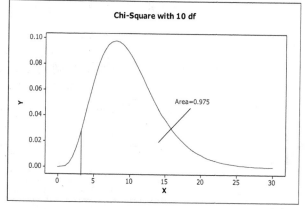

12.9 The χ^2-value having area 0.05 to its left has 0.95 to its right. Table VII gives values with specified areas to their right. If df = 26, $\chi^2_{0.95} = 15.379$

Exercises 12.2

12.11 The term "goodness-of-fit" is used to describe the type of hypothesis test considered in this section because the test is carried out by determining how well the observed frequencies match or fit the expected frequencies.

12.13 The assumptions are satisfied. The expected frequencies are np = 65, 30, and 5. All are 1 or more, and none are less than 5.

12.15 The assumptions are satisfied. The expected frequencies are np = 10, 10, 12.5, 15, and 2.5. All are 1 or more, and exactly 20% are less than 5.

12.17 The assumptions are not satisfied. The expected frequencies are np = 11, 11, 12.5, 15, and 0.5. One of them is not 1 or more.

12.19 (a) The population consists of occupied housing units built after 2000.

Copyright © 2012 Pearson Education, Inc. Publishing as Addison-Wesley.

The variable under consideration is the primary heating fuel of each unit.

(b) The assumptions are not satisfied when n = 200. The expected frequencies are np = 103, 19.6, 61.4, 11.4, 3.8, 0.8. One of the frequencies is less than 1, and 33% are less than 5.
The assumptions are not satisfied when n = 250. The expected frequencies are np = 128.75, 24.5, 76.75, 14.25, 4.75, 1.0. None of the frequencies is less than 1, but two (33%) are less than 5.
The assumptions are satisfied when n = 300. The expected frequencies are np = 154.5, 29.4, 92.1, 17.1, 5.7, 1.2. None of the frequencies is less than 1 and only one (17%) is less than 5.

(c) We need the count for Wood and other fuel to be at least 5. Thus we must have N(0.019) > 5. This implies that N must be greater than 5/0.019 = 263.15. Thus the smallest possible value of N is 264.

12.21 The procedure is summarized in the following table.

Distribution	Observed Frequency O	Expected Frequency E	Difference O - E	Square of Difference $(O - E)^2$	Chi-square subtotal $(O - E)^2/E$
0.2	85	100	-15	225	2.250
0.4	215	200	15	225	1.125
0.3	130	150	-20	400	2.667
0.1	70	50	20	400	8.000
	500				14.042

df = 3; From Table VII, critical Value = 7.815. χ^2 = 14.0422, which is greater than the critical value, so reject the null hypothesis. The data provide sufficient evidence that the variable differs from the given distribution.

12.23 The procedure is summarized in the following table.

Distribution	Observed Frequency O	Expected Frequency E	Difference O - E	Square of Difference $(O - E)^2$	Chi-square subtotal $(O - E)^2/E$
0.2	9	10	-1	1	0.100
0.1	7	5	2	4	0.800
0.1	1	5	-4	16	3.200
0.3	12	15	-3	9	0.600
0.3	21	15	6	36	2.400
	50				7.100

df = 4; From Table VII, critical Value = 7.779. χ^2 = 7.100, which is less than the critical value, so do not reject the null hypothesis. The data do not provide sufficient evidence that the variable differs from the given distribution.

12.25 The procedure is summarized in the following table.

Distribution	Observed Frequency O	Expected Frequency E	Difference O - E	Square of Difference $(O - E)^2$	Chi-square subtotal $(O - E)^2/E$
0.5	147	175	-28	784	4.480
0.3	115	105	10	100	0.952
0.2	88	70	18	324	4.629
	350				10.061

Copyright © 2012 Pearson Education, Inc. Publishing as Addison-Wesley.

df = 2; From Table VII, critical Value = 9.210. χ^2 = 10.061, which is greater than the critical value, so reject the null hypothesis. The data provide sufficient evidence that the variable differs from the given distribution.

12.27 (a) The population consists of this year's incoming college freshmen in the U.S. The variable under consideration is their political view.

(b)

Political View	Distribution	Observed Frequency O	Expected Frequency E	Difference O - E	Square of Difference $(O - E)^2$	Chi-square subtotal $(O - E)^2/E$
Liberal	0.277	160	138.500	21.500	462.250	3.338
Moderate	0.519	246	259.500	-13.500	182.250	0.702
Conservative	0.204	94	102.000	-8.000	64.000	0.627
		500				4.667

Step 1: H_0: The distribution of political views for this year's incoming freshmen is the same as the 2000 distribution.

H_a: The distribution of political views for this year's incoming freshmen is different from the 2000 distribution.

Step 2: α = 0.05

Step 3: χ^2 = 4.667 (See column 7 of the table.)

Step 4: df = 2; From Table VII, critical value = 5.991

For the P-value approach, $0.05 < P(\chi^2 > 4.667) < 0.10$

Step 5: Since 4.667 < 5.991, do not reject H_0. Since the P-value is greater than the significance level, do not reject H_0.

Step 6: The data do not provide sufficient evidence at the 5% level to conclude that the distribution of political views for this year's incoming freshmen is different from the 2000 distribution.

(c) Since the P-value is less than 0.10, reject H_0 at the 10% significance level. The data do provide sufficient evidence at the 10% level to conclude that the distribution of political views for this year's incoming freshmen is different from the 2000 distribution.

12.29

Color	Distribution	Observed Frequency O	Expected Frequency E	Difference O - E	Square of Difference $(O - E)^2$	Chi-square subtotal $(O - E)^2/E$
Brown	0.30	152	152.7	-0.3	0.49	0.003
Yellow	0.20	114	101.8	12.2	148.84	1.462
Red	0.20	106	101.8	4.2	17.64	0.173
Orange	0.10	51	50.9	0.1	0.01	0.000
Green	0.10	43	50.9	-7.9	62.41	1.226
Blue	0.10	43	50.9	-7.9	62.41	1.226
		509				4.091

Step 1: H_0: The color distribution of M&Ms is the same as that reported by M&M/Mars consumer affairs.

H_a: The color distribution of M&Ms is different from that reported by M&M/Mars consumer affairs.

Step 2: α = 0.05

Step 3: χ^2 = 4.091 (See column 7 of the table.)

Step 4: df = 5; From Table VII, the critical value = 11.070

Copyright © 2012 Pearson Education, Inc. Publishing as Addison-Wesley.

For the P-value approach, $P(\chi^2 > 4.091) > 0.10$

Step 5: Since $4.091 < 11.070$, do not reject H_0. Since the P-value is larger than the significance level, do not reject H_0.

Step 6: There is not sufficient evidence to conclude that the color distribution of M&Ms is different from that reported by M&M/Mars consumer affairs.

12.31

Number	Distribution	Observed Frequency O	Expected Frequency E	Difference O - E	Square of Difference $(O - E)^2$	Chi-square subtotal $(O - E)^2/E$
1	0.167	23	25	-2	4	0.160
2	0.167	26	25	1	1	0.040
3	0.167	23	25	-2	4	0.160
4	0.167	21	25	-4	16	0.640
5	0.167	31	25	6	36	1.440
6	0.167	26	25	1	1	0.040
		150				2.480

Step 1: H_0:   The die is not loaded.
         H_a: The die is loaded.

Step 2: $\alpha = 0.05$

Step 3: $\chi^2 = 2.480$ (See column 7 of the table.)

Step 4: df = 5; From Table VII, the critical value = 11.071

For the P-value approach, $P(\chi^2 > 2.480) > 0.10$.

Step 5: Since $2.480 < 11.071$, do not reject H_0. Since the P-value is larger than the significance level, do not reject H_0.

Step 6: The data do not provide sufficient evidence to conclude that the die is loaded.

12.33 (a) Using Minitab, choose **Stat/Tables/Chi-Square Goodness-of-Fit Test**, select the **Observed Counts** option button, enter O in the **Observed counts** text box, enter GAMES in the **Category names** text box, select the **Specific proportions** option button from the **Test** list, enter P in the **Specific proportions** text box, and click **OK**. The results are

Using category names in GAMES

Category	Observed	Test Proportion	Expected	Contribution to Chi-Sq
4	20	0.1250	12.50	4.500
5	23	0.2500	25.00	0.160
6	22	0.3125	31.25	2.738
7	35	0.3125	31.25	0.450

N	DF	Chi-Sq	P-Value
100	3	7.848	0.049

The P-value = 0.049, which is smaller than the significance level 0.05, so reject H_0. The data do provide sufficient evidence to conclude that World Series teams are not evenly matched.

(b) The usual assumption is that the data comprise a random sample from some population about which we want to make some inference. In this case, the data include the entire population of World Series records. We can think of these records as a sample from the population of all World Series that have been played and have yet to be played, but they are not a random sample. Nevertheless, each World Series is independent of the others, and the procedure does allow us to gain some insight into whether the overall records of the number of games played

Copyright © 2012 Pearson Education, Inc. Publishing as Addison-Wesley.

in each series follow a theoretical model based on the teams being evenly matched.

12.35 Using Minitab, choose **Stat/Tables/Chi-Square Goodness-of-Fit Test**, select the **Observed Counts** option button, enter O in the **Observed counts** text box, enter REASON in the **Category names** text box, select the **Specific proportions** option button from the **Test** list, enter P in the **Specific proportions** text box, and click **OK**. The results are

Chi-Square Goodness-of-Fit Test for Observed Counts in Variable: O

Using category names in REASON

Category	Observed	Test Proportion	Expected	Contribution to Chi-Sq
Help from friends/relatives	43	0.062	31.0	4.6452
Industry/business	108	0.178	89.0	4.0562
Job assignment	23	0.072	36.0	4.6944
Job transfer	20	0.048	24.0	0.6667
Joining family	45	0.068	34.0	3.5588
Marriage	205	0.368	184.0	2.3967
Other	9	0.035	17.5	4.1286
Study/training	47	0.169	84.5	16.6420

N	DF	Chi-Sq	P-Value
500	7	40.7886	0.000

The P-value = 0.000, which is smaller than the significance level 0.01, so reject H_0. The data provide sufficient evidence to conclude that distribution of reasons for migration between provinces in China differs from that for migration within provinces.

12.37 (a) When c =2, the chi-square goodness-of-fit test is equivalent to the one-sample z-test for one proportion (Procedure 12.2) since when there are only two categories in the chi-square test, one of them can be thought of as a success and the other as a failure.

(b) The values of p in the chi-square test correspond to p and (1 - p) in the z-test. The two observed frequencies in the chi-square test correspond to x and   n - x in the z-test. In other words, knowing one of the probabilities in the chi-square test when c = 2 allows us to know the other one. The same is true for the observed frequencies and for the expected frequencies. Finally, we compare the z and chi-square statistics when c = 2 and $p = p_0$.

$$\chi^2 = \sum \frac{(O-E)^2}{E} = \frac{(x-np)^2}{np} + \frac{[(n-x)-n(1-p)]^2}{n(1-p)}$$

$$= \frac{(x-np)^2}{np} + \frac{(np-x)^2}{n(1-p)} = (x-np)^2[\frac{1}{np} + \frac{1}{n(1-p)}]$$

$$= (x-np)^2[\frac{1}{np(1-p)}] = \frac{(x-np)^2}{np(1-p)}$$

Note that since $z = \dfrac{\dfrac{x}{n} - p}{\sqrt{\dfrac{p(1-p)}{n}}} = \dfrac{x-np}{\sqrt{np(1-p)}}$, $z^2 = \dfrac{(x-np)^2}{np(1-p)} = \chi^2$.

Thus the computed value of the χ^2 statistic is the same as the square

Copyright © 2012 Pearson Education, Inc. Publishing as Addison-Wesley.

of the z-test statistic. Since the χ^2 statistic is always positive, either a positive or a negative z-value will lead to the same χ^2 value, making the two tests equivalent when the alternative hypothesis for the z-test is two tailed.

Exercises 12.3

12.39 Cells

12.41 To obtain the total number of observations of bivariate data in a contingency table, one can sum the individual cell frequencies, sum the row subtotals, or sum the column subtotals.

12.43 Yes. If there were no association between the gender of the physician and specialty of the physician, then the same percentage of male and female physicians would choose internal medicine. Since different percentages of male and female physicians chose internal medicine, there is an association between the variables "gender" and "specialty." In other words, knowing the gender of a physician imparts information about the likelihood that the physician specialized in internal medicine.

12.45 (a)

	M	F	Total
BUS	2	7	9
ENG	10	2	12
LIB	3	1	4
Total	15	10	25

(b)

	BUS	ENG	LIB	Total
M	0.222	0.833	0.750	0.600
F	0.778	0.167	0.250	0.400
Total	1.000	1.000	1.000	1.000

(c)

	M	F	Total
BUS	0.133	0.700	0.360
ENG	0.667	0.200	0.480
LIB	0.200	0.100	0.160
Total	1.000	1.000	1.000

(d) Yes. The conditional distributions of gender are different within each college.

12.47 (a)

	Fresh	Soph	Jun	Sen	Total
Dem	2	6	8	4	20
Rep	3	9	12	6	30
Other	1	3	4	2	10
Total	6	18	24	12	60

(b)

	Fresh	Soph	Jun	Sen
Dem	0.333	0.333	0.333	0.333
Rep	0.500	0.500	0.500	0.500
Other	0.167	0.167	0.167	0.167
Total	1.000	1.000	1.000	1.000

(c) There is no association between party affiliation and class level. All of the conditional distributions of political party within class level are identical.

Copyright © 2012 Pearson Education, Inc. Publishing as Addison-Wesley.

(d)   The marginal distribution of party affiliation will be identical to each of the conditional distributions, that is

Party	Frequency
Dem	0.333
Rep	0.500
Other	0.167
Total	1.000

(e)   True. Since party affiliation and class level are not associated, all of the conditional distributions of class level within political party will be identical and will be the same as the marginal distribution of class level.

12.49 (a)   6

(b)   First complete the first row, then the first column total, then the second row total, then the Total for other, then complete the Female column.

	Male	Female	Total
White	10,563	1,971	12,534
Black	14,247	7,196	21,443
Other	6,471	2,048	8,519
Total	31,281	11,215	42,496

(c)   42,496   (d)   21,443   (e)   31,281   (f)   1,971

12.51 (a)   1,056,500   (b)   48,600   (c)   10,600

(d)   88,800 + 226,500 - 10,600 = 304,700

(e)   51,900   (f)   51,900   (g)   1,648,900 - 88,800 = 1,560,100

12.53 (a)   Complete the first column, then the second column, the third row, and then the remaining totals.

	Full Owner	Part Owner	Tenant	Total
Under 50	**639**	64	41	**744**
50 - 179	487	131	41	659
180 - 499	203	**153**	**33**	389
500 - 999	54	91	17	162
1000 & Over	46	112	18	176
Total	1429	551	**150**	**2130**

(b)   15   (c)   744,000   (d)   150,000   (e)   91,000   (f)   701,000

(g)   68,000

12.55 The table from Exercise 12.49 was first transposed so that the columns became the rows and vice versa. Then each cell entry was divided by the column total to produce the following table.

	White	Black	Other	Total
Male	0.843	0.664	0.760	0.736
Female	0.157	0.336	0.240	0.264
Total	1.000	1.000	1.000	1.000

(a)   The conditional distributions of gender within each race are given in columns 2, 3, and 4 of the above table. The conditional distributions of gender by race indicate that in all populations, the largest percentage of AIDS cases is male, but there are differences in the percentages.

(b)   The marginal distribution of gender is given in the last column of the above table. The largest percentage of AIDS cases, regardless of race,

Copyright © 2012 Pearson Education, Inc. Publishing as Addison-Wesley.

is male.

(c) Yes. The conditional distributions for gender are different for the three race categories.

(d) 26.4% of the AIDS cases were females.

(e) 15.7% of the AIDS cases among whites were females.

(f) True. Since there is an association between gender and race category, the conditional distributions of race by gender cannot be identical (If they were identical, there would be no association between the two variables.).

(g) Directly from the table in Exercise 12.49, divide each cell entry by the column total below it to obtain the table below.

	Male	Female	Total
White	0.338	0.176	0.295
Black	0.455	0.642	0.505
Other	0.207	0.183	0.200
Total	1.000	1.000	1.000

The conditional distributions of race by gender are given in columns 2 and 3 of the table. The marginal distribution of race is given in the last column of the table. The conditional distributions of race by gender indicate that among both males and females with AIDS, the largest percentage of cases is black. Other interpretations are possible as well.

12.57 (a)

	State	Federal	Local	Total
8th grade or less	0.142	0.119	0.131	0.137
Some high school	0.255	0.145	0.334	0.273
GED	0.285	0.226	0.141	0.238
High school diploma	0.205	0.270	0.259	0.225
Post secondary	0.090	0.158	0.103	0.098
College grad or more	0.024	0.081	0.032	0.029
Total	1.001	0.999	1.000	1.000

(b) Yes. The conditional distributions of educational attainment within the facility categories (columns 2, 3, and 4 of the table in part a) are not identical.

(c) The marginal distribution of educational attainment is given in the last column of the table in part (a).

(d)

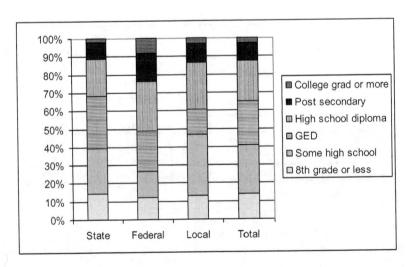

If the conditional distributions of educational attainment within the

Copyright © 2012 Pearson Education, Inc. Publishing as Addison-Wesley.

facility categories (first three bars) had been identical, each of the bars would have been segmented identically and would be identical to the fourth bar, the marginal distribution of educational attainment. The fact that they are not segmented identically means that there is an association between educational attainment and type of facility.

(e) False. Since educational attainment and type of facility are associated, the conditional distributions of facility type within educational attainment will be different.

(f) We interchanged the rows and columns of the table in Exercise 12.51 and then divided each cell entry by its associated column total to obtain the following table.

	8th grade or less	Some high school	GED	High school diploma	Post secondary	College grad or more	Total
State	0.662	0.598	0.768	0.584	0.590	0.521	0.641
Federal	0.047	0.029	0.051	0.065	0.087	0.148	0.054
Local	0.291	0.374	0.181	0.352	0.323	0.331	0.305
Total	1.000	1.001	1.000	1.001	1.000	1.000	1.000

The conditional distributions of facility type within educational attainment are given in columns 2, 3, 4, 5, 6, and 7. The marginal distribution of facility type is given by the last column.

(g) From the table in part (f), 5.4% of the prisoners are in federal facilities.

(h) From the table in part (f), 4.7% of the prisoners with at most an eighth grade education are in federal facilities.

(i) From the table in part (a), 11.9% of the prisoners in federal facilities have at most an 8th grade education.

12.59 Using Minitab, choose **Stat ▶ Tables ▶ Cross Tabulation and Chi-Square**, enter REGION in the **For rows** text box, enter PARTY in the **For columns** text box, check the **Display** box for **Counts** and click **OK**. The following table is produced.

```
Rows: REGION    Columns: PARTY

            Democrat  Republican  All

Midwest        7          5       12
Northeast      6          3        9
South          9          7       16
West           6          7       13
All           28         22       50
```

(b) Repeat the procedure in part (a), but check the box for **Column percents** instead of **Counts**.

```
Rows: REGION    Columns: PARTY

            Democrat  Republican    All

Midwest       25.00     22.73      24.00
Northeast     21.43     13.64      18.00
South         32.14     31.82      32.00
West          21.43     31.82      26.00
All          100.00    100.00     100.00
```

(c) Repeat the procedure in part (a), but check the box for **Row percents** instead of **Counts**.

```
Rows: REGION    Columns: PARTY
```

Copyright © 2012 Pearson Education, Inc. Publishing as Addison-Wesley.

	Democrat	Republican	All
Midwest	58.33	41.67	100.00
Northeast	66.67	33.33	100.00
South	56.25	43.75	100.00
West	46.15	53.85	100.00
All	56.00	44.00	100.00

(d) Yes. The conditional distributions of party within each region are not identical.

12.61 Using Minitab, choose **Stat ▶ Tables ▶ Cross Tabulation and Chi-Square**, enter <u>PARTY</u> in the **For rows** text box, enter <u>CLASS</u> in the **For columns** text box, check the **Display** box for **Counts** and click **OK**. The following table is produced.

Rows: PARTY   Columns: CLASS

	I	II	III	All
Democratic	22	20	16	58
Independent	2	0	0	2
Republican	9	13	18	40
All	33	33	34	100

(b) Repeat the procedure in part (a), but check the box for **Column percents** instead of **Counts**.

Rows: PARTY   Columns: CLASS

	I	II	III	All
Democratic	66.67	60.61	47.06	58.00
Independent	6.06	0.00	0.00	2.00
Republican	27.27	39.39	52.94	40.00
All	100.00	100.00	100.00	100.00

(c) Repeat the procedure in part (a), but check the box for **Row percents** instead of **Counts**.

Rows: PARTY   Columns: CLASS

	I	II	III	All
Democratic	37.93	34.48	27.59	100.00
Independent	100.00	0.00	0.00	100.00
Republican	22.50	32.50	45.00	100.00
All	33.00	33.00	34.00	100.00

(d) Yes. The conditional distributions of party within each class are not identical.

12.63 (a) If there were no association between age group and gender, the conditional distribution of age group within gender would be the same as the marginal distribution of age group. Therefore 7.3% of the resident population would be in the age group 20-24 years.

(b) If there were no association between age group and gender, the conditional distribution of age group within gender would be the same for each gender. Therefore 7.3% of the female residents would be in the age group 20-24 years.

(c) 7.3% of 153.0 million, i.e., 11,169,000, females would be in the age group 20-24 years.

(d) The number of female residents in the age group 20-24 years is not what we would expect IF there were NO association between age group and gender. Therefore there must be an association between age group and gender.

Exercises 12.4

Copyright © 2012 Pearson Education, Inc. Publishing as Addison-Wesley.

12.65 H_0: The two variables under consideration are statistically independent.
H_a: The two variables under consideration are statistically dependent.

12.67 The degrees of freedom equal $(r - 1)(c - 1) = (6 - 1)(4 - 1) = 15$.

12.69 If two variables are not associated, then their frequencies do not rise and fall together, whether by causation or as the result of some third variable. If the variables did have a causal relationship, then they would be associated.

12.71 Step 1: H_0: Siskel's ratings and Ebert's ratings of movies are not associated.
H_a: Siskel's ratings and Ebert's ratings of movies are associated.

Step 2: $\alpha = 0.01$

Step 3: Calculate the expected frequencies using the formula $E = RC/n$ where R = row total, C = column total, and n = sample size. The results are shown in the following table.

Ebert's rating

		Thumbs down	Mixed	Thumbs up	Total
Siskel's Rating	Thumbs down	11.8	8.4	24.8	**45.0**
	Mixed	8.4	6.0	17.6	**32.0**
	Thumbs up	21.8	15.6	45.6	**83.0**
	Total	**42.0**	**30.0**	**88.0**	**160.0**

Compute the value of the test statistic, $\chi^2 = \sum \dfrac{(O-E)^2}{E}$

where O and E represent the observed and expected frequencies respectively. We show the contributions to this sum from each of the cells of the contingency table in the following table.

	Thumbs down	Mixed	Thumbs up
Thumbs down	12.574	0.023	5.578
Mixed	0.019	8.167	2.475
Thumbs up	6.377	2.767	7.376

The total of the 9 table entries above is the value of the chi-square statistic, that is, $\chi^2 = 45.357$. Depending on rounding, your answer could differ slightly.

Step 4: The degrees of freedom are $(3 - 1)((3 - 1) = 4$, so the critical value from Table VII is $\chi^2_{0.01} = 13.277$.

For the P-value approach, P-value < 0.005

Step 5: Since $45.357 > 13.277$, we reject the null hypothesis. Since the P-value is less than the significance level, we reject the null hypothesis

Step 6: We conclude that there is an association between Siskel's ratings and Ebert's ratings of movies.

12.73 (a) The expected frequencies are shown below the observed frequencies in the following contingency table.

Social Class	A few	Some	Lots	Total
	4	13	15	32
	5	11	18	34
Total	9	24	33	66

None of the expected values is less than one, but two of the six (33%)

Copyright © 2012 Pearson Education, Inc. Publishing as Addison-Wesley.

are less than 5. Thus Assumptions 1 and 2 are not both satisfied. The chi-square independence test should not be used.

(b)

Social Class	Never	Sometimes	Often	Total
Middle	2 6.303	8 8.727	22 16.970	32
Working	11 6.697	10 9.273	13 18.030	34
Total	13	18	35	66

None of the expected values is less than one, and none of the six are less than 5. Thus Assumptions 1 and 2 are both satisfied. The chi-square independence test may be used.

The following table gives the contributions from each cell to the chi-square statistic.

Row, Column	O	E	$(O - E)^2/E$
1,1	2	6.303	2.9376
1,2	8	8.727	0.0606
1,3	22	16.970	1.4909
2,1	11	6.697	2.7648
2,2	10	9.273	0.0570
2,3	13	18.030	1.4033
	66		8.7142

Step 1: H_0: The frequency with which parents play "I Spy" games with their children is independent of their economic class.

H_a: The frequency with which parents play "I Spy" games with their children is dependent of their economic class.

Step 2: $\alpha = 0.05$

Step 3: Observed and expected frequencies are presented in the frequency contingency table. Each expected frequency is placed below its corresponding observed frequency. Expected frequencies are calculated using the formula E = (R x C)/n.

The same information about the Os and Es is presented in the table below the contingency table.

$\chi^2 = 8.7142$ (See column 4 of the table below the contingency table.)

Step 4: Critical value = 5.991

For the P-value approach, $0.01 < P(\chi^2 > 8.7142) < 0.025$.

Step 5: Since 8.7142 > 5.991, reject H_0. Since the P-value is less than the significance level, reject H_0.

Step 6: The data do provide sufficient evidence to conclude that the frequency with which parents play "I Spy" with their children is dependent on economic class.

12.75   Size of City

Copyright © 2012 Pearson Education, Inc. Publishing as Addison-Wesley.

Status in Practice	Less than 250,000	250,000-499,999	500,000 or more	Total
Government	12 15.7	4 4.2	14 10.1	30
Judicial	8 5.8	1 1.5	2 3.7	11
Private Practice	122 116.4	31 31.1	69 74.5	222
Salaried	19 23.1	7 6.2	18 14.8	44
Total	161	43	103	307

Step 1: H_0: Size of city and status in practice for lawyers are independent.

H_a: Size of city and status in practice for lawyers are dependent. Observed and expected frequencies are presented in the frequency contingency table. Each expected frequency is placed below its corresponding observed frequency. Expected frequencies are calculated using the formula $E = (R \times C)/n$.

Assumption 2 is violated since 25% (3/12) of the expected frequencies are less than 5. Thus, we do not proceed with the test.

12.77 The expected frequencies are shown below the observed frequencies in the following contingency table.

	Depressed	Not Depressed	Total
Osteoporitic	3 3.21	35 34.79	38
Low BMD	69 50.89	533 551.11	602
Normal	97 114.89	1262 1244.11	1359
Total	169	1830	1999

The table below gives the contributions from each cell to the chi-square statistic.

Row, Column	O	E	$(O - E)^2/E$
1,1	3	3.21	0.014
1,2	35	34.79	0.001
2,1	69	50.89	6.441
2,2	533	551.11	0.595
3,1	97	114.89	2.787
3,2	1262	1244.11	0.257
	1999	1999.00	10.095

Step 1: H_0: Depression and one mineral density in elderly Asian men are statistically independent.

H_a: Depression and one mineral density in elderly Asian men are statistically dependent.

Step 2: $\alpha = 0.01$

Step 3: Observed and expected frequencies are presented in the frequency contingency table. Each expected frequency is placed below its corresponding observed frequency. Expected frequencies are calculated using the formula $E = (R \cdot C)/n$.

Copyright © 2012 Pearson Education, Inc. Publishing as Addison-Wesley.

The same information about the Os and Es is presented in the table below the contingency table.

$\chi^2 = 10.095$ (See column 4 of the last table above.)

Note: Answers may vary slightly due to rounding.

Step 4: Critical value = 9.210

For the P-value approach, $0.005 < P(\chi^2 > 91.258) < 0.01$.

Step 5: Since $10.095 > 9.210$, reject H_0. Since the P-value is smaller than the significance level, reject H_0.

Step 6: At the 1% significance level, we conclude that depression and one mineral density in elderly Asian men are statistically dependent.

12.79 Using Minitab, the data are grouped in columns in order of the seven pairs of variables, with the responses "on the Job", "Off the Job", and "Don't Know" in rows 1, 2, and 3. The procedure is the same for each part. Choose

Stat ▶ Tables ▶ Chi-Square Test (Table in Worksheet), enter the necessary columns for each part of the exercise [e.g., MALE FEMALE for part (a)] in the **Columns containing the table:** text box, and click **OK**. The results are shown below using the P-value approach.

(a)   Expected counts are printed below observed counts
        Chi-Square contributions are printed below expected counts

	Male	Female	Total
1	77	77	154
	82.49	71.51	
	0.366	0.422	
2	263	215	478
	256.05	221.95	
	0.189	0.218	
3	28	27	55
	29.46	25.54	
	0.072	0.084	
Total	368	319	687

Chi-Sq = 1.350, DF = 2, P-Value = 0.509

Since the P-value = 0.509, which is greater than the significance level 0.05, do not reject H_0, the data do not provide evidence that gender and response are associated.

(b)   Expected counts are printed below observed counts
        Chi-Square contributions are printed below expected counts

	18-29 years	30-49 years	50-64 years	65 and older	Total
1	33	77	35	9	154
	39.12	84.53	24.73	5.62	
	0.957	0.671	4.265	2.032	
2	136	274	56	11	477
	121.16	261.83	76.60	17.41	
	1.816	0.566	5.539	2.359	
3	5	25	19	5	54
	13.72	29.64	8.67	1.97	
	5.539	0.727	12.302	4.656	
Total	174	376	110	25	685

Chi-Sq = 41.430, DF = 6, P-Value = 0.000
1 cells with expected counts less than 5.

Copyright © 2012 Pearson Education, Inc. Publishing as Addison-Wesley.

Since the P-value = 0.000, which is less than the significance level 0.05, reject H_0, the data provide evidence that age and response are associated.

(c)   Expected counts are printed below observed counts
        Chi-Square contributions are printed below expected counts

		Urban	Suburban	Rural	Total
1		62	47	42	151
		60.44	53.80	36.75	
		0.040	0.860	0.749	
2		197	171	109	477
		190.94	169.96	116.10	
		0.192	0.006	0.435	
3		14	25	15	54
		21.62	19.24	13.14	
		2.683	1.724	0.262	
Total		273	243	166	682

Chi-Sq = 6.952, DF = 4, P-Value = 0.138

Since the P-value = 0.138, which is greater than the significance level 0.05, do not reject H_0, the data do not provide evidence that type of community and response are associated.

(d)   Expected counts are printed below observed counts
        Chi-Square contributions are printed below expected counts

	Postgraduate	College graduate	Some college	No college	Total
1	23	41	20	68	152
	18.69	41.17	29.60	62.54	
	0.992	0.001	3.113	0.477	
2	51	126	108	192	477
	58.66	129.20	92.89	196.25	
	1.001	0.079	2.459	0.092	
3	10	18	5	21	54
	6.64	14.63	10.52	22.22	
	1.699	0.778	2.893	0.067	
Total	84	185	133	281	683

Chi-Sq = 13.651, DF = 6, P-Value = 0.034

Since the P-value = 0.034, which is less than the significance level 0.05, reject H_0, the data provide evidence that educational attainment and response are associated.

(e)   Expected counts are printed below observed counts
        Chi-Square contributions are printed below expected counts

	Under $20,000	$20,000-$29,999	$30,000-$49,999	$50,000 and over	Total
1	40	32	41	34	147
	29.18	34.52	40.09	43.21	
	4.014	0.184	0.021	1.963	
2	80	116	131	138	465
	92.30	109.20	126.82	136.68	
	1.638	0.423	0.138	0.013	
3	11	7	8	22	48
	9.53	11.27	13.09	14.11	
	0.228	1.620	1.980	4.413	
Total	131	155	180	194	660

Chi-Sq = 16.634, DF = 6, P-Value = 0.011

Copyright © 2012 Pearson Education, Inc. Publishing as Addison-Wesley.

Since the P-value = 0.011, which is less than the significance level 0.05, reject H_0, the data provide evidence that income and response are associated.

(f)
```
        Expected counts are printed below observed counts
        Chi-Square contributions are printed below expected counts

            Private  Government   Self  Total
        1       67          23      61    151
              93.26       25.76   31.98
              7.397       0.295  26.343

        2      326          82      69    477
             294.62       81.37  101.01
              3.343       0.005  10.145

        3       27          11      14     52
              32.12        8.87   11.01
              0.815       0.511   0.811
     Total     420         116     144    680
        Chi-Sq = 49.665, DF = 4, P-Value = 0.000
```
Since the P-value = 0.000, which is less than the significance level 0.05, reject H_0, the data provide evidence that type of employer and response are associated.

12.81 The expected frequencies are shown below the observed frequencies in the following contingency table.

	PhD	MA/Other	Total
Mail	65	21	86
	58.95	27.05	
Email	166	73	239
	163.84	75.16	
Both	84	28	112
	76.78	35.22	
N/A	73	56	129
	88.43	40.57	
Total	388	178	566

PhD	MA	Chi-Square
0.620	1.352	
0.029	0.062	
0.679	1.481	
2.693	5.869	
		12.785

Step 1: H_0: Degree and preference are non-associated.

H_a: Degree and preference are associated.

Step 2: $\alpha = 0.05$

Step 3: Observed and expected frequencies are presented in the frequency contingency table (left hand table above). Each expected frequency is placed below its corresponding observed frequency. Expected frequencies are calculated using the formula E = (R · C)/n. The right hand table contains each cell's contribution to the Chi-Square value shown in the lower right hand corner.

$\chi^2 = 12.785$

Step 4: Critical value = 7.815

For the P-value approach: $0.005 <$ P-value < 0.01

Step 5: Since 12.785 > 7.815, we reject the null hypothesis. Since the P-value is less than the significance level, reject the null hypothesis

Step 6: At the 5% significance level, there is sufficient evidence to conclude that degree and preference are associated.

Exercises 12.5

12.83 If populations are *homogeneous* with respect to a variable, it means that the distribution of the variable is similar among the different populations. If the populations are *nonhomogeneous*, it means the distribution of the variable is different among the different populations.

Copyright © 2012 Pearson Education, Inc. Publishing as Addison-Wesley.

12.85 If a variable has only two possible values, the chi-square homogeneity test provides a procedure for comparing several population <u>proportions.</u>

12.87 If the variable has five possible values among six populations, the degrees of freedom for the chi-square statistic will be $(5-1) \cdot (6-1) = 20$.

12.89 The expected frequencies are shown below the observed frequencies in the following contingency table.

Race

Region	White	Black	Other	Total
Northeast	93	14	6	113
	92.47	14.50	6.03	
Midwest	118	14	4	136
	111.29	17.45	7.25	
South	167	42	7	216
	176.76	27.72	11.52	
West	113	7	15	135
	110.48	17.33	7.20	
Total	491	77	32	600

The table below gives the contributions from each cell to the chi-square statistic.

Row, Column	O	E	$(O - E)^2/E$
1,1	93	92.47	0.003
1,2	14	14.50	0.017
1,3	6	6.03	0.000
2,1	118	111.29	0.405
2,2	14	17.45	0.682
2,3	4	7.25	1.457
3,1	167	176.76	0.539
3,2	42	27.72	7.356
3,3	7	11.52	1.773
4,1	113	110.48	0.057
4,2	7	17.33	6.157
4,3	15	7.20	8.450
	600	600.00	26.896

Step 1: H_0: The distribution of race is the same among the four U.S. regions.

   H_a: There is a difference in the distributions of race among the four U.S. regions.

Step 2: $\alpha = 0.01$

Step 3: Observed and expected frequencies are presented in the frequency contingency table. Each expected frequency is placed below its corresponding observed frequency. Expected frequencies are calculated using the formula E = (R · C)/n.

   The same information about the Os and Es is presented in the table below the contingency table.

   $\chi^2 = 26.896$ (See column 4 of the last table above.)

   Note: Answers may vary slightly due to rounding.

Step 4: Critical value = 16.812

   For the P-value approach, $P(\chi^2 > 26.896) < 0.005$.

Step 5: Since 26.896 > 16.812, reject H_0. Since the P-value is smaller than the significance level, reject H_0.

Step 6: At the 1% significance level, we conclude that there is a difference in race distributions among the four U.S. regions.

Copyright © 2012 Pearson Education, Inc. Publishing as Addison-Wesley.

12.91 The expected frequencies are shown below the observed frequencies in the following contingency table.

	Year		
Age	1996	2002	Total
17 or younger	5 5.52	6 5.48	11
18 – 24	77 77.78	78 77.22	155
25 – 34	88 95.85	103 95.15	191
35 – 44	72 69.25	66 68.75	138
45 – 54	28 22.58	17 22.42	45
55 or older	6 5.02	4 4.98	10
Total	276	274	550

The table below gives the contributions from each cell to the chi-square statistic.

Row, Column	O	E	$(O - E)^2/E$
1,1	5	5.52	0.049
1,2	6	5.48	0.049
2,1	77	77.78	0.008
2,2	78	77.22	0.008
3,1	88	95.85	0.642
3,2	103	95.15	0.647
4,1	72	69.25	0.109
4,2	66	68.75	0.110
5,1	28	22.58	1.300
5,2	17	22.42	1.310
6,1	6	5.02	0.192
6,2	4	4.98	0.193
	550	550.00	4.617

Step 1: H_0: In the two years, jail inmates are homogeneous with respect to age
H_a: In the two years, jail inmates are nonhomogeneous with respect to age.

Step 2: $\alpha = 0.05$

Step 3: Observed and expected frequencies are presented in the frequency contingency table. Each expected frequency is placed below its corresponding observed frequency. Expected frequencies are calculated using the formula $E = (R \cdot C)/n$.
The same information about the Os and Es is presented in the table below the contingency table.

$\chi^2 = 4.617$ (See column 4 of the last table above.)

Note: Answers may vary slightly due to rounding.

Step 4: Critical value = 11.070

For the P-value approach, $P(\chi^2 > 4.617) > 0.10$.

Step 5: Since 4.617 < 11.070, do not reject H_0. Since the P-value is greater than the significance level, do not reject H_0.

Step 6: At the 5% significance level, we do not conclude that jail inmates are nonhomogeneous with respect to age, in the two years.

Copyright © 2012 Pearson Education, Inc. Publishing as Addison-Wesley.

12.93 Using Minitab, choose **Stat ▶ Tables ▶ Chi-Square Test (Table in Worksheet)**, enter 'Not failure' 'Failure' in the **Columns containing the table:** text box, and click **OK**. The results are shown below using the P-value approach.

Expected counts are printed below observed counts

	Not failure	Failure	Total
1	94	17	111
	73.35	37.65	
	5.812	11.323	
2	24	22	46
	30.40	15.60	
	1.347	2.624	
3	71	58	129
	85.25	43.75	
	2.381	4.640	
Total	189	97	286

Chi-Sq = 28.128, DF = 2, P-Value = 0.000

We are testing the null hypothesis that the failure rate is the same for each of the three treatments. Since the P-value = 0.000, which is less than the significance level of 0.05, we reject that hypothesis. At 5% significance level, the data provide sufficient evidence that the failure rate is not the same for each of the three treatments.

12.95 (a) Population 1: February 2009 U.S. adults, $\hat{p}_1 = 0.54$

Population 2: March 2009 U.S. adults, $\hat{p}_2 = 0.61$

$\hat{p}_P = (543.78 + 614.27)/(1007 + 1007) = 0.575$

Step 1: $H_0: p_1 = p_2$, $H_a: p_1 \neq p_2$

Step 2: $\alpha = 0.05$

Step 3: $z = \dfrac{0.54 - 0.61}{\sqrt{0.575(1-0.575)[(1/1007)+(1/1007)]}} = -3.177$

Step 4: Since $\alpha = 0.05$, the critical values are $\pm z_{\alpha/2} = \pm 1.96$

For the P-value approach, $2 * P(z < -3.18) = 0.0014$

Step 5: Since $-3.18 < -1.96$, reject H_0. P-value $< \alpha$. Thus, reject H_0.

Step 6: The test results are significant at the 5% level; that is, at the 5% significance level, the data provide sufficient evidence to conclude that the approval percentages of all U.S. adults are different between the two months.

(b) The expected frequencies are shown below the observed frequencies in the following contingency table.

Month	Approval Yes	No	Total
February 2009	543.78	463.22	1007
	579	428	
March 2009	614.27	392.73	1007
	579	428	
Total	1158	856	2014

The table below gives the contributions from each cell to the chi-square statistic.

Copyright © 2012 Pearson Education, Inc. Publishing as Addison-Wesley.

Row, Column	O	E	$(O - E)^2/E$
1,1	543.78	579	2.142
1,2	463.22	428	2.898
2,1	614.27	579	2.148
2,2	392.73	428	2.906
	2014	2014.00	10.094

Step 1: H_0: The distribution of approval percentages is the same between the two months.

H_a: The distribution of approval percentages is different between the two months.

Step 2: $\alpha = 0.05$

Step 3: Observed and expected frequencies are presented in the frequency contingency table. Each expected frequency is placed below its corresponding observed frequency. Expected frequencies are calculated using the formula $E = (R \cdot C)/n$.

The same information about the Os and Es is presented in the table below the contingency table.

$\chi^2 = 10.094$ (See column 4 of the last table above.)

Note: Answers may vary slightly due to rounding.

Step 4: Critical value = 3.841

For the P-value approach, $P(\chi^2 > 10.094) < 0.005$.

Step 5: Since $10.094 > 3.841$, reject H_0. Since the P-value is smaller than the significance level, reject H_0.

Step 6: At the 5% significance level, we conclude that there is a difference in approval percentages of all U.S. adults between the two months.

(c) Both the homogeneity test and the two proportions z – test resulted in the same conclusion and had similar P-values.

(d) The homogeneity test is the same as a two-tailed two proportion z-test if there are only two populations.

12.97 In Exercise 12.95, the χ^2 test statistic for the homogeneity test is 10.094 and the z test statistic for the two proportions z-test is -3.177. If you square -3.177, the result is $(-3.177)(-3.177) = 10.093$, which is close to our χ^2 test statistic.

12.99 We will show that z^2 and χ^2 are the same when there are two populations, each with two outcomes. Using the notation from Chapter 12, let n_1 and n_2 be the number of trials for the two samples and let x_1 and x_2 be the number of successes in each sample. To simplify the algebra, we will let $N = n_1 + n_2$ and $X = x_1 + x_2$. First we will simplify the expression for z^2 and then for χ^2.

Copyright © 2012 Pearson Education, Inc. Publishing as Addison-Wesley.

$$The\ pooled\ \hat{p} = \frac{x_1 + x_2}{n_1 + n_2}\ and\ 1 - \hat{p} = 1 - \frac{x_1 + x_2}{n_1 + n_2} = \frac{n_1 + n_2 - (x_1 + x_2)}{n_1 + n_2}$$

$$Since\ z = \frac{\left(\dfrac{x_1}{n_1} - \dfrac{x_2}{n_2}\right)}{\sqrt{\hat{p}(1 - \hat{p})\left(\dfrac{1}{n_1} + \dfrac{1}{n_2}\right)}},$$

$$z^2 = \frac{\left(\dfrac{x_1}{n_1} - \dfrac{x_2}{n_2}\right)^2}{\hat{p}(1 - \hat{p})\left(\dfrac{1}{n_1} + \dfrac{1}{n_2}\right)} = \frac{\dfrac{(n_2\,x_1 - n_2\,x_1)^2}{(n_1\,n_2)^2}}{\left(\dfrac{x_1 + x_2}{n_1 + n_2}\right)\left(\dfrac{n_1 + n_2 - (x_1 + x_2)}{n_1 + n_2}\right)\left(\dfrac{1}{n_1} + \dfrac{1}{n_2}\right)}$$

$$= \frac{\dfrac{(n_2\,x_1 - n_2\,x_1)^2}{(n_1\,n_2)^2}}{\left(\dfrac{x_1 + x_2}{n_1 + n_2}\right)\left(\dfrac{n_1 + n_2 - (x_1 + x_2)}{n_1 + n_2}\right)\left(\dfrac{n_2 + n_1}{n_1\,n_2}\right)} = \frac{(n_2\,x_1 - n_2\,x_1)^2(n_1 + n_2)}{(n_1\,n_2)\left[(x_1 + x_2)n_1 + n_2 - (x_1 + x_2)\right]}$$

$$= \frac{(n_2\,x_1 - n_2\,x_1)^2\,N}{(n_1\,n_2)(X)(N - X)}$$

Arranging the data in a 2 x 2 table we have the observed values as

	Successes	Failures	Total
Sample 1	x_1	n_1-x_1	n_1
Sample 2	x_2	n_2-x_2	n_2
Total	$x_1 + x_2$	$(n_1 + n_2)-(x_1 + x_2)$	$n_1 + n_2$

Now the expected values for each cell are

$$E_{11} = \frac{(n_1(x_1 + x_2))}{n_1 + n_2} = \frac{(n_1(x_1 + x_2))}{N};\ E_{12} = \frac{(n_1[(n_1 + n_2)-(x_1 + x_2)])}{n_1 + n_2} = \frac{(n_1[N-(x_1 + x_2)])}{N}$$

$$E_{21} = \frac{(n_2(x_1 + x_2))}{n_1 + n_2} = \frac{(n_2(x_1 + x_2))}{N};\ E_{22} = \frac{(n_2[(n_1 + n_2)-(x_1 + x_2)])}{n_1 + n_2} = \frac{(n_2[N-(x_1 + x_2)])}{N}$$

$$O_{11}-E_{11} = x_1 - \frac{(n_1(x_1 + x_2))}{n_1 + n_2} = \frac{(n_1 + n_2)x_1}{n_1 + n_2} - \frac{(n_1(x_1 + x_2))}{n_1 + n_2} = \frac{n_2\,x_1 - n_1\,x_2}{n_1 + n_2}$$

All four O − E differences are the same in absolute value, so the square of each one equals

$$(O_{11}-E_{11})^2 = \left(\frac{n_2\,x_1 - n_1\,x_2}{n_1 + n_2}\right)^2 = \frac{(n_2\,x_1 - n_1\,x_2)^2}{N^2}$$

Now we will use the above facts to show that $\chi^2 = z^2$.

Copyright © 2012 Pearson Education, Inc. Publishing as Addison-Wesley.

$$\chi^2 = \sum \frac{(O-E)^2}{E} = \frac{(n_2\,x_1 - n_1\,x_2)^2}{N^2}\left[\frac{1}{E_{11}} + \frac{1}{E_{12}} + \frac{1}{E_{21}} + \frac{1}{E_{22}}\right]$$

$$\frac{(n_2\,x_1 - n_1\,x_2)^2}{N^2}\left[\frac{N}{n_1\,(x_1 + x_2)} + \frac{N}{n_1\,(N-(x_1 + x_2))} + \frac{N}{n_2\,(x_1 + x_2)} + \frac{N}{n_2\,[N-(x_1 + x_2)]} + \right]$$

$$= \frac{(n_2\,x_1 - n_1\,x_2)^2}{N}\left[\frac{1}{n_1 X} + \frac{1}{n_1\,(N-X)} + \frac{1}{n_2 X} + \frac{1}{n_2\,(N-X)} + \right]$$

$$= \frac{(n_2\,x_1 - n_1\,x_2)^2}{N}\left[\frac{n_2\,[N-X] + n_2\,X + n_1\,(N-X) + n_1\,X}{n_1\,n_2\,X\,(N-X)}\right]$$

$$= \frac{(n_2\,x_1 - n_1\,x_2)^2}{N}\left[\frac{(n_1 + n_2)N}{n_1\,n_2\,X(N-X)}\right] = \frac{(n_2\,x_1 - n_1\,x_2)^2}{N}\left[\frac{N \cdot N}{n_1\,n_2\,X(N-X)}\right]$$

$$= \frac{(n_2\,x_1 - n_1\,x_2)^2}{1}\left[\frac{N}{n_1\,n_2\,X(N-X)}\right] = \frac{(n_2\,x_1 - n_1\,x_2)^2}{n_1\,n_2\,X(N-X)} = z^2$$

12.101 We will go through the steps of the Chi-square homogeneity test for comparing two population proportions and the steps of the two-tailed two-proportions z-test and show that they are equivalent.

Step 1 for the homogeneity test:

H_0: The distribution of a variable is homogeneous between two population proportions.

H_a: The distribution of a variable is nonhomogeneous between two population proportions.

Step 1 for the two-tailed two-proportions z-test

H_0: $\quad p_1 = p_2$

H_a: $\quad p_1 \neq p_2$

Step 1 is the same for the two tests because when you say that the proportions are homogeneous, it is another way of saying that they are equivalent. Nonhomogeneous means that the two proportions would be different.

Step 2 is the same for the two tests. You are choosing the significance level.

Step 3 is the same for the two tests as shown in Exercise 12.99.

Step 4 is the same for the two tests as shown in Exercise 12.100.

Step 5 is the same for the two tests. If the test statistic falls in the rejection region, reject the null hypothesis.

Step 6 is the same for the two tests. Both tests will result in the same conclusion.

Review Problems For Chapter 12

1. The distributions and curves are distinguished by their numbers of degrees of freedom.

2. (a) zero   (b)   skewed right   (c)   normal curve

3. (a) No. The degrees of freedom for the χ-square goodness-of-fit test depends on the number of categories, not the number of observations.

   (b) No. The degrees of freedom for the χ-square independence test is $(r-1)(c-1)$ where r and c are the number of rows and columns, respectively, in the contingency table.

Copyright © 2012 Pearson Education, Inc. Publishing as Addison-Wesley.

(c) No. The degrees of freedom for the χ-square homogeneity test is (r -1)(c-1) where r and c are the number of rows and columns, respectively, in the contingency table.

4. Values of the test statistic near zero arise when the observed and expected frequencies are in close agreement. It is only when these frequencies differ enough to produce large values of the test statistic that the null hypothesis is rejected. These values are in the right tail of the χ-square distribution.

5. The value would be zero since O - E = 0 for pair of expected and observed frequencies.

6. (a) All expected frequencies are 1 or greater, and at most 20% of the expected frequencies are less than 5.
    (b) Very important. If either of these assumptions are not met, the test should not be carried out by these procedures.

7. (a) 7.8%
    (b) 7.8% of 24.7 million, or 1.93 million
    (c) Since the observed number of Midwest Hispanic households is not equal to the number expected if there were no association between race of the householder and area of residence, we conclude that there *is* an association between the race of the householder and area of residence.

8. (a) Compare the conditional distributions of one of the variables within categories of the other variable. If all of the conditional distributions are identical, there is no association between the variables; if not, there is an association.
    (b) No. Since the data are for an entire population, we are not making an inference from a sample to the population. The association (or non-association) is a fact.

9. (a) Perform a chi-square test of independence. If the null hypothesis (of non-association) is rejected, we conclude that there is an association between the variables.
    (b) Yes. It is possible (with probability α) that we could reject the null hypothesis when it is, in fact, true. It is also possible, that, even though there is actually an association between the variables, the evidence is not strong enough to draw that conclusion. Either one of these types of errors is due to randomness in selecting a sample which does not exactly reflect the characteristics of the population.

10. For df = 17:
    (a) $\chi^2_{0.99} = 6.408$ (b) $\chi^2_{0.01} = 33.409$ (c) $\chi^2_{0.05} = 27.587$
    (d) $\chi^2_{0.95} = 8.672$ (e) $\chi^2_{0.975} = 7.564$ and $\chi^2_{0.025} = 30.191$

11.

Highest level	O	p	E = np	$(O-E)^2/E$
Not HS graduate	84	0.158	79.0	0.316
HS graduate	160	0.332	166.0	0.217
Some college	88	0.176	88.0	0.000
Associate's degree	32	0.078	39.0	1.256
Bachelor's degree	87	0.170	85.0	0.047
Advanced degree	49	0.086	43.0	0.837
Total	500		500.0	2.673

Step 1: H_0: The educational attainment distribution for adults 25 years old and over this year is the same as in 2000.
H_a: The educational attainment distribution for adults 25 years old and over this year differs from that of 2000.

Step 2: $\alpha = 0.05$

Step 3: $\chi^2 = 2.673$ (See column 5 of the table.)

Copyright © 2012 Pearson Education, Inc. Publishing as Addison-Wesley.

Step 4: Critical value = 11.070

For the P-value approach, $0.10 < P(\chi^2 > 2.673) < 0.90$ (It's closer to 0.90 than to 0.10.)

Step 5: Since $2.673 < 11.070$, do not reject H_0.

Since the P-value is larger than the significance level, do not reject H_0.

Step 6: At the 5% significance level, data provides evidence that the educational attainment distribution for adults 25 years old and over this year does not differ from that of 2000.

12. (a) The population consists of U.S. presidents.
    (b) The two variables are the presidents' region of birth and their political party.
    (c)

	D	DR	F	R	U	W	TOTAL
NE	7	1	1	5	0	1	15
MW	1	0	0	10	0	0	11
SO	6	3	1	2	1	3	16
WE	1	0	0	1	0	0	2
TOTAL	15	4	2	18	1	4	44

13. (a) Dividing each cell entry by the column total in the table in part (a), we obtain the conditional distributions of birth region for each party and in the right hand column, the marginal distribution of birth region.

	D	DR	F	R	U	W	TOTAL
NE	0.467	0.250	0.500	0.277	0.000	0.250	0.341
MW	0.067	0.000	0.000	0.556	0.000	0.000	0.250
SO	0.400	0.750	0.500	0.111	1.000	0.750	0.364
WE	0.000	0.000	0.000	0.056	0.000	0.000	0.045
TOTAL	1.000	1.000	1.000	1.000	1.000	1.000	1.000

(b) Dividing each cell entry by the row total in the table in part (a), we obtain the conditional distributions of party for each birth region and in the bottom row, the marginal distribution of party.

	D	DR	F	R	U	W	TOTAL
NE	0.467	0.067	0.067	0.333	0.000	0.067	1.000
MW	0.091	0.000	0.000	0.909	0.000	0.000	1.000
SO	0.375	0.188	0.063	0.125	0.063	0.188	1.000
WE	0.500	0.000	0.000	0.500	0.000	0.000	1.000
TOTAL	0.341	0.093	0.046	0.409	0.023	0.091	1.000

(c) Yes. The conditional distributions of birth region by party are not all the same.
(d) 40.9% of Presidents are Republicans.
(e) 40.9%
(f) 12.5% of Presidents born in the South are Republicans.
(g) 36.4% of presidents were born in the South.
(h) 36.4%
(i) 11.1% of Republican presidents were born in the South.

14. (a) 2046 (b) 737 (c) 266 (d) 3046 (e) 6580 - 1167 = 5413
    (f) 5403 + 1167 - 660 = 5910

15. (a)

	General	Psychiatric	Chronic	Tuberculosis	Other	Total
GOV	0.314	0.361	0.808	0.750	0.144	0.311
PROP	0.122	0.486	0.038	0.000	0.361	0.177
NP	0.564	0.153	0.154	0.250	0.495	0.512
Total	1.000	1.000	1.000	1.000	1.000	1.000

Copyright © 2012 Pearson Education, Inc. Publishing as Addison-Wesley.

The conditional distributions of control type with facility type are given in columns 2, 3, 4, 5, and 6 of the table.

(b) Yes. The conditional distributions of control type within facility type are not all identical.

(c) The marginal distribution of control type is given by the last column of the table above.

(d)

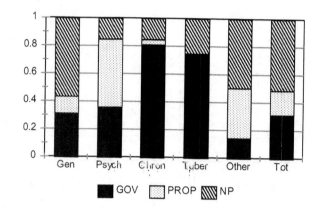

The conditional distributions of control type within facility type are shown be the first five bars of the graph. The marginal distribution of control type is given by the last bar. Since the bars are not identically shaded, control type and facility type are associated variables.

(e) False. Since we have established that facility type and control type are associated, the conditional distributions of facility type within control types will not be identical.

(f) After interchanging the rows and columns of the table given in Problem 13, we divided each cell entry by its associated column total to obtain the table below.

	GOV	PROP	NP	Total
General	0.829	0.566	0.905	0.821
Psychiatric	0.130	0.307	0.034	0.112
Chronic	0.010	0.001	0.001	0.004
Tuberculosis	0.001	0.000	0.000	0.001
Other	0.029	0.127	0.060	0.062
Total	1.000	1.000	1.000	1.000

The conditional distributions of facility type within control type are given in columns 2, 3, and 4 of the table. The marginal distribution of facility type is given by the last column of the table.

(g) From the table in part (a), 17.7% of the hospitals are under proprietary control.

(h) From the table in part (a), 48.6% of the psychiatric hospitals are under proprietary control.

(i) From the table in part (f), 30.7% of the hospitals under proprietary control are psychiatric hospitals.

16.

	Positive	Partial	None	Total
Lymphocyte depletion	18	10	44	72
Lymphocyte dominance	74	18	12	104
Mixed cellularity	154	54	58	266
Nodular sclerosis	68	16	12	96
Total	314	98	126	538

Copyright © 2012 Pearson Education, Inc. Publishing as Addison-Wesley.

Row, Column	O	E	$(O - E)^2/E$
1,1	18	42.02	13.732
1,2	10	13.12	0.740
1,3	44	16.86	43.674
2,1	74	60.70	2.915
2,2	18	18.94	0.047
2,3	12	24.36	6.269
3,1	154	155.25	0.010
3,2	54	48.45	0.635
3,3	58	62.30	0.296
4,1	68	56.03	2.557
4,2	16	17.49	0.126
4,3	12	22.48	4.888
	538	538.00	75.889

Step 1: H_0: Treatment response and histological type are statistically independent.

H_a: Treatment response and histological type are statistically dependent.

Step 2: $\alpha = 0.01$

Step 3: Observed frequencies are presented in the frequency contingency table. Expected frequencies are calculated using the formula $E = (R \cdot C)/n$ and are included in the third column of the second table.

$\chi^2 = 75.889$ (See column 4 of the table below the contingency table.)

Note: Answers may vary slightly due to rounding.

Step 4: Critical value = 16.812

For the P-value approach, $P(\chi^2 > 77.693) < 0.005$.

Step 5: Since $75.889 > 16.812$, reject H_0. Since the P-value is smaller than the significance level, reject H_0.

Step 6: At the 1% significance level, the data provide sufficient evidence to conclude that treatment response and histological type are associated.

17. (a) The populations are they types of residence of people. The can reside inside principal cities, outside principal cities but within metropolitan areas, or outside metropolitan areas.

(b) The variable is income level.

(c)

	IPC	OPC	OMA	Total
Under $5,000	30	49	18	97
$5,000 - $9,999	36	45	20	101
$10,000 - $14,999	41	57	27	125
$15,000 - $24,999	82	122	46	250
$25,000 - $34,999	69	108	41	218
$35,000 - $49,999	73	126	40	239
$50,000 - $74,999	67	135	34	236
$75,000 & over	68	146	20	234
Total	466	788	246	1500

Copyright © 2012 Pearson Education, Inc. Publishing as Addison-Wesley.

Row, Column	O	E	$(O - E)^2/E$
1,1	36	30.13	0.001
1,2	49	50.96	0.075
1,3	18	15.91	0.275
2,1	36	31.38	0.681
2,2	45	53.06	1.224
2,3	20	16.56	0.713
3,1	41	38.83	0.121
3,2	57	65.67	1.144
3,3	27	20.50	2.061
4,1	82	77.67	0.242
4,2	122	131.33	0.663
4,3	46	41.00	0.610
5,1	69	67.73	0.024
5,2	108	114.52	0.372
5,3	41	35.75	0.770
6,1	73	74.25	0.021
6,2	126	125.55	0.002
6,3	40	39.20	0.016
7,1	67	73.32	0.544
7,2	135	123.98	0.980
7,3	34	38.70	0.572
8,1	68	72.70	0.303
8,2	146	122.93	4.330
8,3	20	38.38	8.799
	1500	1500.00	24.543

Step 1: H_0: People residing in the three types of residence are homogeneous with respect to income level.
H_a: People residing in the three types of residence are nonhomogeneous with respect to income level.

Step 2: α = 0.05

Step 3:   Observed frequencies are presented in the frequency contingency table. Expected frequencies are calculated using the formula E = (R $\cdot$ C)/n and are included in the third column of the second table.

χ^2 = 24.543   (See column 4 of the table below the contingency table.)

Note: Answers may vary slightly due to rounding.

Step 4: Critical value = 23.685

For the P-value approach, $0.025 < P(\chi^2 > 24.543) < 0.05$.

Step 5: Since 24.543 > 23.685, reject H_0. Since the P-value is smaller than the significance level, reject H_0.

Step 6:   At the 5% significance level, the data provide sufficient evidence to conclude that people residing in the three types of residence are nonhomogeneous with respect to income level.

18.

	Democrat	Republican	Independent	Total
Yes	528	231	472	1231
No	100	240	174	514
Total	628	471	646	1745

Copyright © 2012 Pearson Education, Inc. Publishing as Addison-Wesley.

Row, Column	O	E	$(O - E)^2/E$
1,1	528	433.02	16.301
1,2	231	332.26	30.862
1,3	472	455.72	0.582
2,1	100	184.98	39.041
2,2	240	138.74	73.913
2,3	174	190.28	1.393
	1745	1745.00	162.093

Step 1: H_0: The percentages of Democrats, Republicans, and Independents who thought the U.S. economy was in a recession in May 2008 are not deferent

H_a: There is a difference in the percentages of Democrats, Republicans, and Independents who thought the U.S. economy was in a recession in May 2008.

Step 2: $\alpha = 0.01$

Step 3:   Observed frequencies are presented in the frequency contingency table. Expected frequencies are calculated using the formula $E = (R \cdot C)/n$ and are included in the third column of the second table.

$\chi^2 = 162.093$ (See column 4 of the table below the contingency table.)

Note: Answers may vary slightly due to rounding.

Step 4:   Critical value = 9.210

For the P-value approach, $P(\chi^2 > 162.093) < 0.005$.

Step 5:   Since $162.093 > 9.210$, reject H_0. Since the P-value is smaller than the significance level, reject H_0.

Step 6:   At the 1% significance level, the data provide sufficient evidence to conclude that there is a difference in the percentages of Democrats, Republicans, and Independents who thought the U.S. economy was in a recession in May 2008.

19.   (a) Using Minitab, **Stat ▶ Tables ▶ Cross Tabulation and Chi-Square**, enter LOT SIZE in the **For rows:** text box and HOUSE SIZE in the **For columns:** text box, click on the **Counts** box under **Display** to check it. Click **OK**. The results are

```
Rows: LOT SIZE    Columns: HOUSE SIZE

        H1  H2  H3  All

L1      18   3   6   27
L2       6   4   1   11
L3       2   2   1    5
L4       2   2   1    5
All     28  11   9   48
```

(b) Repeat the steps above, except click on the **Column Percents** box under **Display**. Click **OK**. The results are

Copyright © 2012 Pearson Education, Inc. Publishing as Addison-Wesley.

```
Rows: LOT SIZE    Columns: HOUSE SIZE

            H1       H2       H3      All
   L1     64.29    27.27    66.67    56.25
   L2     21.43    36.36    11.11    22.92
   L3      7.14    18.18    11.11    10.42
   L4      7.14    18.18    11.11    10.42
   All   100.00   100.00   100.00   100.00
```

The marginal distribution for lot size is given in the last column above.

(c) Repeat the steps above, except click on the **Row Percents** box under **Display**. Click **OK**. The results are

```
Rows: LOT SIZE    Columns: HOUSE SIZE

            H1       H2       H3      All
   L1     66.67    11.11    22.22   100.00
   L2     54.55    36.36     9.09   100.00
   L3     40.00    40.00    20.00   100.00
   L4     40.00    40.00    20.00   100.00
   All    58.33    22.92    18.75   100.00
```

The marginal distribution for House Size is given in the last row above.

(d) By looking at the distributions, you could say that an association exists between the lot size and the house size, because the conditional distributions of house size by lot size are different. The assumptions are not met to conduct a chi-square independence test because too many of the expected counts are less than 5.

20. Using Minitab, **Stat ▶ Tables ▶ Cross Tabulation and Chi-Square**, enter REGION in the **For rows:** text box and AGE in the **For columns:** text box, click on the **Counts** box under **Display** to check it. Click on the Chi-square button, check the box for Chi-Square analysis and click **OK** twice. The results are

```
Rows: OPINION    Columns: EDUCATION

                                        Not
                                         HS
              College    Some
                grad   college  HS grad  grad    All

Favor           264      205      461     290    1220
Oppose           17       26       81      81     205
No opinion        6        7       34      56     103
All             287      238      576     427    1528

Cell Contents:      Count


Pearson Chi-Square = 77.837, DF = 6, P-Value = 0.000
Likelihood Ratio Chi-Square = 79.327, DF = 6, P-Value = 0.000
```

Since the Pearson Chi-Square P-value = 0.000, which is less than the significance level 0.01, reject the null hypothesis. The data provide sufficient evidence to conclude that opinion on this issue and educational level are associated.

Copyright © 2012 Pearson Education, Inc. Publishing as Addison-Wesley.

CHAPTER 13 ANSWERS

Exercises 13.1

13.1 We state the two numbers of degrees of freedom.

13.3 $F_{0.05}$, $F_{0.025}$, F_α

13.5 (a) The first number in parentheses—12—is the number of degrees of freedom for the numerator.

(b) The second number in parentheses—7—is the number of degrees of freedom for the denominator.

13.7 (a) $F_{0.05} = 1.89$ (b) $F_{0.01} = 2.47$ (c) $F_{0.025} = 2.14$

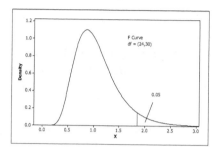 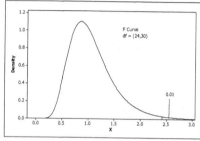 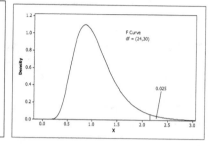

13.9 (a) $F_{0.01} = 2.88$ (b) $F_{0.05} = 2.10$ (c) $F_{0.10} = 1.78$

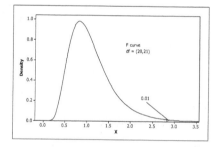 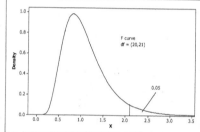 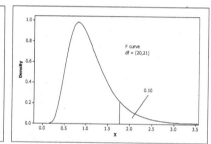

13.11 A straightforward method for finding $F_{0.05}$ for df = (25, 20) using Table VIII is by interpolation. Specifically, proceed along the top row of the table until locating the numbers 24 and 30, between which is 25, the desired number of degrees of freedom for the numerator. Notice that 25 is not reported in this row but is 1/6 of the way in moving from 24 to 30; i.e., (25 − 24)/(30 − 24) = 1/6.

The outside columns of the table give the degrees of freedom for the denominator, which is 20 in this case. Go down either of the outside columns to the row labeled "20." Then go across that row until you are under the columns headed "24" and "30." The numbers in the body of the table intersecting these row and column positions are the F-values 2.08 and 2.04. Thus, 24 is matched with 2.08, and 30 is matched with 2.04.

Now, proceed with interpolation. With 25 known to be 1/6 of the distance between 24 and 30, we likewise want to find the F-value that is 1/6 of the distance between 2.08 and 2.04. We do this by taking (1/6) of (2.04 − 2.08) = −0.0066. Adding −0.0066 to 2.08 gives us the desired F-value, which is 2.0734, or 2.07.

Thus, an approximation to $F_{0.05}$ for df = (25, 20) using Table VIII is 2.07. The limitation of this approximation is that it uses a linear technique on values that are related in a nonlinear fashion. However, because the difference between 2.08 and 2.04 is so small in the first place, the approximation errors will likewise be small.

Copyright © 2012 Pearson Education, Inc. Publishing as Addison-Wesley.

Exercises 13.2

13.13 The pooled-t procedure (Procedure 10.1) is a method for comparing the means of two populations. One-way ANOVA is a procedure for comparing the means of several populations. Thus, one-way ANOVA is a generalization of the pooled-t procedure.

13.15 The reason for the word "variance" in the phrase "analysis of variance" is because the analysis-of-variance procedure for comparing means involves analyzing the *variation* in the sample data.

13.17 (a) MSTR (or SSTR) is a statistic that measures the variation among the sample means for a one-way ANOVA.

(b) MSE (or SSE) is a statistic that measures the variation within the samples for a one-way ANOVA.

(c) F = MSTR/MSE is a statistic that compares the variation among the sample means to the variation within the samples.

13.19 The term *one-way* is used because the type of ANOVA compares the means of a variable for populations that result from a classification by *one* other variable, call the factor.

13.21 No, because the variation among the sample means is not large relative to the variation within the samples.

13.23 In a one-way ANOVA, the residual of an observation is the difference between an observation and the mean of the sample containing it.

13.25

$$\bar{x}_1 = 18/3 = 6; \ \bar{x}_2 = 6/3 = 2; \ \bar{x}_3 = 16/4 = 4$$

$$\bar{x} = \sum x_i / n = (18 + 6 + 16)/(3 + 3 + 4) = 40/10 = 4$$

$$(n_1 - 1)s_1^2 = (8-6)^2 + (4-6)^2 + (6-6)^2 = 8$$

$$(n_2 - 1)s_2^2 = (2-2)^2 + (1-2)^2 + (3-2)^2 = 2$$

$$(n_3 - 1)s_3^2 = (4-4)^2 + (3-4)^2 + (6-4)^2 + (3-4)^2 = 6$$

(a) $SSTR = n_1(\bar{x}_1 - \bar{x})^2 + n_2(\bar{x}_2 - \bar{x})^2 + n_3(\bar{x}_3 - \bar{x})^2 = 3(6-4)^2 + 3(2-4)^2 + 4(4-4)^2 = 24$

(b) $MSTR = SSTR/(k-1) = 24/(3-1) = 12$

(c-d)

$$SSE = (n_1 - 1)s_1^2 + (n_2 - 1)s_2^2 + (n_3 - 1)s_3^2 = 8 + 2 + 6 = 16$$

$$MSE = SSE/(n-k) = 16/(10-3) = 2.286$$

(e) $F = MSTR/MSE = 12/2.286 = 5.25$

13.27

$$\bar{x}_1 = 20/4 = 5; \ \bar{x}_2 = 18/3 = 6; \ \bar{x}_3 = 30/5 = 6; \ \bar{x}_4 = 21/5 = 5; \ \bar{x}_5 = 27/3 = 9$$

$$\bar{x} = \sum x_i / n = (20 + 18 + 30 + 21 + 27)/(4 + 3 + 5 + 5 + 3) = 120/20 = 6$$

$$(n_1 - 1)s_1^2 = (7-5)^2 + (4-5)^2 + (5-5)^2 + (4-5)^2 = 6$$

$$(n_2 - 1)s_2^2 = (5-6)^2 + (9-6)^2 + (4-6)^2 = 14$$

$$(n_3 - 1)s_3^2 = (6-6)^2 + (7-6)^2 + (5-6)^2 + (4-6)^2 + (8-6)^2 = 10$$

$$(n_4 - 1)s_4^2 = (3-5)^2 + (7-5)^2 + (7-5)^2 + (4-5)^2 + (4-5)^2 = 14$$

$$(n_5 - 1)s_5^2 = (7-9)^2 + (9-9)^2 + (11-9)^2 = 8$$

$$SSTR = n_1(\bar{x}_1 - \bar{x})^2 + n_2(\bar{x}_2 - \bar{x})^2 + n_3(\bar{x}_3 - \bar{x})^2 + n_4(\bar{x}_4 - \bar{x})^2 + n_5(\bar{x}_5 - \bar{x})^2$$

(a)
$$= 4(5-6)^2 + 3(6-6)^2 + 5(6-6)^2 + 5(5-6)^2 + 3(9-6)^2 = 36$$

(b) $MSTR = SSTR/(k-1) = 36/(5-1) = 9$

Copyright © 2012 Pearson Education, Inc. Publishing as Addison-Wesley.

(c-d)

$$SSE = (n_1 - 1)s_1^2 + (n_2 - 1)s_2^2 + (n_3 - 1)s_3^2 + (n_4 - 1)s_4^2 + (n_5 - 1)s_5^2$$
$$= 6 + 14 + 10 + 14 + 8 = 52$$

$$MSE = SSE/(n-k) = 52/(20-5) = 3.467$$

(e) $F = MSTR/MSE = 9/3.467 = 2.600$

13.29

$$\bar{x}_1 = 24/3 = 8; \quad \bar{x}_2 = 15/3 = 5; \quad \bar{x}_3 = 36/3 = 12; \quad \bar{x}_4 = 9/3 = 3$$

$$\bar{x} = \sum x_i / n = (24 + 15 + 36 + 9)/(3 + 3 + 3 + 3) = 84/12 = 7$$

$$(n_1 - 1)s_1^2 = (11-8)^2 + (6-8)^2 + (7-8)^2 = 14$$

$$(n_2 - 1)s_2^2 = (9-5)^2 + (2-5)^2 + (4-5)^2 = 26$$

$$(n_3 - 1)s_3^2 = (16-12)^2 + (10-12)^2 + (10-12)^2 = 24$$

$$(n_4 - 1)s_4^2 = (5-3)^2 + (1-3)^2 + (3-3)^2 = 8$$

(a)
$$SSTR = n_1(\bar{x}_1 - \bar{x})^2 + n_2(\bar{x}_2 - \bar{x})^2 + n_3(\bar{x}_3 - \bar{x})^2 + n_4(\bar{x}_4 - \bar{x})^2 + n_5(\bar{x}_5 - \bar{x})^2$$
$$= 3(8-7)^2 + 3(5-7)^2 + 3(12-7)^2 + 3(3-7)^2 = 138$$

(b) $MSTR = SSTR/(k-1) = 138/(4-1) = 46$

(c-d)

$$SSE = (n_1 - 1)s_1^2 + (n_2 - 1)s_2^2 + (n_3 - 1)s_3^2 + (n_4 - 1)s_4^2 + (n_5 - 1)s_5^2$$
$$= 14 + 26 + 24 + 8 = 72$$

$$MSE = SSE/(n-k) = 72/(12-4) = 9$$

(e) $F = MSTR/MSE = 46/9 = 5.11$

13.31 If the hypothesis test situation abides by the characteristics outlined in this exercise, we can use the pooled-t test or we can use the one-way ANOVA test discussed in this section using k = 2.

Exercises 13.3

13.33 A small value of F results when MSTR is small compared to MSE, i.e., when the variation between sample means is small compared to the variation within samples. This describes what should happen when the null hypothesis is true; thus it does not comprise evidence that the null hypothesis is false. Only when the variation between sample means is large compared to the variation within samples, i.e., when F is large, do we have evidence that the null hypothesis is not true.

13.35 SST = SSTR + SSE. This means that the total variation can be partitioned into a component representing variation among the sample means and another component representing variation within the samples.

13.37 (a) One-way ANOVA

(b) Two-way ANOVA

13.39 Since 0.708 = 2.124/df(Treatment), we have df(Treatment) = 3
It follows that df (Total) = 3 + 20 = 23.
Since 0.75 = 0.708/MSE, we have MSE = 0.708/0.75 = 0.944.
Then since 0.944 = SSE/20, we have SSE = 0.944(20) = 18.88.
Finally SST = SSTR + SSE = 2.124 + 18.880 = 21.004
Thus, the completed table is

Copyright © 2012 Pearson Education, Inc. Publishing as Addison-Wesley.

Source	df	SS	MS=SS/df	F-statistic
Treatment	3	2.124	0.708	0.75
Error	20	18.880	0.944	
Total	23	21.004		

13.41 Since the Total df is 16, the df for Treatment is 14 - 12 = 2.
Since MSTR = SSTR/2, SSTR = 2(MSTR) = 2(1.4) = 2.8
Similarly, SSE = 12(MSE) = 12(0.9) = 10.8
Now SST = SSTR + SSE = 2.8 + 10.8 = 13.6
Finally, F = MSTR/MSE = 1.4/0.9 = 1.556
Thus, the completed table is

Source	df	SS	MS=SS/df	F-statistic
Treatment	2	2.8	1.4	1.556
Error	12	10.8	0.9	
Total	14	13.6		

13.43 We have the following:
$k = 3$
$n_1 = 3 \qquad n_2 = 3, \qquad n_3 = 4$
$T_1 = 18 \qquad T_2 = 6 \qquad T_3 = 16$
$n = \sum n_j = 3 + 3 + 4 = 10$
$\sum x_i = \sum T_j = 18 + 6 + 16 = 40$
$\sum x_i^2 = (8)^2 + (4)^2 + (6)^2 + ... + (6)^2 + (3)^2 = 200$

(a) Consequently,

$$SST = \sum x_i^2 - \left(\sum x_i\right)^2 / n = 200 - 40^2/10 = 200 - 160 = 40$$

$$SSTR = \sum(T_j^2 / n_j) - \left(\sum x_i\right)^2 / n = (18)^2/3 + (6)^2/3 + (16)^2/4 - (40)^2/10 = 184 - 160 = 24$$

$$SSE = SST - SSTR = 40 - 24 = 16$$

(b) The results are the same as in Exercise 13.25.

(c)

Source	df	SS	MS=SS/df	F-statistic
Treatment	2	24	12.00	5.25
Error	7	16	2.29	
Total	9	40		

(d) The df are (2,7). The critical value is 4.74. The P-value is 0.025 <
P < 0.50. Since 5.25 > 4.74 and since the p-value is less than the
significance level, we reject the null hypothesis that the population
means are equal.

13.45 We have the following:
$k = 5$
$n_1 = 4 \qquad n_2 = 3 \qquad n_3 = 5 \qquad n_4 = 5 \qquad n_5 = 3$
$T_1 = 20 \qquad T_2 = 18 \qquad T_3 = 30 \qquad T_4 = 25 \qquad T_5 = 27$
$n = \sum n_j = 4 + 3 + 5 + 5 + 3 = 20$
$\sum x_i = \sum T_j = 20 + 18 + 30 + 25 + 27 = 120$
$\sum x_i^2 = (7)^2 + (4)^2 + (5)^2 + ... + (9)^2 + (11)^2 = 808$

Copyright © 2012 Pearson Education, Inc. Publishing as Addison-Wesley.

(a) Consequently,

$$SST = \sum x_i^2 - \left(\sum x_i\right)^2 / n = 808 - 120^2 / 20 = 808 - 720 = 88$$

$$SSTR = \sum (T_j^2 / n_j) - \left(\sum x_i\right)^2 / n$$

$$= (20)^2 / 4 + (18)^2 / 3 + (30)^2 / 5 + (25)^2 / 5 + (27)^2 / 3 - (120)^2 / 20 = 756 - 720 = 36$$

$$SSE = SST - SSTR = 88 - 36 = 52$$

(b) The results are the same as in Exercise 13.27.

(c)

Source	df	SS	MS=SS/df	F-statistic
Treatment	4	36	9.00	2.60
Error	15	52	3.47	
Total	19	88		

(d) The df are (4,15). The critical value is 3.06. The P-value is 0.05 < P < 0.10. Since 2.60 > 3.06 and since the P-value is greater than the significance level, we do not reject the null hypothesis that the population means are equal.

13.47 We have the following:

k = 4

$n_1 = 3$ $n_2 = 3$ $n_3 = 3$ $n_4 = 3$
$T_1 = 24$ $T_2 = 15$ $T_3 = 36$ $T_4 = 9$

$n = \sum n_j = 3 + 3 + 3 + 3 = 12$

$\sum x_i = \sum T_j = 24 + 15 + 36 + 9 = 84$

$\sum x_i^2 = (11)^2 + (6)^2 + (7)^2 + \dots + (1)^2 + (3)^2 = 798$

(a) Consequently,

$$SST = \sum x_i^2 - \left(\sum x_i\right)^2 / n = 798 - 84^2 / 12 = 798 - 588 = 210$$

$$SSTR = \sum (T_j^2 / n_j) - \left(\sum x_i\right)^2 / n$$

$$= (24)^2 / 3 + (15)^2 / 3 + (36)^2 / 3 + (9)^2 / 3 - (84)^2 / 12 = 726 - 588 = 138$$

$$SSE = SST - SSTR = 210 - 138 = 72$$

(b) The results are the same as in Exercise 13.29.

(c)

Source	df	SS	MS=SS/df	F-statistic
Treatment	3	138	46.0	5.11
Error	8	72	9.0	
Total	11	210		

(d) The df are (3,8). The critical value is 4.07. The P-value is 0.025 < P < 0.05. Since 5.11 > 4.07 and since the P-value is less than the significance level, we reject the null hypothesis that the population means are equal.

13.49 The total number of populations being sampled is k = 3. Let the subscripts 1, 2, and 3 refer to diatoms, bacteria, and macroalgae, respectively. The total number of observations is n = 12. Also, $n_1 = 4$, $n_2 = 4$, and $n_3 = 4$.

Step 1: $H_0: \mu_1 = \mu_2 = \mu_3$ (population means are equal)

H_a: population means are different

Step 2: $\alpha = 0.05$

Copyright © 2012 Pearson Education, Inc. Publishing as Addison-Wesley.

Step 3: Let T_1, T_2 and T_3 refer to the sum of the data values in each of the three samples, respectively. Thus:

$T_1 = 1828$ $T_2 = 1225$ $T_3 = 1175$.

Also, the sum of all the data values is $\sum x_i = 4228$, and their sum of squares is $\sum x_i^2 = 1,561,154$. Then,

$$SST = \sum x_i^2 - \left(\sum x_i\right)^2 / n = 1561.154 - 4228^2 / 12 = 71488.67$$

$$SSTR = \frac{T_1^2}{n_1} + \frac{T_2^2}{n_2} + \frac{T_3^2}{n_3} - \frac{\left(\sum x_i\right)^2}{n} = \frac{1828^2}{4} + \frac{1225^2}{4} + \frac{1175^2}{4} - \frac{4228^2}{12} = 66043.17$$

and SSE = SST - SSTR = 71488.67 - 66043.17 = 5445.50

MSTR = SSTR/(k - 1) = 66043.17/(3 - 1) = 33021.585

MSE = SSE/(n - k) = 5445.50/(12 - 3) = 605.056

F = MSTR/MSE = 33021.585/605.056 = 54.98

The one-way ANOVA table is

Source	Df	SS	MS = SS/df	F-statistic
Treatment	2	66043.17	33021.585	54.58
Error	9	5445.50	605.056	
Total	11	71488.67		

Step 4: df = (k - 1, n - k) = (2, 9); critical value = 4.26
For the p-value approach, P(F > 54.58) < 0.005

Step 5: Since 54.58 > 4.26, we reject the null hypothesis. Since P-value < 0.05, we reject the null hypothesis.

Step 6: At the 5% significance level, we reject the null hypothesis and conclude that there is a difference in the mean number of copepods among the three different diets.

13.51 The number of populations being sampled is k = 5. Let the subscripts 1-5 refer to strains A-E, respectively, The total number observations is n = 25. Also, $n_1 = n_2 = n_3 = n_4 = n_5 = 5$.

Step 1: H_0: $\mu_1 = \mu_2 = \mu_3 = \mu_4 = \mu_5$ (population means are equal)

H_a: Not all of the population means are equal

Step 2: $\alpha = 0.05$

Step 3: Let $T_1 - T_5$ refer to the sum of the data values in each of the five samples, respectively. Thus,

$T_1 = 104$ $T_2 = 129$ $T_3 = 185$ $T_4 = 98$ $T_5 = 194$.

Also, the sum of all the data values is $\sum x_i = 710$, and their sum of squares is $\sum x_i^2 = 25424$.

Consequently, $SST = \sum x_i^2 - \left(\sum x_i\right)^2 / n = 25424 - 710^2 / 25 = 5260$

$$SSTR = \frac{T_1^2}{n_1} + \frac{T_2^2}{n_2} + \frac{T_3^2}{n_3} + \frac{T_4^2}{n_4} + \frac{T_5^2}{n_5} - \frac{\left(\sum x_i\right)^2}{n}$$

$$= \frac{104^2}{5} + \frac{129^2}{5} + \frac{185^2}{5} + \frac{98^2}{5} + \frac{194^2}{5} - \frac{710^2}{25} = 1620$$

and SSE = SST - SSTR = 5260 - 1620 = 3640.

MSTR = SSTR/(k - 1) = 1620/(5 - 1) = 405

MSE = SSE/(n - k) = 3640/(25 - 5) = 182

F = MSTR/MSE = 405/182 = 2.23

Copyright © 2012 Pearson Education, Inc. Publishing as Addison-Wesley.

The one-way ANOVA table is:

Source	df	SS	MS = SS/df	F-statistic
Treatment	4	1620	405	2.23
Error	20	3640	182	
Total	24	5260		

Step 4: df = (k - 1, n - k) = (4, 20); critical value = 2.87
For the P-value approach, $P(F > 2.23) > 0.10$.

Step 5: From Step 3, F = 2.23. From Step 4, $F_{0.05} = 2.87$. Since 2.23 < 2.87, do not reject H_0.
Since the P-value is larger than the significance level, do not reject H_0.

Step 6: At the 5% significance level, the data do not provide sufficient evidence to conclude that the population means are different.

13.53 The total number of populations being sampled is k = 3. Let the subscripts 1, 2, and 3 refer to 2000, 2001, and 2002, respectively. The total number of observations is n = 34. Also, $n_1 = 11$, $n_2 = 13$, $n_3 = 10$.

Step 1: H_0: $\mu_1 = \mu_2 = \mu_3$ (mean resistance to *Stagonospora nororum* among the three years of wheat harvests are equal)
H_a: Not all the means are equal.

Step 2: $\alpha = 0.05$

Step 3: Let T_1, T_2, and T_3, refer to the sum of the data values in each of the four samples, respectively. Thus,
$T_1 = 55.0$ $T_2 = 84.2$ $T_3 = 46.0$.
Also, the sum of all the data values is $\sum x_i = 185.2$, and their sum of squares is $\sum x_i^2 = 1108.48$.

Consequently, $SST = \sum x_i^2 - (\sum x_i)^2 / n = 1108.48 - 185.2^2 / 34 = 99.685$

$$SSTR = \frac{T_1^2}{n_1} + \frac{T_2^2}{n_2} + \frac{T_3^2}{n_3} + \frac{T_4^2}{n_4} - \frac{(\sum x_i)^2}{n}$$

$$= \frac{55^2}{11} + \frac{84.2^2}{13} + \frac{46^2}{10} - \frac{185.2^2}{34} = 23.162$$

and SSE = SST - SSTR = 99.685 - 23.162 = 76.523
MSTR = SSTR/(k - 1) = 23.162/(3 - 1) = 11.581
MSE = SSE/(n - k) = 76.523/(34 - 3) = 2.468
F = MSTR/MSE = 11.581/2.468 = 4.692

The one-way ANOVA table is:

Source	df	SS	MS = SS/df	F-statistic
Treatment	2	23.162	11.581	4.692
Error	31	76.523	2.468	
Total	33	99.685		

Step 4: df = (k - 1, n - k) = (2, 31); critical value = 3.32
For the p-value approach, $0.01 < P(F > 4.692) < 0.025$.

Step 5: From Step 3, F = 4.692. From Step 4, $F_{0.05} = 3.32$. Since 4.692 > 3.32, reject H_0.
Since the p-value is smaller than the significance level, reject H_0.

Copyright © 2012 Pearson Education, Inc. Publishing as Addison-Wesley.

Step 6:   At the 5% significance level, the data provide sufficient evidence to conclude that a difference exists in mean resistance to *Stagonospora nororum* among the three years of wheat harvests.

13.55 (a) To carry out the Analysis of variance and residual analysis, we will use the data in a single column, which is named RENT. In the next column, which is named REGION, are the names of each of the regions corresponding to each data value in RENT. The residual analysis consists of plotting the residuals against the means (or fits) and constructing a normal probability plot. This is done at the same time as the analysis of variance in Minitab. To do this, we choose **Stat ▶**

Anova ▶One-way..., select <u>RENT</u> for the **Response:** text box and <u>REGION</u> in the **Factor:** text box. Then click on the **Graphs** button and click to place check marks in the boxes for **Normal plot of residuals** and **Residuals versus fits**, and click **OK**. The printed results are

One-way ANOVA: RENT versus REGION

```
Source   DF      SS      MS     F      P
REGION    3   400513  133504  7.54  0.002
Error    16   283265   17704
Total    19   683778

S = 133.1   R-Sq = 58.57%   R-Sq(adj) = 50.81%
```

```
                              Individual 99% CIs For Mean Based on
                              Pooled StDev
Level       N    Mean   StDev   -+---------+---------+---------+--------
Midwest     6   743.0    92.1   (-------*-------)
Northeast   5  1055.2   150.2                    (--------*-------)
South       4   854.5   168.0        (---------*--------)
West        5  1064.4   128.4                   (-------*--------)
                                -+---------+---------+---------+--------
                               600       800      1000      1200
```

Pooled StDev = 133.1

(b) The F-value of 7.54 is significant at the 0.05 level (P-Value = 0.002). If the assumptions of normal populations and equal standard deviations are met, we reject the null hypothesis of equal population means and conclude that there is a difference in the mean monthly rents among newly completed apartments in the four U.S. regions.

(c) The graphs produced by the procedure in part (a) are shown following.

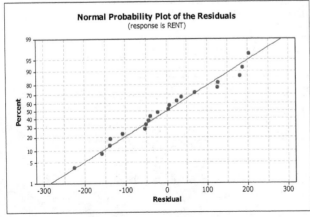

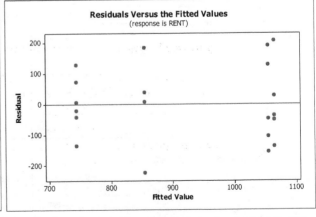

From the printed output, the ratio of the largest standard deviation to the smallest is 168.0/92.1 = 1.82. This is less than 2, indicating

Copyright © 2012 Pearson Education, Inc. Publishing as Addison-Wesley.

that the rule of 2 is not violated, leading us to conclude that the assumption of equal standard deviations is reasonable. This can also be seen in the right-hand graph above. There are no outliers in the normal probability plot and the points are linear, indicating that there are no problems with the normality assumption as well. [A Ryan-Joiner normality test of the residuals has a P-value of >0.10, leading us to conclude that the normality assumption is reasonable.]

13.57 (a) To carry out the Analysis of variance and residual analysis, we will use the data in the single column named RATE. In the next column, which is named BREAST, are names of each of the three feather treatments corresponding to each data value in RATE. The residual analysis consists of plotting the residuals against the means (or fits) and constructing a normal probability plot. This is done at the same time as the analysis of variance in Minitab. To do this, we choose

Stat ▶ Anova ▶One-way..., select RATE for the **Response:** text box and BREAST in the **Factor:** text box. Enter 99 in the **Confidence level** box. Then click on the **Graphs** button and click to place check marks in the boxes for **Normal plot of residuals** and **Residuals versus fits**, and click **OK.** The printed results are

One-way ANOVA: RATE versus BREAST

```
Source  DF      SS      MS      F       P
BREAST   2   960.3   480.1   6.09   0.008
Error   21  1655.3    78.8
Total   23  2615.6

S = 8.878   R-Sq = 36.71%   R-Sq(adj) = 30.69%
```

```
                                   Individual 99% CIs For Mean Based on
                                   Pooled StDev
Level      N    Mean    StDev   ----+---------+---------+---------+-----
Control    8  14.838    6.556                (--------*--------)
Enlarged   8  19.775   13.048                    (--------*--------)
Reduced    8   4.588    4.821   (--------*-------)
                                ----+---------+---------+---------+-----
                                    0        10        20        30

Pooled StDev = 8.878
```

(b) The F-value of 6.09 is significant at the 0.01 level (P-Value = 0.008). If the assumptions of normal populations and equal standard deviations are met, we reject the null hypothesis of equal population means. There is sufficient evidence to conclude that there is a difference in the mean singing rates among male rock sparrows exposed to the three types of breast treatments.

(c) The graphical results are

Copyright © 2012 Pearson Education, Inc. Publishing as Addison-Wesley.

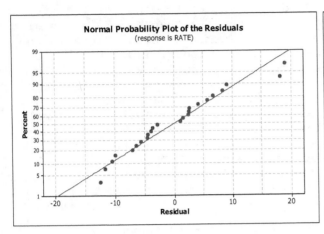

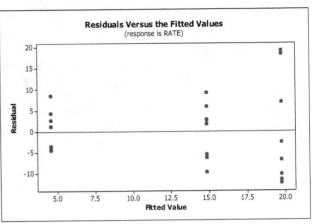

From the printed output, the ratio of the largest standard deviation to the smallest is $13.048/4.821 = 2.71$. This is greater than 2, indicating that the rule of 2 is violated and leading us to conclude that the assumption of equal standard deviations may not be reasonable. The small sample sizes may be partly responsible for this occurrence, which can also be seen the right-hand graph above. The points in the probability plot are linear, indicating no problems with the normality assumption. [A Ryan-Joiner normality test of the residuals has a P-value of >0.10, leading us to conclude that the normality assumption is reasonable.]

13.59 (a) To carry out the analysis of variance and residual analysis, we will use the data in the single column named VOLUME. In the next column, which is named MATERIAL, are names of each of the three feather treatments corresponding to each data value in VOLUME. The residual analysis consists of plotting the residuals against the means (or fits) and constructing a normal probability plot. This is done at the same time as the analysis of variance in Minitab. To do this, we choose

Stat ▶ Anova ▶One-way..., select <u>VOLUME</u> for the **Response:** text box and <u>MATERIAL</u> in the **Factor:** text box. Enter <u>95</u> in the **Confidence level** box. Then click on the **Graphs** button and click to place check marks in the boxes for **Normal plot of residuals** and **Residuals versus fits,** and click **OK.** The printed results are

One-way ANOVA: HARDNESS versus MATERIAL

```
Source    DF      SS      MS      F       P
MATERIAL   2   805.31  402.66  114.71  0.000
Error     15    52.65    3.51
Total     17   857.97

S = 1.874   R-Sq = 93.86%   R-Sq(adj) = 93.04%
```

```
                                  Individual 95% CIs For Mean Based on
                                  Pooled StDev
Level       N    Mean   StDev    -----+---------+---------+---------+----
Duracross   6  39.633   3.120                                  (--*---)
Duradent    6  24.017   0.615    (--*--)
Endura      6  27.533   0.647           (--*--)
                                  -----+---------+---------+---------+----
                                     25.0      30.0      35.0      40.0

Pooled StDev = 1.874
```

Copyright © 2012 Pearson Education, Inc. Publishing as Addison-Wesley.

(b) The F-value of 114.71 is significant at the 0.05 level (P-Value = 0.000). If the assumptions of normal populations and equal standard deviations are met, we reject the null hypothesis of equal population means. There is sufficient evidence to conclude that there is a difference in the mean hardness among the three materials for making artificial teeth.

(c) The graphical results are

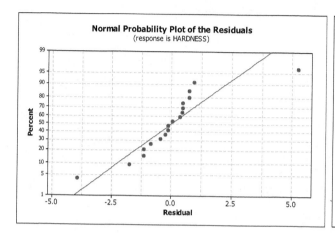
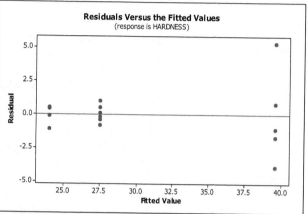

From the printed output, the ratio of the largest standard deviation to the smallest is 3.120/0.615 = 5.07. This is greater than 2. The rule of 2 is violated, leading us to conclude that the assumption of equal standard deviations may not be reasonable. The small sample sizes may be partly responsible for this occurrence, which can also be seen the right-hand graph above. The points in the probability plot are not linear, indicating that there may also be problems with the normality assumption. [A Ryan-Joiner normality test of the residuals has a P-value of < 0.010, so we conclude that the normality assumption is not reasonable.] Although the F value is so large that it is unlikely that either violation would have any serious effect on our conclusions, these data should probably be analyzed with a more robust method.

13.61 The total number of populations being sampled is k = 3. The total number of observations is n = 300. Also, $n_1 = 90$, $n_2 = 17$, $n_3 = 193$.

Step 1: H_0: $\mu_1 = \mu_2 = \mu_3$ (mean IQ for the three groups are the same)

H_a: Not all the means are equal.

Step 2: $\alpha = 0.01$

Step 3: We can determine the mean of all the observations by the following

$$\bar{x} = \frac{n_1\bar{x}_1 + n_2\bar{x}_2 + n_3\bar{x}_3}{n_1 + n_2 + n_3} = \frac{90(92.8) + 17(94.8) + 193(103.7)}{90 + 17 + 193} = 99.926$$

Then, use the defining formulas to get the SSTR and SSE.

$$SSTR = \sum n_i(\bar{x}_i - \bar{x})^2$$

$$= 90(92.8 - 99.926)^2 + 17(94.8 - 99.926)^2 + 193(103.7 - 99.926)^2$$

$$= 7765.792$$

$$SSE = \sum (n_i - 1)s_i^2 = 89(15.2)^2 + 16(19.0)^2 + 192(15.3)^2$$

$$= 71283.840$$

and SST = SSE + SSTR = 71283.840 + 7765.792 = 79049.632

MSTR = SSTR/(k - 1) = 7765.792/(3 - 1) = 3882.896

MSE = SSE/(n - k) = 71283.840/(300 - 3) = 240.013

F = MSTR/MSE = 3882.896/240.013 = 16.178

Copyright © 2012 Pearson Education, Inc. Publishing as Addison-Wesley.

The one-way ANOVA table is:

Source	df	SS	MS = SS/df	F- statistic
Treatment	2	7765.792	3882.896	16.178
Error	297	71283.840	240.013	
Total	299	79049.632		

Step 4: $df = (k - 1, n - k) = (2, 297)$; From the table in the problem, critical value = 4.68
For the p-value approach, $P(F > 16.178) < 0.005$.

Step 5: From Step 3, $F = 16.178$. From Step 4, $F_{0.01} = 4.68$. Since $16.178 > 4.68$, reject H_0.
Since the p-value is smaller than the significance level, reject H_0.

Step 6: At the 1% significance level, the data provide sufficient evidence to conclude that a difference exists in mean IQ at age 7 ½ - 8 years for preterm children among the three groups.

13.63 The total number of populations being sampled is $k = 6$. The total number of observations is $n = 160$.

Step 1: H_0: $\mu_1 = \mu_2 = \mu_3 = \mu_4 = \mu_5 = \mu_6$ (mean starting salaries are the same)
H_a: Not all the means are equal.

Step 2: $\alpha = 0.01$

Step 3: We can determine the mean of all the observations by the following

$$\overline{x} = \frac{n_1\overline{x}_1 + n_2\overline{x}_2 + n_3\overline{x}_3 + n_4\overline{x}_4 + n_5\overline{x}_5 + n_5\overline{x}_5}{n_1 + n_2 + n_3 + n_4 + n_5 + n_6}$$

$$= \frac{46(56.5) + 11(52.4) + 30(35.9) + 11(43.7) + 44(59.1) + 18(48.9)}{46 + 11 + 30 + 11 + 44 + 18} = 51.336$$

Then, use the defining formulas to get the SSTR and SSE.

$$SSTR = \sum n_i(\overline{x}_i - \overline{x})^2$$

$$= 46(56.5 - 51.336)^2 + 11(52.4 - 51.336)^2 + 30(35.9 - 51.336)^2$$

$$+ 11(43.7 - 51.336)^2 + 44(59.1 - 51.336)^2 + 18(48.9 - 51.336)^2$$

$$= 11787.747$$

$$SSE = \sum (n_i - 1)s_i^2 = 45(5.6)^2 + 10(4.7)^2 + 29(4.0)^2 + 10(5.0)^2 + 43(5.7)^2 + 17(4.8)^2$$

$$= 4134.850$$

and $SST = SSE + SSTR = 4134.850 + 11787.747 = 15922.597$
$MSTR = SSTR/(k - 1) = 11787.747/(6 - 1) = 2357.549$
$MSE = SSE/(n - k) = 4134.850/(160 - 6) = 26.850$
$F = MSTR/MSE = 2357.549/26.850 \approx 87.804$

The one-way ANOVA table is:

Source	df	SS	MS = SS/df	F- statistic
Treatment	5	11787.747	2357.549	87.804
Error	154	4134.850	26.850	
Total	159	15922.597		

Step 4: $df = (k - 1, n - k) = (5, 154)$; From the table in the problem, critical value = 3.14
For the p-value approach, $P(F > 87.804) < 0.005$.

Copyright © 2012 Pearson Education, Inc. Publishing as Addison-Wesley.

Step 5: From Step 3, F = 87.804. From Step 4, $F_{0.01}$ = 3.14. Since 87.804 > 3.14, reject H_0.
Since the p-value is smaller than the significance level, reject H_0.

Step 6: At the 1% significance level, the data provide sufficient evidence to conclude that a difference exists in mean starting salaries among bachelors degree candidates in the six fields.

13.65 (a) We will obtain the plots and standard deviations using the normal probability plot in Minitab. Choose **Graph ▶ Probability plot...**, select the **Simple** version, enter WEIGHT Lahna, WEIGHT Siika, WEIGHT Saerki, and WEIGHT Parkki in the **Variables** text box, Click on the **Multiple graphs** button and select **In separate panels of the same graph**, and click **OK**. Repeat this process with LENGTH Lahna, LENGTH Siika, LENGTH Saerki, and LENGTH Parkki. The two sets of graphs are

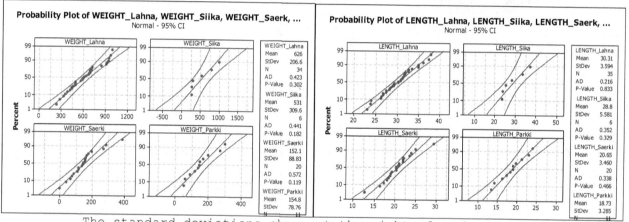

The standard deviations shown at the right of each set of graphs are for the weights: 206.6, 309.6, 88.83, and 78.76. For the lengths they are 3.594, 5.581, 3.460, and 3.285.

(b) To carry out the residual analysis (and possibly the ANOVA), we will use the data in the single columns named WEIGHT and LENGTH. In the next column, which is named SPECIES, are names of each of the four species corresponding to each data value in WEIGHT and LENGTH. The residual analysis consists of plotting the residuals against the means (or fits) and constructing a normal probability plot. This is done at the same time as the analysis of variance in Minitab. To do this, we choose **Stat ▶ Anova ▶ One-way...**, select WEIGHT for the **Response:** text box and SPECIES in the **Factor:** text box. Enter 95 in the **Confidence level** box. Then click on the **Graphs** button and click to place check marks in the boxes for **Normal plot of residuals** and **Residuals versus fits**, and click **OK** twice. Repeat the process for LENGTH. The residual analysis results for WEIGHT and LENGTH are

Copyright © 2012 Pearson Education, Inc. Publishing as Addison-Wesley.

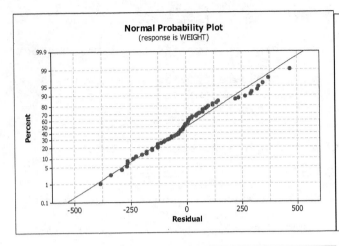

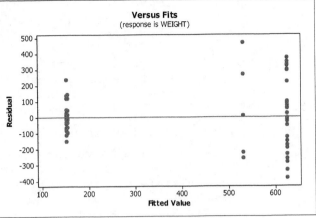

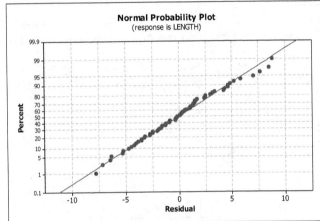

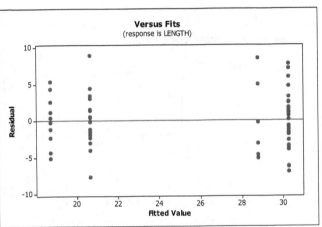

(c) For the weights, the ratio of the largest to smallest standard deviation is 309.6/78.76 = 3.93. The rule of 2 is violated for the weights. Conducting a one-way ANOVA test on the weights data does not appear to be reasonable. Parts (d)-(e) are omitted for the weights. For the lengths, the ratio of the largest to smallest standard deviation is 5.581/3.285 = 1.70. The rule of 2 is not violated for the lengths. The normal probability plots of the individual samples and of the residuals are approximately linear in every plot. Conducting a one-way ANOVA test on these data does appear to be reasonable. Parts (d)-(e) will be performed for the lenths data.

(d) The printed portion of the output from the procedure in part (b) is

One-way ANOVA: LENGTH versus SPECIES

```
Source    DF       SS      MS      F      P
SPECIES    3   1845.9   615.3  44.98  0.000
Error     68    930.2    13.7
Total     71   2776.1

S = 3.699   R-Sq = 66.49%   R-Sq(adj) = 65.02%

                               Individual 95% CIs For Mean Based on
                               Pooled StDev
Level     N    Mean   StDev  ---------+---------+---------+---------+
Lahna    35  30.306   3.594                              (--*--)
Siika     6  28.800   5.581                        (-------*-------)
Saerki   20  20.645   3.460        (----*---)
Parkki   11  18.727   3.285  (-----*----)
                             ---------+---------+---------+---------+
                                 20.0      24.0      28.0      32.0
```

Copyright © 2012 Pearson Education, Inc. Publishing as Addison-Wesley.

```
Pooled StDev = 3.699
```
Since the F-value of 44.98 has a P-value of 0.000, we reject the null hypothesis of equal population means.
(e) We conclude that at the 5% significance level, the data provide sufficient evidence that the mean lengths of the four species of fish caught in Lake Laengelmaevesi are different. Note: we determined in part (c) that it was not reasonable to perform the one-way ANOVA analysis on the weight data.

13.67 (a) We will obtain the plots and standard deviations using the normal probability plot in Minitab. Choose **Graph ▶ Probability plot...**, select the **Simple** version, enter Meadow Pipit, Tree Pipit, and Hedge Sparrow in the **Variables** text box, Click on the **Multiple graphs** button and select **In separate panels of the same graph**, and click **OK**. Then repeat with Robin, Pied Wagtail, and Wren. The graphs are

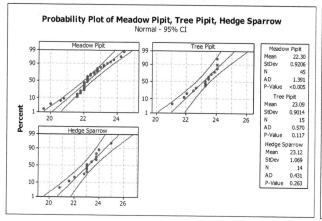 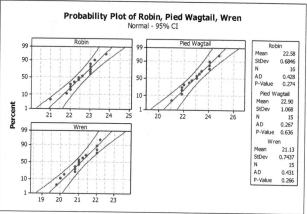

The standard deviations for the six birds are, respectively, 0.9206, 0.9014, 1.069, 0.6846, 1.068, and 0.7437.

(b) To carry out the residual analysis (and possibly the ANOVA), we will use the data in the single column named LENGTH. In the next column, which is named SPECIES, are the names of each of the six species corresponding to each data value in LENGTH. The residual analysis consists of plotting the residuals against the means (or fits) and constructing a normal probability plot. This is done at the same time as the analysis of variance in Minitab. To do this, we choose **Stat ▶ Anova ▶One-way...**, select LENGTH for the **Response:** text box and SPECIES in the **Factor:** text box. Enter 95 in the **Confidence level** box. Then click on the **Graphs** button and click to place check marks in the boxes for **Normal plot of residuals** and **Residuals versus fits**, and click **OK**. The residual analysis results are

Copyright © 2012 Pearson Education, Inc. Publishing as Addison-Wesley.

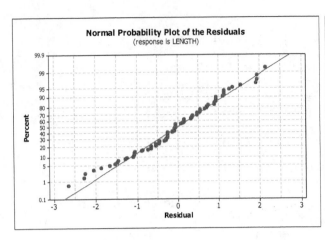

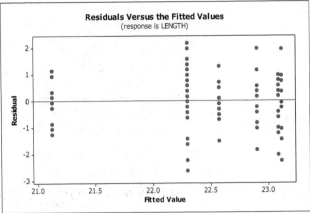

(c) The ratio of the largest to smallest standard deviation is 1.069/0.7437 = 1.44. The rule of 2 is not violated. This is also seen in the graph at the right above. The normal probability plots of the individual samples and of the residuals are all linear except for that of the Meadow Pipit, which appears to have two possible outliers with low values. Conducting a one-way ANOVA test on these data appears to be reasonable, although a more robust alternative procedure may be preferable.

(d) The printed portion of the output from the procedure in part (b) is

One-way ANOVA: LENGTH versus SPECIES

```
Source    DF      SS     MS      F      P
SPECIES    5  42.940  8.588  10.39  0.000
Error    114  94.248  0.827
Total    119 137.188

S = 0.9093   R-Sq = 31.30%   R-Sq(adj) = 28.29%
```

```
                                   Individual 95% CIs For Mean Based on
                                   Pooled StDev
Level           N    Mean   StDev  --+---------+---------+---------+-------
Hedge Sparrow  14  23.121   1.069                        (-----*-----)
Meadow Pipit   45  22.299   0.921             (---*--)
Pied Wagtail   15  22.903   1.068                   (-----*-----)
Robin          16  22.575   0.685                 (----*-----)
Tree Pipit     15  23.090   0.901                      (-----*-----)
Wren           15  21.130   0.744   (-----*-----)
                                   --+---------+---------+---------+-------
                                   20.80     21.60     22.40     23.20
```

```
Pooled StDev = 0.909
```

Since the F-value of 10.39 has a P-value of 0.000, we reject the null hypothesis of equal population means.

(e) At the 5% significance level, we conclude that the data provide sufficient evidence that the mean egg lengths for the eggs laid by cuckoos in the nests of the six species are different.

13.69 (a) We will obtain the plots and standard deviations using the normal

probability plot in Minitab. Choose **Graph ▶ Probability plot...**, select the **Simple** version, enter Law, Science, Medicine, and Technology in the **Variables** text box, Click on the **Multiple graphs** button and select **In separate panels of the same graph**, and click **OK**. The graphs are

Copyright © 2012 Pearson Education, Inc. Publishing as Addison-Wesley.

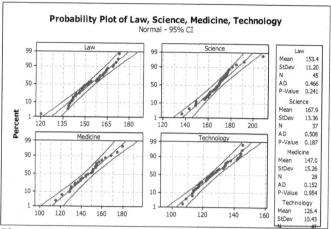

The standard deviations for the four subject areas are, respectively, 11.20, 13.36, 15.26, and 10.43.

(b) To carry out the residual analysis (and possibly the ANOVA), we will use the data in the single column named PRICE. In the next column, which is named SUBJECT, are the names of each of the four subject areas corresponding to each data value in PRICE. The residual analysis consists of plotting the residuals against the means (or fits) and constructing a normal probability plot. This is done at the same time

as the analysis of variance in Minitab. To do this, we choose **Stat** ▶

Anova ▶**One-way...**, select PRICE for the **Response:** text box and SUBJECT in the **Factor:** text box. Enter 95 in the **Confidence level** box. Then click on the **Graphs** button and click to place check marks in the boxes for **Normal plot of residuals** and **Residuals versus fits**, and click **OK**. The residual analysis results are

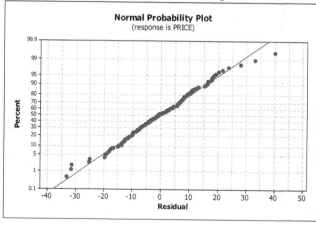
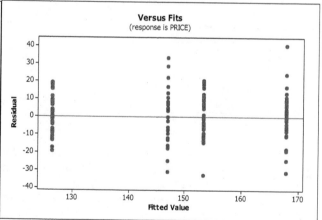

(c) The ratio of the largest to smallest standard deviation is 15.26/10.43 = 1.46. The rule of 2 is not violated. This is also seen in the graph at the right above. The normal probability plots of the individual samples and of the residuals are all linear. Conducting a one-way ANOVA test on these data appears to be reasonable.

(d) The printed portion of the output from the procedure in part (b) is

One-way ANOVA: PRICE versus SUBJECT

Source	DF	SS	MS	F	P
SUBJECT	3	35479	11826	77.13	0.000
Error	148	22692	153		
Total	151	58172			

Copyright © 2012 Pearson Education, Inc. Publishing as Addison-Wesley.

```
          S = 12.38   R-Sq = 60.99%   R-Sq(adj) = 60.20%

                                   Individual 95% CIs For Mean Based on
                                   Pooled StDev
          Level        N   Mean   StDev  --------+---------+---------+---------+-
          Law         45  153.37  11.20                      (-*--)
          Medicine    28  146.97  15.26                (--*--)
          Science     37  167.88  13.36                              (--*--)
          Technology  42  126.43  10.43  (-*--)
                                         --------+---------+---------+---------+-
                                               135       150       165       180

          Pooled StDev = 12.38
```

Since the F-value of 77.13 has a P-value of 0.000, we reject the null hypothesis of equal population means.

(e) At 5% significance, we conclude that the data provide sufficient evidence that the mean prices of hardcover books in the four subject areas are different.

13.71 (a) We will obtain the plots and standard deviations using the normal probability plot in Minitab. Choose **Graph ▶ Probability plot...**, select the **Simple** version, enter HB SC, HB SS, and HB ST in the **Variables** text box, click on the **Multiple graphs** button and select **In separate panels of the same graph**, and click **OK**. The graphs are

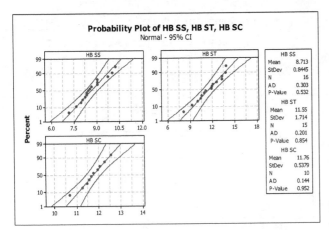

The standard deviations for the three types of sickle cell disease are, respectively, 0.8445, 1.714, and 0.5379.

(b) To carry out the residual analysis (and possibly the ANOVA), we will use the data in the single column named HEMOLEVEL. In the next column, which is named TYPE, are the names of each of the three sickle cell types corresponding to each data value in HEMOLEVEL. The residual analysis consists of plotting the residuals against the means (or fits) and constructing a normal probability plot. This is done at the same time as the analysis of variance in Minitab. To do this, we choose

Stat ▶ Anova ▶One-way..., select HEMOLEVEL for the **Response:** text box and TYPE in the **Factor:** text box. Enter 95 in the **Confidence level** box. Then click on the **Graphs** button and click to place check marks in the boxes for **Normal plot of residuals** and **Residuals versus fits**, and click **OK**. The residual analysis results are

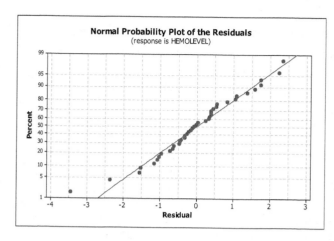

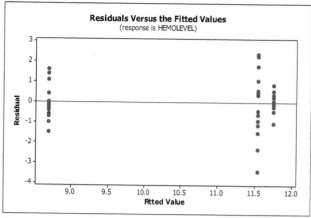

(c) The ratio of the largest to smallest standard deviation is 1.714/0.5379
= 3.19. The rule of 2 is violated. This is also seen in the graph at
the right above. The normal probability plots of the individual
samples and of the residuals are all linear. Conducting a one-way
ANOVA test on these data may be questionable. We will proceed with the
ANOVA and later we will compare the results with those of a more robust
procedure.

(d) The printed portion of the output from the procedure in part (b) is

One-way ANOVA: HEMOLEVEL versus TYPE

```
Source   DF      SS     MS      F      P
TYPE      2   83.43  41.71  29.13  0.000
Error    38   54.42   1.43
Total    40  137.85
S = 1.197    R-Sq = 60.52%    R-Sq(adj) = 58.44%
```

```
                                Individual 95% CIs For Mean Based on
                                Pooled StDev
Level    N    Mean   StDev  --+---------+---------+---------+-------
HB SC   10  11.760   0.538                          (-----*-----)
HB SS   16   8.713   0.844  (----*----)
HB ST   15  11.547   1.714                  (----*----)
                            --+---------+---------+---------+-------
                            8.4       9.6      10.8      12.0
```

Pooled StDev = 1.197

Since the F-value of 29.13 has a P-value of 0.000, we reject the null
hypothesis of equal population means, assuming that the assumptions for
an ANOVA test are met.

(e) We conclude that the data provide sufficient evidence that the mean
hemoglobin levels in the three types of sickle cell disease are
different.

13.73 (a) If you are given summary statistics and wish to conduct a one-way
ANOVA, you can use the following formula to obtain the mean of all the
observations.

$$\bar{x} = \frac{n_1\bar{x}_1 + n_2\bar{x}_2 + \cdots + n_k\bar{x}_k}{n_1 + n_2 + \cdots + n_k}$$

To verify this formula, recall that for the individual means for the

samples from the separate populations, $\bar{x}_j = \dfrac{\sum x_{ji}}{n_j}$. Let the notation

Copyright © 2012 Pearson Education, Inc. Publishing as Addison-Wesley.

$\sum x_{1i}$ mean to sum up all of the individual observations in the sample from the first population. We can substitute this formula for the individual sample means into the formula for the overall sample mean.

$$\bar{x} = \frac{n_1 \dfrac{\sum x_{1i}}{n_1} + n_2 \dfrac{\sum x_{2i}}{n_2} + \cdots + n_k \dfrac{\sum x_{ki}}{n_k}}{n_1 + n_2 + \cdots + n_k}.$$

The sample sizes cancel out of the numerator and we are left with

$$\bar{x} = \frac{\sum x_{1i} + \sum x_{2i} + \cdots + \sum x_{ki}}{n_1 + n_2 + \cdots + n_k} = \frac{\sum x_i}{n}.$$

The numerator of this expression means that we will add up all of the observations from each of the individual samples. In other words, add up all of the observations. The denominator of the expression means that we will add up all of the individual sample sizes, which is the overall sample size. Therefore, we have verified the formula for obtaining the mean of all the observations from the summary statistics.

(b) If all of the sample sizes are the same then we can simplify the formula by the following:

$$\bar{x} = \frac{n_1 \bar{x}_1 + n_2 \bar{x}_2 + \cdots + n_k \bar{x}_k}{n_1 + n_2 + \cdots + n_k} = \frac{n\bar{x}_1 + n\bar{x}_2 + \cdots + n\bar{x}_k}{kn}.$$

We can factor an n out of the numerator which will then cancel out with the n in the denominator. We are left with the following:

$$\bar{x} = \frac{\bar{x}_1 + \bar{x}_2 + \cdots + \bar{x}_k}{k}.$$

This means that, if all the sample sizes are equal, the mean of all the observations is found by adding up all of the sample means and dividing by the number of populations. In other words, it is just the mean of the sample means.

(c) To get the value of the F-statistic from the summary statistics, we use the defining formulas for the SSTR and SSE calculations. First, calculate SSTR. $SSTR = \sum n_i (\bar{x}_i - \bar{x})^2$. For each of the sample means, find its deviation from the overall sample mean found using the formula in part (a), square this deviation, multiply it by the respective sample size, and find the sum of all these terms for each of the samples. For SSE, $SSE = \sum (n_i - 1)s_i^2$, square each sample standard deviation, multiply it by the sample size minus one, and find the sum of all these terms for each of the samples. Then, the MSTR value is found by dividing the SSTR term by the degrees of freedom for the treatment: (k – 1). The MSE term is found by dividing the SSE term by the degrees of freedom for the error: (n – k). Then, finally, the F- statistic is found by dividing MSTR by MSE.

(iv) The standard deviations of the four populations are equal.

13.75 Suppose that we define the sample events A and B as follows:

A: the interval constructed around the difference $\mu_1 - \mu_2$

B: the interval constructed around the difference $\mu_1 - \mu_3$.

The 95% confidence interval for each event above is defined as:

$$P(A) = 0.95 \text{ and } P(B) = 0.95.$$

The probability of both A and B occurring simultaneously is written P(A and B). Since A and B are realistically not independent, the general multiplication rule applies; i.e., P(A and B) = P(A)•P(B|A).

Copyright © 2012 Pearson Education, Inc. Publishing as Addison-Wesley.

A difficulty arises in calculating a precise number for P(A and B) because we need additional information about P(B|A), which we do not have. However, P(B|A) is by no means equal to 1.00, which results in the product of P(A) and P(B|A) being less than 0.95 (because 0.95 times a probability less than 1.00 is clearly less than 0.95). Thus, the probability of both A and B occurring simultaneously is not 0.95.

Review Problems For Chapter 13

1. One-way ANOVA is used to compare means of a variable for populations that result from a classification by one other variable called the factor.
2. (i) Simple random samples: Check by carefully studying the way that the sampling was done.
   (ii) Independent samples: Check by carefully studying the way that the sampling was done.
   (iii) Normal populations; Check with normal probability plots, histograms, dotplots.
   (iv) All populations have equal standard deviations; this is a reasonable assumption if the ratio of the largest standard deviation to the smallest one is less than 2.
3. F distribution
4. There are n = 17 observations and k = 3 samples. The degrees of freedom are (k - 1, n - k) = (2, 14).
5. (a) Variation among sample means is measured by the mean square for treatments, MSTR = SSTR/(k - 1).
   (b) The variation within samples is measured by the error mean square, MSE = SSE/(n - k).
6. (a) SST is the total sum of squares. It measures the total variation among all of the sample data. $SST = \sum(x - \bar{x})^2$

   SSTR is the treatment sum of squares. It measures the variation among the sample means. $SSTR = \sum n_i(\bar{x}_i - \bar{x})^2$

   SSE is the error sum of squares. It measures the variation within the samples. $SSE = \sum(x - \bar{x}_i)^2 = \sum(n_i - 1)s_i^2$

   (b) SST = SSTR + SSE. This means that the total variation in the sample can be broken down into two components, one representing the variation between the sample means and one representing the variation within the samples.
7. (a) One purpose of a one-way ANOVA table is to organize and summarize the quantities required for ANOVA.
   (b)

Source	df	SS	MS=SS/df	F-statistic
Treatment	k - 1	SSTR	MSTR=SSTR/(k - 1)	F=MSTR/MSE
Error	n - k	SSE	MSE=SSE/(n - k)	
Total	n - 1	SST		

8. (a) 2 (b) 14 (c) 3.74 (d) 6.51 (e) 3.74
9. (a) The total number of populations being sampled is k = 3. Let the subscripts 1, 2, and 3 refer to A, B, and C, respectively. The total number of pieces of sample data is n = 12. Also, $n_1 = 3$, $n_2 = 5$, $n_3 = 4$. The following statistics for each sample are: $\bar{x}_1 = 3$, $\bar{x}_2 = 3$, and $\bar{x}_3 = 6$. Also, $s_1 = 2.000$, $s_2 = 2.449$, and

Copyright © 2012 Pearson Education, Inc. Publishing as Addison-Wesley.

$s_3 = 4.243$. Note, the mean of all the sample data is $\bar{x} = 4$. We use this information as follows:

(b) SST $= \sum (x - \bar{x})^2 = (1 - 4)^2 + \ldots + (3 - 4)^2 = 110.0$

SSTR $= n_1(\bar{x}_1 - \bar{x})^2 + n_2(\bar{x}_2 - \bar{x})^2 + n_3(\bar{x}_3 - \bar{x})^2$
$= 3(3 - 4)^2 + 5(3 - 4)^2 + 4(6 - 4)^2 = 24.0$

SSE $= (n_1 - 1)s_1^2 + (n_2 - 1)s_2^2 + (n_3 - 1)s_3^2$
$= (3 - 1)(2.000)^2 + (5 - 1)(2.449)^2 + (4 - 1)(4.243)^2 = 86.0$

SST = SSTR + SSE since $110.0 = 24.0 + 86.0$

(c) Let T_1, T_2, and T_3 refer to the sum of the data values in each of the three samples, respectively. Thus: $T_1 = 9$ $T_2 = 15$ $T_3 = 24$. Also, the sum of all the data values is $\sum x = 48$, and their sum of squares is $\sum x^2 = 302$.

Consequently: $SST = \sum x^2 - \dfrac{(\sum x)^2}{n} = 302 - \dfrac{48^2}{12} = 110$

$$SSTR = \dfrac{T_1^2}{n_1} + \dfrac{T_2^2}{n_2} + \dfrac{T_3^2}{n_3} - \dfrac{(\sum x)^2}{n} = \dfrac{9^2}{3} + \dfrac{15^2}{5} + \dfrac{24^2}{4} - \dfrac{48^2}{12} = 24$$

and SSE = SST - SSTR = 110 - 24 = 86 .

(d)

Source	Df	SS	MS=SS/df	F-statistic
Treatment	2	24	12.000	1.255
Error	9	86	9.556	
Total	11	110		

10. (a) MSTR is a measure of the variation between the sample mean losses for highway robberies, gas station robberies, and convenience store robberies.

(b) MSE is a measure of the variation within the three samples.

(c) The four assumptions for one-way ANOVA, given in Key Fact 13.2: simple random samples, independent samples, normal populations, and equal standard deviations. Assumptions 1 and 2 on simple random independent samples are absolutely essential to the one-way ANOVA procedure. Assumption 3 on normality is not too critical as long as the populations are not too far from being normally distributed. Assumption 4 on equal standard deviations is also not that important provided the sample sizes are roughly equal.

11. (a) The samples and their normal scores are shown in the table below along with the residuals which are $x_i - \bar{x}_j$.

	Hwy	w	Residual	Gas	w	Residual	Conv. St.	w	Residual
	952	-1.18	-27.8	1298	-1.28	129.2	844	-1.28	82.8
	996	-0.50	16.2	1195	-0.64	26.2	921	-0.64	159.8
	839	0.00	-140.8	1174	-0.20	5.2	880	-0.20	118.8
	1088	0.50	108.2	1113	0.20	-55.8	706	0.20	-55.2
	1024	1.18	44.2	953	0.64	-215.8	602	0.64	-159.2
				1280	1.28	111.2	614	1.28	-147.2
Total	4899			7013			4567		
Mean	979.8			1168.8			761.2		
St.Dev.	92.90			126.1			139.0		

We will plot the normal scores w against the data for each type of store. The graphs have been produced using Minitab, but you can easily do them by hand. The standard deviations for each type of robbery are shown in the above table.

Copyright © 2012 Pearson Education, Inc. Publishing as Addison-Wesley.

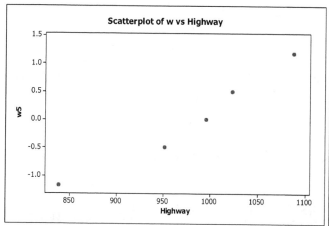

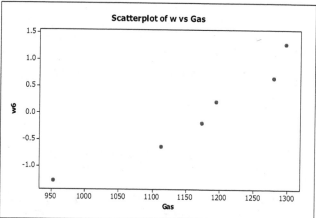

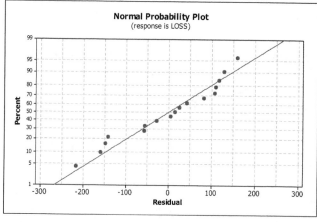

(b) Now order all of the residuals in the above table from smallest to largest and plot them against the normal scores in Table III for n = 17. Also plot the residuals against the means for each type of robbery. The results are

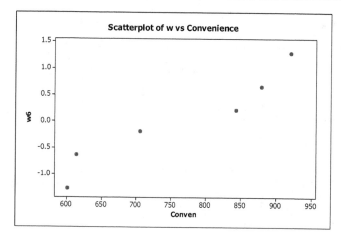

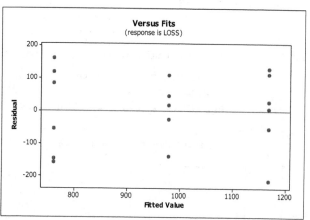

(c) All three of the normal probability plots are close to linear, and the ratio of the largest to smallest standard deviation is 1.50. The residual analysis shows that the normal probability plot of the residuals is close to linear and the last graph shows that the spreads of the residuals are about the same for each type of robbery. Thus the ANOVA assumptions of normality and equal standard deviations appear to be reasonable.

Copyright © 2012 Pearson Education, Inc. Publishing as Addison-Wesley.

12. The total number of populations being sampled is k = 3. Let the subscripts 1, 2, and 3 refer to Highway, Gas station, and Convenience store, respectively. The total number of observations is n = 17. Also, $n_1 = 5$, $n_2 = 6$, and $n_3 = 6$.

Step 1: H_0: $\mu_1 = \mu_2 = \mu_3$ (population means are equal)

H_a: Not all the means are equal.

Step 2: $\alpha = 0.05$

Step 3: The sums of squares are

$$SST = \sum x^2 - \frac{(\sum x)^2}{n} = 16,683,857 - \frac{16,479^2}{17} = 709,889.882$$

$$SSTR = \frac{T_1^2}{n_1} + \frac{T_2^2}{n_2} + \frac{T_3^2}{n_3} - \frac{(\sum x)^2}{n} = \frac{4899^2}{5} + \frac{7013^2}{6} + \frac{4567^2}{6} - \frac{16479^2}{17} = 499,349.416$$

and SSE = SST − SSTR = 709,889.882 − 499,349.416 = 210,540.466

The one-way ANOVA table is:

Source	df	SS	MS = SS/df	F-statistic
Treatment	2	499,349.416	249,674.708	16.602
Error	14	210,540.466	15,038.605	
	16	709,889.882		

The F-statistic is defined and calculated as

$$F = \frac{MSTR}{MSE} = \frac{249,674.708}{15,038.605} = 16.602$$

Step 4: df = (k − 1, n − k) = (2, 14); critical value = 3.74
For the P-value approach, P(F > 16.602) < 0.005.

Step 5: From Step 3, F = 16.602. From Step 4, $F_{0.05} = 3.74$. Since 16.602 > 3.74, reject H_0. Since the P-value is smaller than the significance level, reject H_0.

Step 6: At the 5% significance level, the data provide sufficient evidence to conclude that a difference in mean losses exists among the three types of robberies.

13. (a) Using Minitab, choose **Graph ▶ Probability Plot**, select the **Simple** version and click **OK**. Enter Loss, Stable, and Gain in the **Graph Variables** text box, click on the **Multiple Graphs** button and select **In separate panels of the same graph**, and click **OK**. The result is

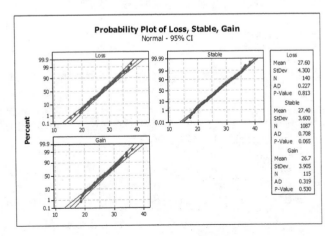

The standard deviations of the three groups are shown in the right hand panel as 4.300, 3.600, and 3.905.

Copyright © 2012 Pearson Education, Inc. Publishing as Addison-Wesley.

(b) To carry out the Analysis of variance and residual analysis, we will use the data in a single column, which is named BMI. In the next column, which is named GROUP, are the names of each of the regions corresponding to each data value in BMI. The residual analysis consists of plotting the residuals against the means (or fits) and constructing a normal probability plot. This is done at the same time

as the analysis of variance in Minitab. To do this, we choose **Stat ▶**

Anova ▶One-way..., select <u>BMI</u> for the **Response:** text box and <u>GROUP</u> in the **Factor:** text box. Enter <u>95</u> in the **Confidence level** box. Then click on the **Graphs** button and click to place check marks in the boxes for **Normal plot of residuals** and **Residuals versus fits**, and click **OK**. Click on the **Comparisons** button and select **Tukey's**, and enter <u>5</u> in the **family error rate** box, and click **OK**. The graphical results are

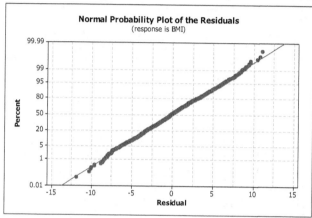
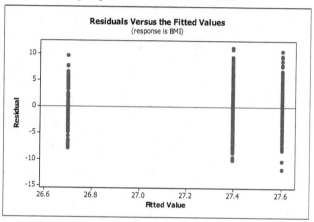

(c) The normal probability plots in part (a) are linear as is the normal probability plot of the residuals in part (b). The plot of the residuals against the fitted values shows approximately equal spreads. The rule of 2 is not violated since the ratio of the largest to smallest standard deviation is 4.3/3.6 = 1.19. A one-way ANOVA is reasonable.

(d) The one-way ANOVA results from the procedure in part (b) are

One-way ANOVA: BMI versus GROUP

Source	DF	SS	MS	F	P
GROUP	2	60.0	30.0	2.18	0.113
Error	1339	18379.7	13.7		
Total	1341	18439.6			

S = 3.705   R-Sq = 0.33%   R-Sq(adj) = 0.18%

```
                               Individual 95% CIs For Mean Based on
                               Pooled StDev
Level      N    Mean   StDev   ------+---------+---------+---------+---
Gain     115  26.700   3.905   (----------*----------)
Loss     140  27.603   4.300                     (--------*---------)
Stable  1087  27.399   3.600                       (---*--)
                               ------+---------+---------+---------+---
                                  26.40     27.00     27.60     28.20
```

(e) The P-value of the test is 0.113, which is larger than the 0.05 significance level. Do not reject the null hypothesis of equal means. The data do not provide sufficient evidence to conclude that there is a

Copyright © 2012 Pearson Education, Inc. Publishing as Addison-Wesley.

difference in body mass index among the three groups of men with different weight losses.

14. (a) Using Minitab, choose **Graph ▶ Probability Plot**, select the **Simple** version and click **OK**. Enter <u>Loss</u>, <u>Stable</u>, and <u>Gain</u> in the **Graph Variables** text box, click on the **Multiple Graphs** button and select **In separate panels of the same graph**, and click **OK**. The result is

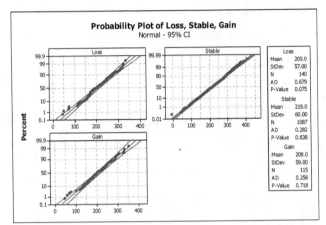

The standard deviations of the three groups are shown in the right hand panel as 57, 60, and 59.

(b) To carry out the Analysis of variance and residual analysis, we will use the data in a single column, which is named POWER. In the next column, which is named GROUP, are the names of each of the regions corresponding to each data value in POWER. The residual analysis consists of plotting the residuals against the means (or fits) and constructing a normal probability plot. This is done at the same time as the analysis of variance in Minitab. To do this, we choose **Stat ▶ Anova ▶One-way...**, select <u>POWER</u> for the **Response:** text box and <u>GROUP</u> in the **Factor:** text box. Enter <u>95</u> in the **Confidence level** box. Then click on the **Graphs** button and click to place check marks in the boxes for **Normal plot of residuals** and **Residuals versus fits**, and click **OK**. Click on the **Comparisons** button and select **Tukey's**, and enter <u>5</u> in the **family error rate** box, and click **OK**. The graphical results are

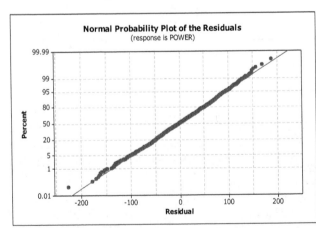
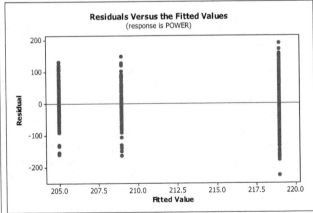

(c) The normal probability plots in part (a) are linear as is the normal probability plot of the residuals in part (b). The plot of the residuals against the fitted values shows approximately equal spreads.

Copyright © 2012 Pearson Education, Inc. Publishing as Addison-Wesley.

The rule of 2 is not violated since the ratio of the largest to smallest standard deviation is 60/57 = 1.05. A one-way ANOVA is reasonable.

(d) The one-way ANOVA results from the procedure in part (b) are

One-way ANOVA: POWER versus GROUP

Source	DF	SS	MS	F	P
GROUP	2	31742	15871	4.47	0.012
Error	1339	4758204	3554		
Total	1341	4789946			

S = 59.61   R-Sq = 0.66%   R-Sq(adj) = 0.51%

```
                                    Individual 95% CIs For Mean Based on Pooled
                                    StDev
Level      N    Mean   StDev       -+---------+---------+---------+--------
Gain     115  209.00   59.00             (---------------*-------------)
Loss     140  205.00   57.00       (-------------*-------------)
Stable  1087  219.00   60.00                                      (----*----)
                                   -+---------+---------+---------+--------
                                  196.0     203.0     210.0     217.0
```

Pooled StDev = 59.61

(e) The P-value of the test is 0.012, which is smaller than the 0.05 significance level. Reject the null hypothesis of equal means. The data provide sufficient evidence to conclude that there is a difference in leg power among the three groups of men with different weight losses.

15. (a) Using Minitab, choose **Graph ▶ Probability Plot**, select the **Simple** version and click **OK**. Enter 25-34, 35-44, 45-54 and 55-64 in the **Graph Variables** text box, click on the **Multiple Graphs** button and select **In separate panels of the same graph**, and click **OK**. The result is

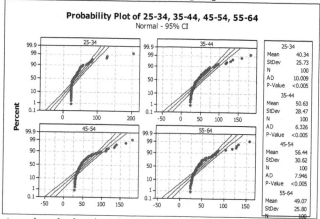

The standard deviations of the incomes in the four age groups are 25.73, 28.47, 30.62, and 25.80.

(b) To carry out the Analysis of variance and residual analysis, we will use the data in a single column, which is named INCOME. In the next column, which is named AGE, are the categories for ages corresponding to each data value in INCOME. The residual analysis consists of plotting the residuals against the means (or fits) and constructing a normal probability plot. This is done at the same time as the analysis of variance in Minitab. To do this, we choose **Stat ▶ Anova ▶ One-way...**, select INCOME for the **Response:** text box and AGE in the **Factor:**

Copyright © 2012 Pearson Education, Inc. Publishing as Addison-Wesley.

text box. Enter 95 in the **Confidence level** box. Then click on the **Graphs** button and click to place check marks in the boxes for **Normal plot of residuals** and **Residuals versus fits**, and click **OK**. Click on the **Comparisons** button and select **Tukey's**, and enter 5 in the **family error rate** box, and click **OK**. The graphical results are

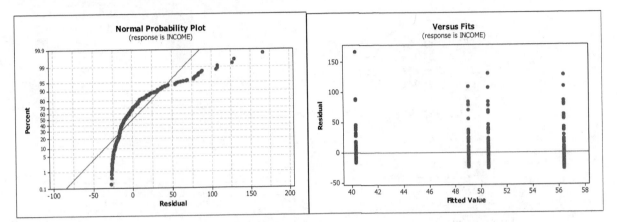

(c) The normal probability plots for each age group are non-linear, as is the residual normal probability plot in part (b). The rule of 2 is not violated as the ratio of the largest to smallest standard deviation is $30.62/25.73 = 1.19$. The normality assumption is not reasonable, so parts (d)-(e) are omitted.

Copyright © 2012 Pearson Education, Inc. Publishing as Addison-Wesley.

CHAPTER 14 ANSWERS

Exercises 14.1

14.1 Conditional distribution, conditional mean, and conditional standard deviation

14.3 (a) population regression line   (b) σ

(c) normal, $\beta_0 + 6\beta_1$, σ

14.5 The sample regression line is the best estimate of the population regression line.

14.7 Residual

14.9 The plot of the residuals against the values of the predictor variable provides the same information as a scatter diagram of the data points. However, it has the advantage of making it easier to spot patterns such as curvature and non-constant standard deviation.

14.11 (a) We first create the table below.

x	y	$\hat{y}$	$e = y - \hat{y}$	e^2
3	-4	-5	1	1
1	0	-1	1	1
2	-5	-3	-2	4
				6

$$s_e = \sqrt{\frac{SSE}{n-2}} = \sqrt{\frac{6}{1}} = 2.449$$

(b-c) For the residual plot, we plot e against x in the graph below at the left. We can use Excel for the normal probability plot. With the values of e in a column named 'e', highlight the data with the mouse, choose **DDXL**, then select **Charts and Plots**, click on the **Function Type** drop down box, and click on **Normal Probability Plot**. Drag the 'e' from the **Names and Columns** box into the **Quantitative Variable** box and click **OK**. The graph is shown below at the right.

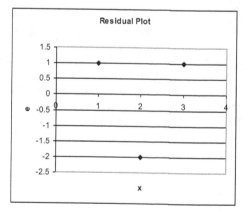
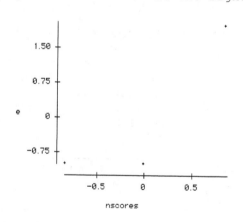

14.13 (a) We first create the table below.

x	y	$\hat{y}$	$e = y - \hat{y}$	e^2
3	4	3	1	1
4	5	5	0	0
1	0	-1	1	1
2	-1	1	-2	4
				6

$$s_e = \sqrt{\frac{SSE}{n-2}} = \sqrt{\frac{6}{2}} = 1.732$$

(b-c) For the residual plot, we plot e against x in the graph following at the left. We can use Excel for the normal probability plot.

Copyright © 2012 Pearson Education, Inc. Publishing as Addison-Wesley.

With the values of e in a column named 'e', highlight the data with the mouse, choose **DDXL**, then select **Charts and Plots**, click on the **Function Type** drop down box, and click on **Normal Probability Plot**. Drag the 'e' from the **Names and Columns** box into the **Quantitative Variable** box and click **OK**. The graph is shown following at the right.

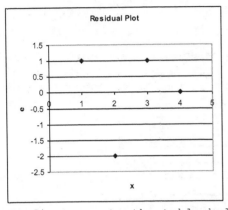

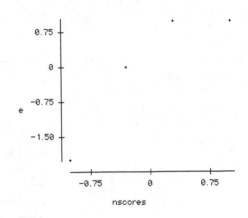

14.15 (a) We first create the table below.

x	y	$\hat{y}$	$e = y - \hat{y}$	e^2
0	4	2.875	1.125	1.265625
2	2	1.625	0.375	0.140625
2	0	1.625	-1.625	2.640625
5	-2	-0.250	-1.750	3.062500
6	1	-0.875	1.875	3.515625
				10.625000

$$s_e = \sqrt{\frac{SSE}{n-2}} = \sqrt{\frac{10.625}{3}} = 1.882$$

(b-c) For the residual plot, we plot e against x in the graph following at the left. We can use Excel for the normal probability plot. With the values of e in a column named 'e', highlight the data with the mouse, choose **DDXL**, then select **Charts and Plots**, click on the **Function Type** drop down box, and click on **Normal Probability Plot**. Drag the 'e' from the **Names and Columns** box into the **Quantitative Variable** box and click **OK**. The graph is shown following at the right.

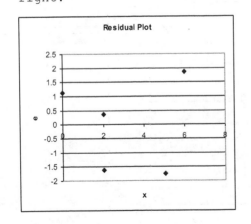

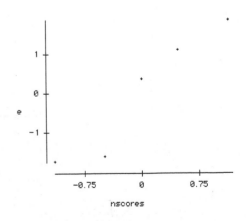

14.17 If the assumptions for regression inferences are satisfied for a model relating a Corvette's age to its price, this means that there are constants

Copyright © 2012 Pearson Education, Inc. Publishing as Addison-Wesley.

β_0, β_1, and σ such that, for each age x, the prices for Corvettes of that age are normally distributed with mean $\beta_0 + \beta_1 x$ and standard deviation σ.

14.19 If the assumptions for regression inferences are satisfied for a model relating the volume of plant emissions of volatile compounds to the weight of the plant, this means that there are constants β_0, β_1, and σ such that, for each weight x, the volumes of emissions y are normally distributed with mean $\beta_0 + \beta_1 x$ and standard deviation σ.

14.21 If the assumptions for regression inferences are satisfied for a model relating the test scores of calculus students to their study-times, this means that there are constants β_0, β_1, and σ such that, for each study-time x, the test scores y are normally distributed with mean $\beta_0 + \beta_1 x$ and standard deviation σ.

14.23 To compute the standard error of the estimate, first retrieve the computation for the error sum of squares (SSE) in part (a) of Exercise 4.89 and then apply the formula for s_e in Definition 14.1 of the text.

(a) In part (a) of Exercise 4.89, n = 10 and SSE was computed as 1,623.7. Applying Definition 14.1, the standard error of the estimate is

$$s_e = \sqrt{\frac{SSE}{n-2}} = \sqrt{\frac{1623.7}{10-2}} = 14.2465$$

Roughly speaking, the predicted prices for the Corvettes differ, on average, from the observed prices by $1424.65 .

(b) Presuming that the variables age (x) and price (y) for Corvettes satisfy Assumptions (1) - (3) for regression inferences, the standard error of the estimate $s_e = 14.2465$ provides an estimate for the common population standard deviation σ of prices for all Corvettes of any given age.

(c) Obtain the necessary normal scores from Table III in the text.

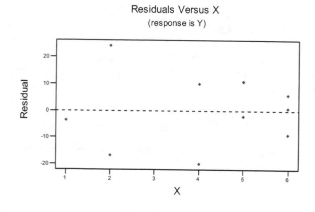

Residuals Versus X
(response is Y)

Age x	Residual e
6	0.82
6	-9.18
6	5.82
2	24.20
2	-16.80
5	-2.09
4	10.01
5	10.91
1	-3.70
4	-19.99

Copyright © 2012 Pearson Education, Inc. Publishing as Addison-Wesley.

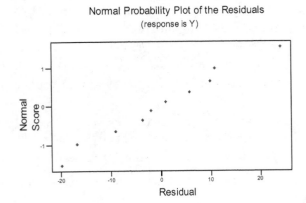

Normal Probability Plot of the Residuals
(response is Y)

Residual e	Normal score n
-19.99	-1.55
-16.80	-1.00
-9.18	-0.65
-3.70	-0.37
-2.09	-0.12
0.82	0.12
5.82	0.37
10.01	0.65
10.91	1.00
24.20	1.55

(d) Taking into account the small sample size, we can say that the residuals fall roughly in a horizontal band centered and symmetric about the x-axis. We can also say that the normal probability plot for residuals is very roughly linear. Therefore, it appears reasonable to consider the assumptions for regression inferences for the variables age and price of Corvettes to be met.

14.25 To compute the standard error of the estimate, first retrieve the computation for the error sum of squares (SSE) in part (a) of Exercise 4.91 and then apply the formula for s_e in Definition 14.1 of the text.

(a) In part (a) of Exercise 4.91, n = 10 and SSE was computed as 88.5. Applying Definition 14.1, the standard error of the estimate is

$$s_e = \sqrt{\frac{SSE}{n-2}} = \sqrt{\frac{264.16}{11-2}} = 5.4177 \text{ hundred nanograms}$$

Roughly speaking, the predicted volumes of volatile emissions from the plants differ, on average, from the observed volumes by 541.77 nanograms.

(b) Presuming that the weight (x) of the potato plants *Solanum tubersom* and volume of volatile emissions (y) from the plants satisfy Assumptions (1) - (3) for regression inferences, the standard error of the estimate $s_e = 5.4177$ hundred nanograms (541.77 nanograms) provides an estimate for the common population standard deviation σ of emission volumes for all plants of any particular weight.

(c)

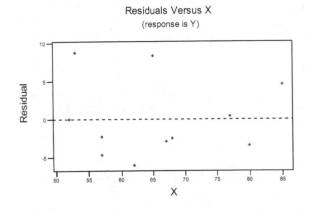

Residuals Versus X
(response is Y)

Weight x	Residual e
57	-4.81
85	4.63
57	-2.31
65	8.39
52	0.01
67	-2.93
62	-6.12
80	-3.55
77	0.44
53	8.85
68	-2.60

Copyright © 2012 Pearson Education, Inc. Publishing as Addison-Wesley.

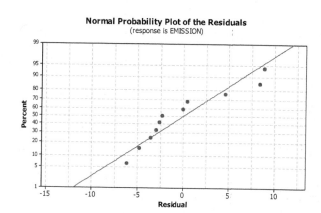

Normal Probability Plot of the Residuals
(response is EMISSION)

Residual e	Normal score n
-6.12	-1.59
-4.81	-1.06
-3.55	-0.73
-2.93	-0.46
-2.60	-0.22
-2.31	0.00
0.01	0.22
0.44	0.46
4.63	0.73
8.39	1.06
8.85	1.59

(d) Taking into account the small sample size, we can say that the residuals fall roughly in a horizontal band centered and symmetric about the x-axis. We can also say that the normal probability plot for residuals is approximately linear.

Therefore, based on the sample data, there are no obvious violations of the assumptions for regression inferences for the variables plant weight and volume of volatile emissions.

14.27 To compute the standard error of the estimate, first retrieve the computation for the error sum of squares (SSE) in part (a) of Exercise 4.93 and then apply the formula for s_e in Definition 14.1 of the text.

(a) In part (a) of Exercise 4.93, n = 8 and SSE was computed as 75.11. Applying Definition 14.1, the standard error of the estimate is

$$s_e = \sqrt{\frac{SSE}{n-2}} = \sqrt{\frac{75.11}{8-2}} = 3.5382$$

Roughly speaking, the predicted test scores differ, on average, from the observed scores by 3.5382 points.

(b) Presuming that the study-time (x) of the calculus students and their test scores (y) satisfy Assumptions (1) - (3) for regression inferences, the standard error of the estimate $s_e = 3.5382$ provides an estimate for the common population standard deviation σ of test scores for all calculus students with any particular study-time.

(c) Obtain the necessary normal scores from Table III in the text.

Residual Plot

Study Time x	Residual e
10	5.59
15	-1.18
12	-0.72
20	-3.95
8	-3.10
16	-1.34
14	0.97
22	3.74

Copyright © 2012 Pearson Education, Inc. Publishing as Addison-Wesley.

Normal Probability Plot of Residuals

Residual e	Normal score n
-3.95	-1.43
-3.10	-0.85
-1.34	-0.47
-1.18	-0.15
-0.72	0.15
0.97	0.47
3.74	0.85
5.59	1.43

(d) Taking into account the small sample size, we can say that the residuals fall roughly in a horizontal band centered and symmetric about the x-axis. We can also say that the normal probability plot for residuals is approximately linear. Therefore, based on the sample data, there are no obvious violations of the assumptions for regression inferences for the variables study-time and test scores of calculus students.

14.29 (a) The assumption of linearity (Assumption 1) may be violated since the band is not horizontal, as may be the assumption of equal standard deviations (Assumption 2) since there is more variation in the residuals for small x than for large x.

(b) It appears that the standard deviation does not remain constant; thus, Assumption 2 is violated.

(c) The graph does not suggest violation of one or more of the assumptions for regression inferences.

(d) The normal probability plot appears to be more curved than linear; thus, the assumption of normality is violated.

14.31 Using Minitab, with the data in columns named INAUGURATION and DEATH, we choose **Stat ▶ Regression ▶ Regression...**, select DEATH in the **Response** text box, and select INAUGURATION in the **Predictors** text box. Click the **Graphs** button, select the **Regular** option button from the **Residuals for Plots** list, select the **Normal plot of residuals** check box from the **Residual Plots** list, click in the **Residuals versus the variables** text box and specify INAUGURATION, click **OK**, and click **OK**. The results are

```
The regression equation is
DEATH = 5.4 + 1.18 INAUGURATION

Predictor       Coef   SE Coef      T      P
Constant        5.37     14.02   0.38  0.704
INAUGURATION  1.1848    0.2537   4.67  0.000

S = 9.70072   R-Sq = 37.1%   R-Sq(adj) = 35.4%

Analysis of Variance

Source           DF      SS      MS      F      P
Regression        1  2051.6  2051.6  21.80  0.000
Residual Error   37  3481.8    94.1
Total            38  5533.4
```

(a) The standard error of the estimate is 9.70072. Roughly speaking, the predicted death ages differ, on average, from the observed death ages by 9.70072 years.

Copyright © 2012 Pearson Education, Inc. Publishing as Addison-Wesley.

(b)

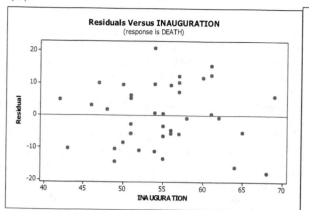
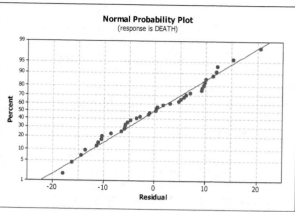

(c)   The residuals fall roughly in a horizontal band centered and symmetric
      about the x-axis. We can also say that the normal probability plot for
      residuals is approximately linear. Therefore, based on the sample
      data, there are no obvious violations of the assumptions for regression
      inferences for the variables inauguration age and death age of U.S.
      Presidents.

14.33 Using Minitab, with the data in columns named LOT SIZE and VALUE, we choose

Stat ▶ Regression ▶ Regression..., select <u>VALUE</u> in the **Response** text box,
and select `'LOT SIZE'` in the **Predictors** text box. Click the **Graphs** button,
select the **Regular** option button from the **Residuals for Plots** list, select
the **Normal plot of residuals** check box from the **Residual Plots** list, click
in the **Residuals versus the variables** text box and specify `'LOT SIZE'`, click
OK, and click **OK**. The results are

```
The regression equation is
VALUE = 292 + 67.1 LOT SIZE

Predictor    Coef   SE Coef     T      P
Constant    292.0     140.3   2.08  0.043
LOT SIZE    67.11      59.87  1.12  0.269

S = 139.012   R-Sq = 2.9%   R-Sq(adj) = 0.6%

Analysis of Variance
Source          DF       SS      MS     F      P
Regression       1    24286   24286  1.26  0.269
Residual Error  42   811619   19324
Total           43   835905
```

(a)   The standard error of the estimate is 139.012 thousand dollars.
      Roughly speaking, the predicted values differ, on average, from the
      observed values by 139.012 thousand dollars.

(b)

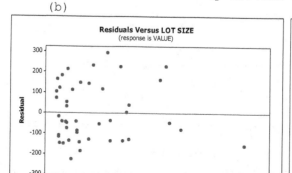
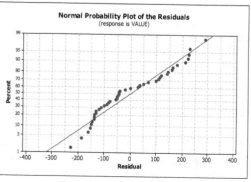

Copyright © 2012 Pearson Education, Inc. Publishing as Addison-Wesley.

(c) The residuals appear to fall roughly in a band centered and symmetric about the x-axis. We can also say that the normal probability plot for residuals deviates from a linear pattern, but it is difficult to say whether the deviation is enough to rule out normality. Therefore, for the time being, based on the sample data, the assumptions for regression inferences seem to be reasonable for the variables lot size and value. More on these data later.

14.35 Using Minitab, with the data in columns named HIGH and LOW, we choose **Stat**

▶ **Regression** ▶ **Regression...**, select <u>LOW</u> in the **Response** text box, and select <u>HIGH</u> in the **Predictors** text box. Click the **Graphs** button, select the **Regular** option button from the **Residuals for Plots** list, select the **Normal plot of residuals** check box from the **Residual Plots** list, click in the **Residuals versus the variables** text box and specify <u>HIGH</u>, click **OK**, and click **OK**. The results are

```
The regression equation is
LOW = - 7.57 + 0.917 HIGH

Predictor       Coef   SE Coef        T       P
Constant      -7.572     1.786    -4.24   0.000
HIGH         0.91685   0.02967    30.91   0.000

S = 4.15111    R-Sq = 95.2%    R-Sq(adj) = 95.1%

Analysis of Variance

Source         DF      SS      MS        F       P
Regression      1   16459   16459   955.17   0.000
Residual Error 48     827      17
Total          49   17286
```

(a) The standard error of the estimate is 4.15111 degrees. Roughly speaking, the predicted low temperatures differ, on average, from the observed low temperatures by 4.15111 degrees.

(b)

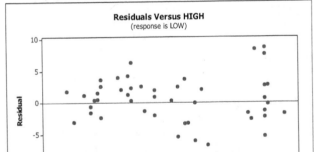
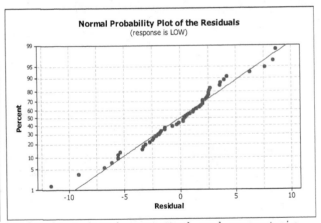

(c) The residuals appear to fall roughly in a band centered and symmetric about the x-axis. We can also say that the normal probability plot for residuals deviates only slightly from a linear pattern. Therefore, based on the sample data, the assumptions for regression inferences seem to be reasonable for the variables average high January temperature and average low January temperature.

14.37 Using Minitab, with the data in columns named DISP and MPG, we choose **Stat**

▶ **Regression** ▶ **Regression...**, select <u>DISP</u> in the **Response** text box, select <u>MPG</u> in the **Predictors** text box, select **Residuals** from the **Storage** check-box list. Click the **Graphs...** button, select the **Regular** option button from the

Copyright © 2012 Pearson Education, Inc. Publishing as Addison-Wesley.

Residuals for Plots list, select the **Normal plot of residuals** check box from the **Residual Plots** list, click in the **Residuals versus the variables** text box and specify <u>DISP</u>, click **OK**, and click **OK**. The results are

(a)   The regression equation is
            MPG = 32.3 - 3.58 DISP

Predictor	Coef	StDev	T	P
Constant	32.2732	0.6704	48.14	0.000
DISP	-3.5762	0.2099	-17.03	0.000

S = 2.248   R-Sq = 70.9%   R-Sq(adj) = 70.7%

Analysis of Variance

Source	DF	SS	MS	F	P
Regression	1	1465.8	1465.8	290.18	0.000
Residual Error	119	601.1	5.1		
Total	120	2067.0			

The standard error of the estimate is the first entry in the sixth line of the computer output. It is reported as s = 2.248. Presuming that the variables DISP and MPG satisfy the assumptions for regression inferences, the standard error of the estimate, 2.248, provides an estimate for the common population standard deviation, σ, of miles per gallon for all cars with a particular engine displacement.

(b)   The two graphs that result follow.

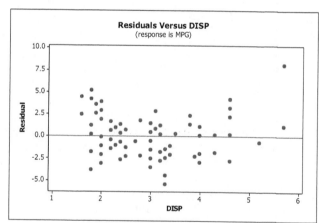
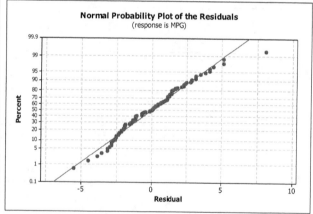

(c)   The residuals plotted against engine displacement display a concave upward curve, indicating that the linear model is not appropriate for the data. Thus, Assumption 1 is violated and the statement made in part (a) interpreting the meaning of the standard error does not hold.

14.39 Using Minitab, with the data in columns named DIAMETER and VOLUME, we choose

Stat ▶ Regression ▶ Regression..., select <u>VOLUME</u> in the **Response** text box, select <u>DIAMETER</u> in the **Predictors** text box, select **Residuals** from the **Storage** check-box list. Click the **Graphs...** button, select the **Regular** option button from the **Residuals for Plots** list, select the **Normal plot of residuals** check box from the **Residual Plots** list, click in the **Residuals versus the variables** text box and specify <u>DIAMETER</u>, click **OK**, and click **OK**. The results are

(a)   The regression equation is
            VOLUME = - 41.6 + 6.84 DIAMETER

Predictor	Coef	StDev	T	P
Constant	-41.568	3.427	-12.13	0.000
DIAMETER	6.8367	0.2877	23.77	0.000

Copyright © 2012 Pearson Education, Inc. Publishing as Addison-Wesley.

```
       S = 9.875        R-Sq = 89.3%      R-Sq(adj) = 89.1%

       Analysis of Variance

       Source          DF         SS         MS        F        P
       Regression       1      55083      55083   564.88    0.000
       Residual Error  68       6631         98
       Total           69      61714
```

The standard error of the estimate is the first entry in the sixth line of the computer output. It is reported as s = 9.875. Presuming that the variables DIAMETER and VOLUME satisfy the assumptions for regression inferences, the standard error of the estimate, 9.875, provides an estimate for the common population standard deviation, σ, of volumes for all trees having a particular diameter at breast height.

(b)   The two graphs that result follow.

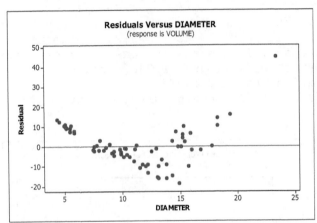
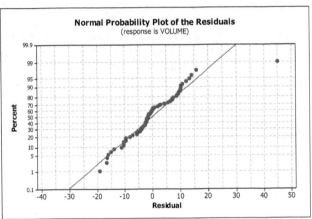

(c)   The residuals plotted against diameter are displayed in a curved concave upward pattern, indicating that the linear model is not appropriate. This is a violation of Assumption 1. In addition, both plots indicate the presence of an outlier. Finally, it appears that there is more variability in the center of the first graph than at the left side, so Assumption 2 is also violated. Removal of the outlier will not change the curved pattern of the other residuals or the unequal variation in the residuals, so this data should not be analyzed using a linear model.

Exercises 14.2

14.41 normal distribution with mean β_1 = -3.5

14.43 We can also use the coefficient of determination, r^2, and the linear correlation coefficient, r, as a basis for a test to decide whether a regression equation is useful for prediction.

14.45 From Exercise 14.11, we have $s_e = 2.44949$.

Also, $\sum x_i = 6$; $\sum x_i^2 = 14$; n=3; $S_{xx} = \sum x_i^2 - \left(\sum x_i\right)^2 / n = 14 - 6^2/3 = 2$

(a) $t = \dfrac{b_1}{s_e / \sqrt{S_{xx}}} = \dfrac{-2}{2.44949/\sqrt{2}} = -1.155$; the critical values with 1 df are

±6.314. Since -6.314 < t < 6.314, we cannot reject $\beta_1 = 0$ and conclude that x is not useful for predicting y.

(b)   The 90% confidence interval for the slope of the regression line is

Copyright © 2012 Pearson Education, Inc. Publishing as Addison-Wesley.

$$b_1 \pm t_{\alpha/2} \frac{s_e}{\sqrt{S_{xx}}} = -2 \pm 6.314 \frac{2.44949}{\sqrt{2}} = (-12.936, 8.936)$$

14.47 From Exercise 14.13, we have $s_e = 1.732$.

Also, $\sum x_i = 10$; $\sum x_i^2 = 30$; n=4; $S_{xx} = \sum x_i^2 - \left(\sum x_i\right)^2 / n = 30 - 10^2/4 = 5$

(a) $t = \dfrac{b_1}{s_e/\sqrt{S_{xx}}} = \dfrac{2}{1.732/\sqrt{5}} = 2.582$; the critical values with 2 df are ± 2.920.

Since $-2.920 < t < 2.920$, we cannot reject $\beta_1 = 0$ and conclude that x is not useful for predicting y.

(b) The 90% confidence interval for the slope of the regression line is

$$b_1 \pm t_{\alpha/2} \frac{s_e}{\sqrt{S_{xx}}} = 2 \pm 2.920 \frac{1.732}{\sqrt{5}} = (-0.260, 4.262)$$

14.49 From Exercise 14.15, we have $s_e = 1.88193$.

Also, $\sum x_i = 15$; $\sum x_i^2 = 69$; n=5; $S_{xx} = \sum x_i^2 - \left(\sum x_i\right)^2 / n = 69 - 15^2/5 = 24$

(a) $t = \dfrac{b_1}{s_e/\sqrt{S_{xx}}} = \dfrac{-0.625}{1.88193/\sqrt{24}} = -1.627$; the critical values with 3 df are

± 2.353. Since $-2.353 < t < 2.353$, we cannot reject $\beta_1 = 0$ and conclude that x is not useful for predicting y.

(b) The 90% confidence interval for the slope of the regression line is

$$b_1 \pm t_{\alpha/2} \frac{s_e}{\sqrt{S_{xx}}} = -0.625 \pm 2.353 \frac{1.88193}{\sqrt{24}} = (-1.512, 0.282)$$

14.51 From Exercise 4.51, $\sum x = 41$, $\sum x^2 = 199$, and $b_1 = -27.9029$. From Exercise 14.23, $s_e = 14.2464$.

Step 1: $H_0: \beta_1 = 0$, $H_a: \beta_1 \neq 0$

Step 2: $\alpha = 0.10$

Step 3: $t = \dfrac{b_1}{s_e/\sqrt{\sum x^2 - \left(\sum x\right)^2/n}} = \dfrac{-27.9029}{14.2465/\sqrt{199 - 41^2/10}} = -10.887$

Step 4: df = n - 2 = 8; critical values = ± 1.860
For the p-value approach, p < 0.01.

Step 5: Since $-10.887 < -1.860$, reject H_0.
Because the p-value is less than the significance level of 0.10, we can reject H_0.

Step 6: At 10% significance level, the data provide sufficient evidence to conclude that the slope of the population regression line is not zero and, hence, that age is useful as a predictor of price for Corvettes.

14.53 From Exercise 4.53, $\sum x = 723$, $\sum x^2 = 48747$, and $b_1 = 0.16285$. From Exercise 14.25, $s_e = 5.418$.

Step 1: $H_0: \beta_1 = 0$, $H_a: \beta_1 \neq 0$

Step 2: $\alpha = 0.05$

Step 3: $t = \dfrac{b_1}{s_e/\sqrt{\sum x^2 - \left(\sum x\right)^2/n}} = \dfrac{0.16285}{5.418/\sqrt{48747 - 723^2/11}} = 1.053$

Step 4: df = n - 2 = 9; critical values = ± 2.262

Copyright © 2012 Pearson Education, Inc. Publishing as Addison-Wesley.

For the p-value approach, p > 0.20.

Step 5: Since 1.053 < 2.262, do not reject H_0.
Because the p-value is greater than the significance level of 0.05, we cannot reject H_0.

Step 6: At the 5% significance level, we conclude that plant weight is not useful as a predictor of volume of volatile emissions.

14.55 From Exercise 4.55, $\sum x = 117$, $\sum x^2 = 1869$, and $b_1 = -0.846$. From Exercise 14.27, $s_e = 3.538$.

Step 1: H_0: $\beta_1 = 0$, H_a: $\beta_1 \neq 0$

Step 2: $\alpha = 0.01$

Step 3: $t = \dfrac{b_1}{s_e / \sqrt{\sum x^2 - (\sum x)^2 / n}} = \dfrac{-0.846}{3.538 / \sqrt{1869 - 117^2 / 8}} = -3.004$

Step 4: df = n - 2 = 6; critical values = ±3.707
For the p-value approach, 0.02 < p < 0.05.

Step 5: Since -3.707 < -3.004 < 3.707, do not reject H_0.
Because the p-value is less than the significance level of 0.01, we cannot reject H_0.

Step 6: At the 1% significance level, we conclude that study-time is not useful as a predictor of test scores of calculus students.

14.57 From Exercise 4.51, $\sum x = 41$, $\sum x^2 = 199$, and $b_1 = -27.9029$. From Exercise 14.23, $s_e = 14.2465$.

Step 1: For a 90% confidence interval, $\alpha = 0.10$. With df = n - 2 = 8, $t_{\alpha/2} = t_{0.05} = 1.860$.

Step 2: The endpoints of the confidence interval for β_1 are

$$b_1 \pm t_{\alpha/2} \cdot s_e / \sqrt{\sum x^2 - (\sum x)^2 / n}$$

$$-27.9029 \pm 1.860 \cdot 14.2465 / \sqrt{199 - 41^2 / 10}$$

$$-27.9029 \pm 4.7670$$

$$-32.6699 \ to \ -23.1359$$

We can be 90% confident that the yearly decrease in mean price for Corvettes is somewhere between $2314 and $3267.

14.59 From Exercise 4.53, $\sum x = 723$, $\sum x^2 = 48747$, and $b_1 = 0.16285$. From Exercise 14.25, $s_e = 5.4177$.

Step 1: For a 95% confidence interval, $\alpha = 0.05$. With df = n - 2 = 9, $t_{\alpha/2} = t_{0.025} = 2.262$.

Step 2: The endpoints of the confidence interval for β_1 are

$$b_1 \pm t_{\alpha/2} \cdot s_e / \sqrt{\sum x^2 - (\sum x)^2 / n}$$

$$0.16285 \pm 2.262 \cdot 5.4177 / \sqrt{48747 - 723^2 / 11}$$

$$0.16285 \pm 0.34997$$

$$-0.187 \ to \ 0.513$$

We can be 95% confident that the increase in mean volatile plant emissions per one gram increase in weight is somewhere between -0.187 and 0.513 hundred nanograms.

14.61 From Exercise 4.55, $\sum x = 117$, $\sum x^2 = 1869$, and $b_1 = -0.846$. From Exercise 14.27, $s_e = 3.53816$.

Copyright © 2012 Pearson Education, Inc. Publishing as Addison-Wesley.

Step 1: For a 99% confidence interval, $\alpha = 0.01$. With df $= n - 2 = 6$, $t_{\alpha/2}$ $= t_{0.005} = 3.707$.

Step 2: The endpoints of the confidence interval for β_1 are

$$b_1 \pm t_{\alpha/2} \cdot s_e / \sqrt{\sum x^2 - \left(\sum x\right)^2 / n}$$

$$-0.846 \pm 3.707 \cdot 3.53816 / \sqrt{1869 - 117^2 / 8}$$

$$-0.846 \pm 1.044$$

$$-1.890 \;\; to \;\; 0.198$$

We can be 99% confident that the change in test score per one hour increase in study-time for calculus students is somewhere between −1.890 and 0.198.

14.63 (a) In Exercise 14.31, part (c), we determined that it was reasonable to consider Assumptions 1-3 for regression inferences met by these variables.

(b) Part of the Minitab output from Exercise 14.31 is reproduced below.

Predictor	Coef	SE Coef	T	P
Constant	5.37	14.02	0.38	0.704
INAUGURATION	1.1848	0.2537	4.67	0.000

The t value and the P-value for the regression t-test are 4.67 and 0.000, respectively. Since 0.000 < 0.05, we conclude that the inauguration ages are useful for predicting the presidents' death ages.

14.65 (a) In Exercise 14.33, part (c), we determined that it was temporarily reasonable to consider Assumptions 1-3 for regression inferences met by these variables, although Assumption 3 was questionable. Note: In Exercise 4.67, we concluded that the observations did not follow a linear pattern. This exercise will help to resolve the conflict.

(b) Part of the Minitab output from Exercise 14.33 is reproduced below.

Predictor	Coef	SE Coef	T	P
Constant	292.0	140.3	2.08	0.043
LOT SIZE	67.11	59.87	1.12	0.269

The t value and the P-value for the regression t-test are 1.12 and 0.269, respectively. Since 0.269 > 0.05, we conclude that the lot size is not useful for predicting the home values.

14.67 (a) In Exercise 14.35, part (c), we determined that it was reasonable to consider Assumptions 1-3 for regression inferences met by these variables.

(b) Part of the Minitab output from Exercise 14.35 is reproduced below.

Predictor	Coef	SE Coef	T	P
Constant	-7.572	1.786	-4.24	0.000
HIGH	0.91685	0.02967	30.91	0.000

The t value and the P-value for the regression t-test are 30.91 and 0.000, respectively. Since 0.000 < 0.05, we conclude that average high temperatures are useful for predicting average low temperatures in January.

14.69 (a) In Exercise 14.37, part (c), we determined that it was not reasonable to consider Assumption 1 for regression inferences met by these variables. The residual plot was convex upward. Therefore, parts (b) and (c) are omitted.

14.71 (a) In Exercise 14.39, part (c), we determined that it was not reasonable to consider Assumptions 1 and 2 for regression inferences met by these variables. Therefore, parts (b) and (c) are omitted.

Exercises 14.3

14.73 $11,443

Copyright © 2012 Pearson Education, Inc. Publishing as Addison-Wesley.

14.75 From Exercise 14.11, $\hat{y} = 1 - 2x$. From Exercise 14.45, $s_e = 2.44949$ and $S_{xx} = 2$.

(a) The point estimate for the conditional mean of y at x = 2 is
$\hat{y}_p = 1 - 2x = 1 - 2(2) = -3$.

(b) The 95% confidence interval for the conditional mean at x = 2 is

$$\hat{y}_p \pm t_{\alpha/2} s_e \sqrt{\frac{1}{n} + \frac{(x_p - \sum x_i / n)^2}{S_{xx}}} = -3 \pm 12.706(2.44949)\sqrt{\frac{1}{3} + \frac{(2 - 6/3)^2}{2}} = -3 \pm 17.97$$

$$= (-20.97, 14.97)$$

(c) The predicted value of y at x = 2 is $\hat{y}_p = 1 - 2x = 1 - 2(2) = -3$.

(d) The 95% prediction interval for y at x = 2 is

$$\hat{y}_p \pm t_{\alpha/2} s_e \sqrt{1 + \frac{1}{n} + \frac{(x_p - \sum x_i / n)^2}{S_{xx}}} = -3 \pm 12.706(2.44949)\sqrt{1 + \frac{1}{3} + \frac{(2 - 6/3)^2}{2}}$$

$$= -3 \pm 35.94 = (-38.94, 32.94)$$

14.77 From Exercise 14.13, $\hat{y} = -3 + 2x$. From Exercise 14.47,
$s_e = 1.73205$ and $S_{xx} = 24$.

(a) The point estimate for the conditional mean of y at x = 4 is
$\hat{y}_p = -3 + 2x = -3 + 2(4) = 5$.

(b) The 95% confidence interval for the conditional mean at x = 4 is

$$\hat{y}_p \pm t_{\alpha/2} s_e \sqrt{\frac{1}{n} + \frac{(x_p - \sum x_i / n)^2}{S_{xx}}} = 5 \pm 4.303(1.73205)\sqrt{\frac{1}{4} + \frac{(4 - 10/4)^2}{5}} = 5 \pm 6.24$$

$$= (-1.24, 11.24)$$

(c) The predicted value of y at x = 4 is $\hat{y}_p = -3 + 2x = -3 + 2(4) = 5$.

(d) The 95% prediction interval for y at x = 4 is

$$\hat{y}_p \pm t_{\alpha/2} s_e \sqrt{1 + \frac{1}{n} + \frac{(x_p - \sum x_i / n)^2}{S_{xx}}} = 5 \pm 4.303(1.73205)\sqrt{1 + \frac{1}{4} + \frac{(4 - 10/4)^2}{5}}$$

$$= 5 \pm 9.72 = (-4.72, 14.72)$$

14.79 From Exercise 14.15, $\hat{y} = 2.875 + 0.625x$. From Exercise 14.49,
$s_e = 1.88193$ and $S_{xx} = 24$.

(a) The point estimate for the conditional mean of y at x = 3 is
$\hat{y}_p = 2.875 + 0.625x = 2.875 + 0.625(3) = 4.75$.

(b) The 95% confidence interval for the conditional mean at x = 3 is

$$\hat{y}_p \pm t_{\alpha/2} s_e \sqrt{\frac{1}{n} + \frac{(x_p - \sum x_i / n)^2}{S_{xx}}} = 4.75 \pm 3.182(1.88193)\sqrt{\frac{1}{5} + \frac{(3 - 15/5)^2}{24}} = 4.75 \pm 2.68$$

$$= (2.07, 7.43)$$

(c) The predicted value of y at x = 3 is
$\hat{y}_p = 2.875 + 0.625x = 2.875 + 0.625(3) = 4.75$.

(d) The 95% prediction interval for y at x = 3 is

Copyright © 2012 Pearson Education, Inc. Publishing as Addison-Wesley.

$$\hat{y}_p \pm t_{\alpha/2} s_e \sqrt{1 + \frac{1}{n} + \frac{\left(x_p - \sum x_i / n\right)^2}{S_{xx}}} = 4.75 \pm 3.182(1.88193)\sqrt{1 + \frac{1}{5} + \frac{(3 - 15/5)^2}{24}}$$

$$= 4.75 \pm 6.56 = (-1.81, 11.31)$$

14.81 From Exercise 4.51, $\hat{y}$ = 456.602 - 27.9029x, $\sum x$ = 41, and $\sum x^2$ = 199. From Exercise 14.23, s_e = 14.2465.

(a) $\hat{y}_p$ =456.602 - 27.9029(4) = 344.9904 = \$34,499.

(b) Step 1:   For a 90% confidence interval, α = 0.10.

With df = n - 2 = 8, $t_{\alpha/2}$ = $t_{0.05}$= 1.860.

Step 2: $\hat{y}_p$ = 344.9904

Step 3:   The endpoints of the confidence interval are

$$\hat{y}_p \pm t_{\alpha/2} \cdot s_e \cdot \sqrt{\frac{1}{n} + \frac{\left(x_p - \sum x / n\right)^2}{\sum x^2 - \left(\sum x\right)^2 / n}}$$

$$= 344.9904 \pm 1.860 \cdot 14.2465 \sqrt{\frac{1}{10} + \frac{(4 - 41/10)^2}{199 - 41^2 / 10}}$$

$$= 344.9904 \pm 8.3931 = (336.60, 353.38)$$

The interpretation of this interval is as follows: We can be 90% confident that the mean price of all four-year-old Corvettes is somewhere between \$33,660 and \$35,338.

(c)   This is the same as the answer in part (a): $\hat{y}_p$ = 344.9904.

(d)   For a 90% prediction interval, Steps 1 and 2 are the same as Steps 1 and 2, respectively, in part (b). Thus, they are not repeated here. Only Step 3 is presented.

Step 3:   The endpoints of the prediction interval are

$$\hat{y}_p \pm t_{\alpha/2} \cdot s_e \cdot \sqrt{1 + \frac{1}{n} + \frac{\left(x_p - \sum x / n\right)^2}{\sum x^2 - \left(\sum x\right)^2 / n}}$$

$$= 344.9904 \pm 1.860 \cdot 14.265 \sqrt{1 + \frac{1}{10} + \frac{(4 - 41/10)^2}{199 - 41^2 / 10}}$$

$$= 344.9904 \pm 27.7959 = (317.20, 372.78)$$

The interpretation of this interval is as follows: We can be 90% certain that the price of a randomly selected four-year-old Corvette will be somewhere between \$31,720 and \$37,278.

(e)

Copyright © 2012 Pearson Education, Inc. Publishing as Addison-Wesley.

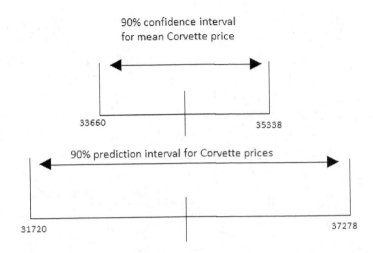

(f) The error in the estimate of the mean price of four-year-old Corvettes is due only to the fact that the population regression line is being estimated by a sample regression line; whereas, the error in the prediction of the price of a randomly selected four-year-old Corvette is due to that fact plus the variation in prices for four-year-old Corvettes.

14.83 From Exercise 4.53, $\hat{y} = 3.523688 + 0.162848x$, $\sum x = 726$, and $\sum x^2 = 48747$. From Exercise 14.25, $s_e = 5.4177$.

(a) $\hat{y}_p = 3.523688 + 0.162848(60) = 13.2946$

(b) Step 1: For a 95% confidence interval, $\alpha = 0.05$.

   With df = n - 2 = 9, $t_{\alpha/2} = t_{0.025} = 2.262$.

   Step 2: $\hat{y}_p = 13.2946$
   Step 3: The endpoints of the confidence interval are

   $$\hat{y}_p \pm t_{\alpha/2} \cdot s_e \cdot \sqrt{\frac{1}{n} + \frac{(x_p - \sum x/n)^2}{\sum x^2 - (\sum x)^2/n}}$$

   $$= 13.2946 \pm 2.262 \cdot 5.4177 \cdot \sqrt{\frac{1}{11} + \frac{(60 - 723/11)^2}{48747 - 723^2/11}}$$

   $$= 813.2946 \pm 4.2036 = (9.0910, 17.4962)$$

The interpretation of this interval is as follows: We can be 95% confident that the mean quantity of volatile emissions of all plants that weigh 60 grams is somewhere between 9.0910 and 17.4982 hundred nanograms.

(c) This is the same as the answer in part (a): $\hat{y}_p = 13.2946$.

(d) For a 95% prediction interval, Steps 1 and 2 are the same as Steps 1 and 2, respectively, in part (b). Thus, they are not repeated here. Only Step 3 is presented.

   Step 3: The endpoints of the prediction interval are

   $$\hat{y}_p \pm t_{\alpha/2} \cdot s_e \cdot \sqrt{1 + \frac{1}{n} + \frac{(x_p - \sum x/n)^2}{\sum x^2 - (\sum x)^2/n}}$$

   $$= 13.2946 \pm 2.262 \cdot 5.4177 \cdot \sqrt{1 + \frac{1}{11} + \frac{(60 - 723/11)^2}{48747 - 723^2/11}}$$

   $$= 813.2946 \pm 12.9557 = (0.3389, 26.2503)$$

Copyright © 2012 Pearson Education, Inc. Publishing as Addison-Wesley.

The interpretation of this interval is as follows: We can be 95% certain that the quantity of volatile emissions of a randomly selected plant that weighs 60 grams is somewhere between 0.3389 and 26.2503 hundred nanograms.

14.85 From Exercise 4.55, $\hat{y} = 94.86698 - 0.84561x$, $\sum x = 117$, and $\sum x^2 = 1869$. From Exercise 14.27, $s_e = 3.5382$.

(a) $\hat{y}_p = 94.86698 - 0.84561(15) = 82.1828$

(b)   Step 1:   For a 99% confidence interval, $\alpha = 0.01$.

With df $= n - 2 = 6$, $t_{\alpha/2} = t_{0.005} = 3.707$.

Step 2: $\hat{y}_p = 82.1828$

Step 3:   The endpoints of the confidence interval are

$$\hat{y}_p \pm t_{\alpha/2} \cdot s_e \cdot \sqrt{\frac{1}{n} + \frac{(x_p - \sum x/n)^2}{\sum x^2 - (\sum x)^2/n}}$$

$$= 82.1828 \pm 3.707 \cdot (3.5382) \cdot \sqrt{\frac{1}{8} + \frac{(15 - 117/8)^2}{1869 - 117^2/8}}$$

$$= 82.1828 \pm 4.6537 = (77.5291, 86.8365)$$

The interpretation of this interval is as follows: We can be 99% confident that the mean calculus test score of students who study 15 hours is somewhere between 77.53 and 86.84 mm.

(c)   This is the same as the answer in part (a): $\hat{y}_p = 82.1828$.

(d)   For a 99% prediction interval, Steps 1 and 2 are the same as Steps 1 and 2, respectively, in part (b). Thus, they are not repeated here. Only Step 3 is presented.

Step 3:   The endpoints of the prediction interval are

$$\hat{y}_p \pm t_{\alpha/2} \cdot s_e \cdot \sqrt{1 + \frac{1}{n} + \frac{(x_p - \sum x/n)^2}{\sum x^2 - (\sum x)^2/n}}$$

$$= 82.1828 \pm 3.707 \cdot (3.5382) \cdot \sqrt{1 + \frac{1}{8} + \frac{(15 - 117/8)^2}{1869 - 117^2/8}}$$

$$= 82.1828 \pm 13.9172 = (68.2656, 96.0100)$$

The interpretation of this interval is as follows: We can be 99% certain that a randomly selected calculus student who studies 125 hours will have a test score somewhere between 68.27 and 96.01 mm.

14.87 (a)   In Exercise 14.31, part (c), we determined that it was reasonable to consider Assumptions 1-3 for regression inferences met by these variables.

(b)   Using Minitab, with the data in columns named INAUGURATION and DEATH,

we choose **Stat ▶ Regression ▶ Regression...**, select <u>DEATH</u> in the

Response text box, and select <u>INAUGURATION</u> in the **Predictors** text box. Click the **Options** button, select the **Regular** option button from the **Residuals for Plots** list, enter <u>53</u> in the **Prediction intervals for new observations** box and <u>95</u> in the **Confidence level** box, click **OK**, and click **OK**. Since much of the output was shown in Exercise 14.31, we show below only that part of the output pertinent to this exercise.
Predicted Values for New Observations

New Obs	Fit	SE Fit	95% CI	95% PI
1	68.16	1.63	(64.87, 71.46)	(48.23, 88.09)

Copyright © 2012 Pearson Education, Inc. Publishing as Addison-Wesley.

The point estimate for the conditional mean of DEATH for inauguration age 53 is found under the word 'Fit' in the output as 68.16. This is the same as the predicted score for inauguration at age 53.

(c) The 95% confidence interval for the conditional mean death age for presidents inaugurated at age 53 is (64.87, 71.46). We can be 95% confident that the mean death age of U.S. Presidents inaugurated at age 53 lies somewhere between 64.87 and 71.46 years.

(d) The predicted death age for Presidents inaugurated at age 53 is found under the word 'Fit' in the output as 68.16.

(e) The 95% prediction interval for the death age of a President inaugurated at age 53 is (48.23, 88.09). We can be 95% certain that the death age of a president inaugurated at age 53 will fall somewhere between 48.23 and 88.09. Note: The lower end of this prediction interval can be safely increased to 53 since a president cannot die prior to being inaugurated.

(f) The confidence interval and the prediction interval are both centered at the predicted value for an inauguration age of 53. The prediction interval is longer because there is variation associated with the additional observation that is not present in the confidence interval for the mean.

14.89 (a) In Exercise 14.33, part (c), we determined that it was reasonable to consider Assumptions 1-3 for regression inferences met by these variables. Note: In Exercise 4.67, we concluded that the pattern of the observations was NOT linear, meaning that Assumption 1 for regressions inferences was not met by these data. In Exercise 14.65, we found that the evidence for a linear pattern was weak or none. Thus we will not continue with parts (b)-(f).

14.91 (a) In Exercise 14.35, part (c), we determined that it was reasonable to consider Assumptions 1-3 for regression inferences met by these variables.

(b) Using Minitab, with the data in columns named HIGH and LOW, we choose

Stat ▶ Regression ▶ Regression..., select LOW in the **Response** text box, and select HIGH in the **Predictors** text box. Click the **Options** button, select the **Regular** option button from the **Residuals for Plots** list, enter 55 in the **Prediction intervals for new observations** box and 95 in the **Confidence level** box, click **OK**, and click **OK**. Since much of the output was shown in Exercise 14.35, we show below only that part of the output pertinent to this exercise.

Predicted Values for New Observations

```
New
Obs    Fit  SE Fit      95% CI            95% PI
  1  42.855  0.590  (41.669, 44.040)  (34.425, 51.285)
```

The point estimate for the conditional mean January low temperature for an average high temperature of 55 degrees is found under the word 'Fit' in the output as 42.855 degrees. This is the same as the predicted value for an average high temperature of 55 degrees.

(c) The 95% confidence interval for the conditional mean low January temperature for cities with mean high temperature of 55 degrees is (41.669, 44.040). We can be 95% confident that the mean low January temperature for cities with a mean January high temperature of 55 degrees lies somewhere between 41.669 and 44.040 degrees.

(d) The predicted mean low January temperature for a city with a mean high January temperature of 55 degrees is found under the word 'Fit' in the output as 42.855 degrees.

(e) The 95% prediction interval for the mean January low temperature for a city with a mean January high temperature is (34.425, 51.285). We can be 95% certain that the mean low January temperature for a city with a

Copyright © 2012 Pearson Education, Inc. Publishing as Addison-Wesley.

mean January high temperature of 55 degrees will lie somewhere between 34.425 and 51.285 degrees.

  (f) The confidence interval and the prediction interval are both centered at the predicted value for a city with a mean January high temperature of 55 degrees. The prediction interval is longer because there is variation associated with the additional observation that is not present in the confidence interval for the mean.

14.93 (a) In Exercise 14.37, part (c), we determined that it was not reasonable to consider Assumption 1 for regression inferences met by these variables. The residual plot was convex upward. Therefore, parts (b)-(f) are omitted.

14.95 (a) In Exercise 14.39, part (c), we determined that it was not reasonable to consider Assumptions 1 and 2 for regression inferences met by these variables. Therefore, parts (b)-(f) are omitted.

14.97 (a)-(b) In Example 14.8 of the text, the 95% confidence interval for the mean price of all Orions of a given age is given by

$$134.69 \pm 2.262 \cdot 12.58 \cdot \sqrt{\frac{1}{11} + \frac{(x_p - 58/11)^2}{326 - 58^2/11}}$$

Substituting the values 2 through 7 into this formula for x_p, we obtain the following confidence intervals and a plot showing the lower and upper confidence limits for the six different values of x_p. The margin of error (half of the difference between the upper and lower confidence limits) is shown in the last column of the table and is plotted against x_p in the second graph.

x_p	Lower Confidence Limit	Upper Confidence Limit	Margin of Error
2	112.25	157.13	22.44
3	117.93	151.45	16.76
4	122.92	146.46	11.77
5	125.94	143.44	8.75
6	124.95	144.43	9.74
7	120.79	148.59	13.90

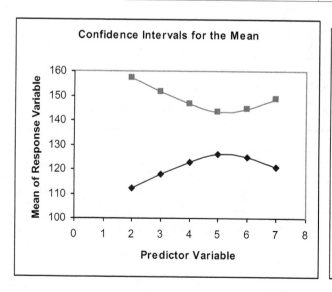

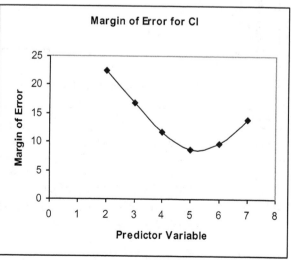

Copyright © 2012 Pearson Education, Inc. Publishing as Addison-Wesley.

We see from the table and the first plot that the confidence intervals get wider as the predictor variable gets farther from the mean of the predictor variable. This is also seen in the second plot, which shows that the margin of error first decreases as x_p increases to the mean 5.27 of the predictor variable and then increases again as x_p moves away from 5.27.

(c) In Example 14.9 of the text, the 95% prediction interval for an Orion of a given age is given by

$$= 134.69 \pm 2.262 \cdot 12.58 \cdot \sqrt{1 + \frac{1}{11} + \frac{(x_p - 58/11)^2}{326 - 58^2/11}}$$

Substituting the values 2 through 7 into this formula for x_p, we obtain the following confidence intervals and a plot showing the lower and upper confidence limits for the six different values of x_p. The margin of error (half of the difference between the upper and lower confidence limits) is shown in the last column of the table and is plotted against x_p in the second graph.

x_p	Lower Prediction Limit	Upper Prediction Limit	Margin of Error
2	98.45	170.93	36.24
3	101.67	167.71	33.02
4	103.89	165.49	30.80
5	104.92	164.46	29.77
6	104.61	164.77	30.08
7	103.02	166.36	31.67

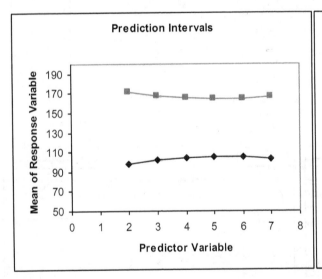
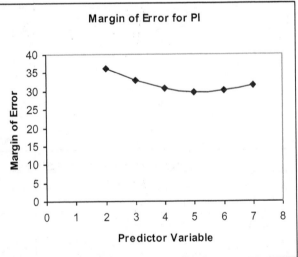

We see from the table and the first plot that the prediction intervals get wider as the predictor variable gets farther from the mean of the predictor variable. This is also seen in the second plot which shows that the margin of error first decreases as x_p increases to the mean 5.27 of the predictor variable and then increases again as x_p moves away from 5.27.

Copyright © 2012 Pearson Education, Inc. Publishing as Addison-Wesley.

Exercises 14.4

14.99   r, the sample linear correlation coefficient

14.101 (a) uncorrelated

(b) increases

(c) negatively

14.103   Step 1 $H_0: \rho = 0$, $H_a: \rho < 0$

Step 2 $\alpha = 0.10$

Step 3 From Exercise 4.117, r = -0.76, n = 3

$$t = \frac{r}{\sqrt{\frac{1-r^2}{n-2}}} = \frac{-0.76}{\sqrt{\frac{1-(-0.76)^2}{1}}} = -1.169$$

Step 4 The critical value is -3.0784

For the P-value approach, P > 0.10.

Step 5 Since -1.169 > -3.078, do not reject H_0.

Since the P-value is greater than 0.10, do not reject H_0.

14.105   Step 1 $H_0: \rho = 0$, $H_a: \rho > 0$

Step 2 $\alpha = 0.10$

Step 3 From Exercise 4.119, r = 0.877, n = 4

$$t = \frac{r}{\sqrt{\frac{1-r^2}{n-2}}} = \frac{0.877}{\sqrt{\frac{1-0.877^2}{2}}} = 2.581$$

Step 4 The critical value is 1.886

For the P-value approach, 0.05 < P < 0.10

Step 5 Since 2.581 > 1.886, reject H_0.

Since the P-value is less than 0.10, reject H_0.

14.107   Step 1 $H_0: \rho = 0$, $H_a: \rho \neq 0$

Step 2 $\alpha = 0.10$

Step 3 From Exercise 4.121, r = -0.685, n = 5

$$t = \frac{r}{\sqrt{\frac{1-r^2}{n-2}}} = \frac{-0.685}{\sqrt{\frac{1-(-0.685)^2}{3}}} = -1.629$$

Step 4 The critical values are ± 2.353

For the P-value approach, P > 0.20.

Step 5 Since -2.353 < -1.629 < 2.353, do not reject H_0.

Since the P-value is greater than 0.10, do not reject H_0.

14.109   From Exercise 4.123, r = - 0.96787 and n = 10.

Step 1: H_0: $\rho = 0$ H_a: $\rho < 0$

Step 2: $\alpha = 0.05$

Step 3: $t = \dfrac{r}{\sqrt{\dfrac{1-r^2}{n-2}}} = \dfrac{-0.96787}{\sqrt{\dfrac{1-(-0.96787)^2}{10-2}}} = -10.8870$

Step 4: The critical value for n - 2 = 8 df is -1.860.

For the p-value approach, p < 0.005.

Step 5: Since -10.8870 < -1.860, reject H_0 and conclude that $\rho < 0$.

Since p < 0.05, reject H_0 and conclude that $\rho < 0$.

14.111   From Exercise 4.125, r = 0.33107 and n = 11.

Step 1: H_0: $\rho = 0$ H_a: $\rho \neq 0$

Step 2: $\alpha = 0.05$

Copyright © 2012 Pearson Education, Inc. Publishing as Addison-Wesley.

Step 3: $t = \dfrac{r}{\sqrt{\dfrac{1-r^2}{n-2}}} = \dfrac{0.33107}{\sqrt{\dfrac{1-(0.33107)^2}{11-2}}} = 1.0526$

Step 4: The critical values for n - 2 = 9 df are $\pm$2.262.
For the p-value approach, p > 0.200.

Step 5: Since -2.262 < 1.0526 < 2.262, do not reject H_0 and conclude that $\rho = 0$ is reasonable.

Since p > 0.05, do not reject H_0 and conclude that $\rho = 0$ is reasonable.

14.113 From Exercise 4.127, r = -0.7749 and n = 8.

Step 1: H_0: $\rho = 0$ H_a: $\rho < 0$

Step 2: $\alpha = 0.01$

Step 3: $t = \dfrac{r}{\sqrt{\dfrac{1-r^2}{n-2}}} = \dfrac{-0.7749}{\sqrt{\dfrac{1-(-0.7749)^2}{8-2}}} = -3.003$

Step 4: The critical value for n - 2 = 6 df is -3.143.
For the p-value approach, 0.010 < p < 0.025.

Step 5: Since -3.003 > 3.143, do not reject H_0. There is not sufficient evidence to conclude that $\rho \neq 0$.

Since p > 0.01, do not reject H_0.

14.115 The population linear correlation coefficient ρ is a parameter (constant) that measures the linear correlation between the population of all data points. The sample linear correlation coefficient r is a statistic (random variable); it measures the linear correlation between a sample of data points, and so its value depends on chance, namely, on which data points are obtained from sampling.

14.117 In Exercise 14.31, part (c), we determined that it was reasonable to consider Assumptions 1-3 for regression inferences met by these variables.

Using Minitab, choose **Stat ▶ Basic Statistics ▶ Correlation...**, select INAUGURATION and DEATH in the **Variables** text box, and click **OK**. The result is
  Pearson correlation of INAUGURATION and DEATH = 0.609
  P-Value = 0.000
Note: The p-value shown is for a two-tailed test. The p-value for the right-tail test would be even smaller. Since the p-value < 0.05, reject H_0 and conclude that the inauguration ages and death ages of U.S. Presidents are positively linearly correlated.

14.119 In Exercise 14.65, part (a), we determined that it was not reasonable to consider Assumptions 1 for regression inferences met by these variables. The correlation t-test should not be used. Over several exercises, there was some controversy regarding this decision. If you choose to continue with the correlation t-test, the results from Minitab are
  Pearson correlation of LOT SIZE and VALUE = 0.170
  P-Value = 0.269
The high P-value confirms our decision not to reject $\rho = 0$.

14.121 In Exercise 14.35, part (c), we determined that it was reasonable to consider Assumptions 1-3 for regression inferences met by these variables.

Copyright © 2012 Pearson Education, Inc. Publishing as Addison-Wesley.

Using Minitab, choose **Stat ▶ Basic Statistics ▶ Correlation...,** select
<u>HIGH</u> and <u>LOW</u> in the **Variables** text box, and click **OK.** The result is
  Pearson correlation of HIGH and LOW = 0.976
  P-Value = 0.000
Since the P-value < 0.05, reject $\rho = o$.

14.123 In Exercise 14.37, part (c), we determined that it was not reasonable
to consider Assumption 1 for regression inferences met by these
variables. The correlation t-test should not be used.

14.125 In Exercise 14.39, part (c), we determined that it was not reasonable
to consider Assumptions 1 and 2 for regression inferences met by these
variables. The correlation t-test should not be used.

Review Problems for Chapter 14

1. (a) conditional
   (b) The four assumptions for regression inferences are:
      (1) *Population regression line:* There is a line $y = \beta_0 + \beta_1 x$ such
          that, for each x-value, the mean of the corresponding population of
          y-values lies on that line.
      (2) *Equal standard deviations:* The standard deviation σ of the
          population of y-values corresponding to a particular x-value is the
          same, regardless of the x-value.
      (3) *Normality:* For each x-value, the corresponding population of y-
          values is normally distributed.
      (4) *Independence:* The observations of the response variable are
          independent of one another.

2. (a) The slope of the sample regression line, b_1
   (b) The y-intercept of the sample regression line, b_0
   (c) s_e

3. We used a plot of the residuals against the values of the predictor variable
   x and a normal probability plot of the residuals. In the first plot, the
   residuals should lie in a horizontal band centered on and symmetric about
   the x-axis. In the second plot, the points should lie roughly in a line.

4. (a) A residual plot showing curvature indicates that the first assumption,
       that of linearity, is probably not valid.
   (b) This type of plot indicates that the second assumption, that of
       constant standard deviation, is probably not valid.
   (c) A normal probability plot with extreme curvature indicates that the
       third assumption, that the conditional distribution of the response
       variable is normally distributed, is not valid.
   (d) A normal probability plot that is roughly linear, but shows outliers
       may be indicating that: the linear model is appropriate for most values
       of x, but not for all; or, that the standard deviation is not constant
       for all values of x, allowing for a few y values to lie far from the
       regression line; or, that a few data values are 'faulty', that is, the
       experimental conditions represented by the value of the x variable were
       not as they should have been or that the y value was in error.

5. If we reject the null hypothesis, we are claiming that β_1 is not zero. This
   means that different values of x will lead to different values of y. Hence,
   the regression equation is useful for making predictions.

6. t (for b_1), r, r^2

7. No. The best estimate of conditional mean of the response variable and the
   best prediction of a single future value of y are the same.

8. A confidence interval estimates the value of a parameter; a prediction
   interval is used to predict a future value of a random variable.

9. ρ

Copyright © 2012 Pearson Education, Inc. Publishing as Addison-Wesley.

10. (a) If $\rho > 0$, the variables are positively correlated, that is, there is a tendency for one variable to increase linearly as the other one increases.

    (b) If we know only that $\rho \neq 0$, the two variables are linearly correlated, that is, there is a linear relationship between them.

    (c) If $\rho < 0$, the variables are negatively correlated, that is, there is a tendency for one variable to decrease linearly as the other one increases.

11. If the regression assumptions are satisfied, then there is a linear relationship between the student/faculty ratio and the graduation rate, the conditional standard deviation of the graduation rate is the same for all values of the student/faculty ratio, the conditional distribution of the graduation rate is a normal distribution for all values of the student/faculty ratio, and the values of the graduation rate are independent of each other.

12. (a) To determine b_0 and b_1 for the regression line, construct a table of values for x, y, x^2, xy and their sums. A column for y^2 is also presented for later use.

x	y	x^2	xy	y^2
16	45	256	720	2025
20	55	400	1100	3025
17	70	289	1190	4900
19	50	361	950	2500
22	47	484	1034	2209
17	46	289	782	2116
17	50	289	850	2500
17	66	289	1122	4356
10	26	100	260	676
18	60	324	1080	3600
173	515	3081	9088	27907

$S_{xx} = 3081 - 173^2/10 = 88.10$

$S_{xy} = 9088 - (173)(515)/10 = 178.50$

$b_1 = 178.50/88.10 = 2.0261$

$b_0 = 515/10 - 2.0261(173/10) = 16.4484$

$\hat{y}_p = 16.448 + 2.026x$ is the equation of the regression line.

(b) $S_{yy} = 27907 - 515^2/10 = 1384.50$

   $SSE = S_{yy} - S^2_{xy}/S_{xx} = 1384.50 - 178.50^2/88.10 = 1022.8400$

$$s_e = \sqrt{\frac{SSE}{n-2}} = \sqrt{\frac{1922.8400}{10-2}} = 11.3073$$

This value of s_e indicates that, roughly speaking, the predicted values of the graduation rate differ from the observed values of the graduation rate by about 11.31.

(c) Presuming that the variables Student/Faculty ratio and Graduation rate satisfy the assumptions for regression inferences, the standard error of the estimate, $s_e = 11.31$, provides an estimate for the common population standard deviation, σ, of graduation rates of entering freshmen at universities with any particular student/faculty ratio.

Copyright © 2012 Pearson Education, Inc. Publishing as Addison-Wesley.

13. For each value of x, we compute $e = y - \hat{y}$ in the following table and plot e against x.

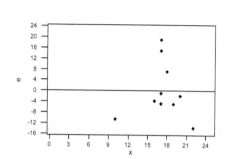

x	e
16	-3.87
20	-1.97
17	19.11
19	-4.94
22	-14.02
17	-4.89
17	-0.89
17	15.11
10	-10.71
18	7.08

Residual	Normal score
-14.02	-1.55
-10.71	-1.00
-4.94	-0.65
-4.89	-0.37
-3.87	-0.12
-1.97	0.12
-0.89	0.37
7.08	0.65
15.11	1.00
19.11	1.55

The small sample size makes it difficult to evaluate the plot of e against x. While there are no obvious patterns in the first plot, there are two points (in the lower left and right corners of the plot) that are cause for concern. The normal probability plot is not far from linear.

14. (a) From Review Problem 11, $\sum x = 173$, $\sum x^2 = 3{,}081$, and $b_1 = 2.0261$. From Review Problem 12, $s_e = 11.3073$.

Step 1: H_0: $\beta_1 = 0$

H_a: $\beta_1 \neq 0$

Step 2: $\alpha = 0.05$

Step 3: $t = \dfrac{b_1}{s_e / \sqrt{\sum x^2 - (\sum x)^2 / n}} = \dfrac{2.0261}{11.3073 / \sqrt{3081 - 173^2 / 10}} = 1.68$

Step 4: df = n - 2 = 8; critical values = ±2.306
For the P-value approach, note that
0.10 < p < 0.20.

Step 5: Since -2.306 < 1.68 < 2.306, do not reject the null hypothesis at the 5% significance level.
Since p > α, do not reject the null hypothesis.

Step 6: The data do not provide sufficient evidence to conclude that the student-to-faculty ratio is useful as a predictor of graduation rate.

(b) Step 1: For a 95% confidence interval, $\alpha = 0.05$. With df = n - 2 = 8, $t_{\alpha/2} = t_{0.025} = 2.306$.

Copyright © 2012 Pearson Education, Inc. Publishing as Addison-Wesley.

Step 2:   The endpoints of the confidence interval for β_1 are

$$b_1 \pm t_{\alpha/2} \cdot s_e / \sqrt{\sum x^2 - \left(\sum x\right)^2 / n}$$

$$2.0261 \pm 2.306 \cdot 11.3073 / \sqrt{3081 - 173^2 / 10}$$

$$2.0261 \pm 2.7780$$

$$-0.7519 \ \textit{to} \ 4.8041$$

We can be 95% confident that the change in graduation rate for universities per 1 unit increase in the student/faculty ratio is somewhere between -0.7519 and 4.8041 percent.

15.   From Review Exercise 11, $\hat{y}$ = 16.4484 + 2.0261x, $\sum x$ = 173, and $\sum x^2$ = 3,081. From Review Exercise 12, s_e = 11.3073.

(a) $\hat{y}_p$ = 16.4484 + 2.0261(17) = 50.89

(b)   Step 1:   For a 95% confidence interval, α = 0.05.

With df = n - 2 = 8, $t_{\alpha/2}$ = $t_{0.025}$= 2.306.

Step 2: $\hat{y}_p$ =50.89

Step 3:   The endpoints of the confidence interval are:

$$\hat{y}_p \pm t_{\alpha/2} \cdot s_e \cdot \sqrt{\frac{1}{n} + \frac{(x_p - \sum x/n)^2}{\sum x^2 - \left(\sum x\right)^2 / n}}$$

$$= 50.89 \pm 2.306 \cdot 11.3073 \cdot \sqrt{\frac{1}{10} + \frac{(17 - 173/10)^2}{3081 - 173^2 / 10}}$$

$$= 50.89 \pm 8.29 = (42.60, \ 59.18)$$

The interpretation of this interval is as follows: We can be 95% confident that the graduation rate of universities with a student/faculty ratio of 17 is somewhere between 42.60% and 59.18%.

(c)   This is the same as the answer in part (a): $\hat{y}_p$ = 50.89.

(d)   For a 95% prediction interval, Steps 1 and 2 are the same as Steps 1 and 2, respectively, in part (b).   Thus, they are not repeated here. Only Step 3 is presented.

Step 3: The endpoints of the prediction interval are:

$$\hat{y}_p \pm t_{\alpha/2} \cdot s_e \cdot \sqrt{1 + \frac{1}{n} + \frac{(x_p - \sum x/n)^2}{\sum x^2 - \left(\sum x\right)^2 n}}$$

$$= 50.89 \pm 2.306 \cdot 11.3073 \cdot \sqrt{1 + \frac{1}{10} + \frac{(17 - 173/10)^2}{3081 - 173^2 / 10}}$$

$$= 50.89 \pm 27.36 = (23.53, \ 78.25)$$

The interpretation of this interval is as follows: We can be 95% certain that the graduation rate of a randomly selected university with a student/faculty ratio of 17 will be somewhere between 23.53% and 78.25%.

(e)   The error in the estimate of the mean graduation rate for universities with a student/faculty ratio of 17 is due only to the fact that the population regression line is being estimated by a sample regression line.   The error in the prediction of the graduation rate of a randomly chosen university with a student/faculty ratio is due to the estimation error mentioned above plus the variation in graduation rates of universities with a student/faculty ratio of 17.

Copyright © 2012 Pearson Education, Inc. Publishing as Addison-Wesley.

16.   From Review Problem 12,

$S_{xx} = 88.10$

$S_{xy} = 178.50$

$S_{yy} = 1384.50$

$$r = \frac{S_{xy}}{\sqrt{S_{xx}S_{yy}}} = \frac{178.50}{\sqrt{(88.10)(1384.50)}} = 0.5111$$

Step 1: $H_0: \rho = 0$, $H_a: \rho > 0$

Step 2: $\alpha = 0.025$

Step 3:

$$t = \frac{r}{\sqrt{\dfrac{1-r^2}{n-2}}} = \frac{0.5111}{\sqrt{\dfrac{1-(0.5111)^2}{10-2}}} = 1.682$$

Step 4:   df = n - 2 = 8; critical value = 2.306
           For the P-value approach, $0.05 < p < 0.10$.

Step 5:   Since 1.682 < 2.306, do not reject H_0.
           Because the p-value is larger than the significance level of
           0.025, we cannot reject H_0.

Step 6:   The data do not provide sufficient evidence to conclude that
           graduation rate and student /faculty ratio are positively
           linearly correlated.

17.   Using Minitab, with the data in columns named IMR and LE, we choose **Stat** ▶

Regression ▶ **Regression...**, select 'LE' in the **Response:** text box, select

IMR in the **Predictors:** text box, click on the **Graphs** button, click in the
Residuals versus the variables text box and select IMR, and click **OK**.
Looking ahead to parts (b-h) and (j), click on the **Options...** button, enter
30 in the **Prediction intervals for new observations:** text box, enter 95 in
the **Confidence level:** text box, and click **OK**. Click on the **Storage...**
button, check the **Residuals** box, click **OK**, and click **OK**. Then choose **Stat**

▶ **Basic statistics** ▶ **Normality test...**, select RESI1 in the **Variable** text
box, select the **Ryan-Joiner** option button from the **Tests for Normality**
field, and click **OK**. We will first show all of the written output and then
answer the questions. The graphical output will be shown in part (k).

           The regression equation is
           LE = 79.4 - 0.354 IMR

```
Predictor       Coef   SE Coef       T      P
Constant      79.396     1.009   78.72  0.000
IMR          -0.35374   0.02601  -13.60  0.000

S = 5.17598    R-Sq = 76.1%    R-Sq(adj) = 75.7%

Analysis of Variance

Source           DF       SS       MS       F      P
Regression        1   4956.2   4956.2  185.00  0.000
Residual Error   58   1553.9     26.8
Total            59   6510.1
```

Copyright © 2012 Pearson Education, Inc. Publishing as Addison-Wesley.

Unusual Observations

Obs	IMR	LE	Fit	SE Fit	Residual	St Resid
11	46	49.890	63.251	0.795	-13.361	-2.61R
25	100	47.430	43.894	1.971	3.536	0.74 X
34	34	44.280	67.418	0.680	-23.138	-4.51R
48	44	50.160	63.828	0.773	-13.668	-2.67R
52	91	43.450	47.365	1.734	-3.915	-0.80 X

R denotes an observation with a large standardized residual.
X denotes an observation whose X value gives it large leverage.

Predicted Values for New Observations

New Obs	Fit	SE Fit	95% CI	95% PI
1	68.784	0.669	(67.445, 70.122)	(58.337, 79.231)

Values of Predictors for New Observations

New Obs	IMR
1	30.0

(a) LE = 79.4 - 0.354 IMR
(b) The standard error of the estimate is 5.17598. Roughly speaking, this indicates how much, on average, the predicted life expectancies differ from the observed values of life expectancies.
(c) The coefficient of determination is 0.761, indicating that 76.1% of the variation in life expectancy is explained by a linear relationship with the infant mortality rate. The t-test for the slope results in t = -13.60 with a P-value of 0.000. Thus, we would reject the null hypothesis that $\beta_1 = 0$. This is a fairly strong relationship and indicates that the IMR may be useful for predicting life expectancy if the assumptions for making such inferences are valid.
(d) The point estimate for the conditional mean of life expectancy for IMR = 30 is 68.784 years. This is the point on the sample regression line that best estimates the corresponding point on the population regression line.
(e) The 95% confidence interval for the conditional mean of life expectancy for IMR = 30 is 67.445 to 70.122 years. We are 95% confident that the mean life expectancy is somewhere between these limits when the IMR is 30.
(f) The predicted value of life expectancy for IMR = 30 is 68.784 years. This is our best estimate of life expectancy for a country with an IMR equal to 30.
(g) The 95% prediction interval for the life expectancy of a country with IMR = 30 is 58.337 to 79.231 years. We are 95% certain that the life expectancy is somewhere between these limits when the IMR is 30.
(h) The error in the estimate of the mean life expectancy of countries with IMR = 30 is due only to the fact that the population regression line is being estimated by a sample regression line; whereas, the error in the prediction of the life expectancy of a single country with an IMR = 30 is due to that fact plus the variation in life expectancies for countries with IMR = 30.

(i) Choose **Stat ▶ Basic Statistics ▶ Correlation**, enter IMR and "LE' in the **Variables** text box, check the box to **Display P-values**, and click **OK**. The result is
Pearson correlation of IMR and LE = -0.873
P-Value = 0.000

Copyright © 2012 Pearson Education, Inc. Publishing as Addison-Wesley.

The P-value for a left-tail test is half of that shown (still 0.000). Since the P-value is less than 0.05, we reject the null hypothesis that $\rho = 0$ and since r is negative, the data provide evidence that there is a negative correlation between IMR and life expectancy assuming that the assumptions for making regression inferences are reasonable.

(j) For the residual analysis, we look at the graph produced by the procedure at the beginning of this solution and also the results of a Ryan-Joiner normality test on the residuals, which were stored earlier.

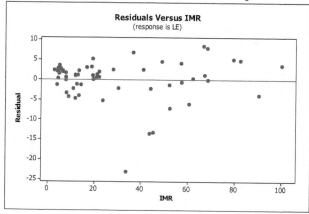

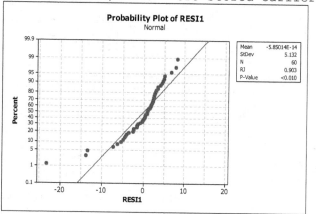

The first graph shows several residuals somewhat farther from the horizontal axis than the rest of the points and there are two residuals at the far right that indicate potential influential observations. This graph does not show a band of residuals centered on and symmetric about the horizontal axis. The normal probability plot shows several points well off the line and the Ryan-Joiner test has a P-value that is less than 0.01. Thus, Assumption 3 for making regression inferences appears to be violated. In addition, the concern about several outliers and potential influential observations indicates that the tests, confidence interval and prediction interval produced earlier may not be valid.

18. Using Minitab, with the data in columns named HIGH and PRECIPITATION, we choose **Stat ▶ Regression ▶ Regression...**, select <u>PRECIPITATION</u> in the **Response:** text box, select <u>HIGH</u> in the **Predictors:** text box, click on the **Graphs** button, click in the **Residuals versus the variables** text box and select <u>HIGH</u>, and click **OK**. Looking ahead to parts (b-h) and (j), click on the **Options...** button, enter <u>83</u> in the **Prediction intervals for new observations:** text box, enter <u>95</u> in the **Confidence level:** text box, and click **OK**. Click on the **Storage...** button, check the **Residuals** box, click

OK, and click **OK**. Then choose **Stat ▶ Basic statistics ▶ Normality test...**, select <u>RESI1</u> in the **Variable** text box, select the **Ryan-Joiner** option button from the **Tests for Normality** field, and click **OK**. We will first show all of the written output and then answer the questions. The graphical output will be shown in part (k).

```
The regression equation is
PRECIPITATION = 2.05 + 0.0067 HIGH

Predictor    Coef    SE Coef     T      P
Constant    2.048     2.170   0.94  0.350
HIGH       0.00667   0.02742  0.24  0.809

S = 2.41336   R-Sq = 0.1%   R-Sq(adj) = 0.0%
```

Copyright © 2012 Pearson Education, Inc. Publishing as Addison-Wesley.

```
Analysis of Variance

Source           DF      SS      MS      F      P
Regression        1    0.344   0.344   0.06   0.809
Residual Error   46  267.919   5.824
Total            47  268.263
```

Unusual Observations

```
Obs  HIGH  PRECIPITATION   Fit   SE Fit  Residual  St Resid
 6    86         8.100   2.621   0.410     5.479    2.30R
10    87         9.600   2.628   0.425     6.972    2.93R
22    45         0.500   2.348   0.972    -1.848   -0.84 X
45    93         7.900   2.668   0.537     5.232    2.22R
```

R denotes an observation with a large standardized residual.
X denotes an observation whose X value gives it large influence.

Predicted Values for New Observations

```
New
Obs    Fit   SE Fit      95% CI            95% PI
 1   2.601   0.373  (1.850, 3.353)   (-2.314, 7.517)
```

Values of Predictors for New Observations

```
New
Obs  HIGH
 1   83.0
```

(a) PRECIPITATION = 2.05 + 0.0067 HIGH
(b) The standard error of the estimate is 2.41336. Roughly speaking, this indicates how much, on average, the predicted average July precipitations differ from the observed values of precipitations.
(c) The coefficient of determination is 0.001, indicating that 0.1% of the variation in average July precipitation is explained by a linear relationship with the average July high temperature. The t-test for the slope results in t = 0.24 with a P-value of 0.809. Thus, we would not reject the null hypothesis that $\beta_1 = 0$. This is a very weak relationship and indicates that the average July high temperature is useless for predicting average July precipitation if the assumptions for making such inferences are valid.
(d) The point estimate for the conditional mean of average July precipitation for an average July high temperature of 83 degrees is 0.373 inches. This is the point on the sample regression line that best estimates the corresponding point on the population regression line.
(e) The 95% confidence interval for the conditional mean of average July precipitation for an average July high temperature of i3 degrees is 1.850 to 3.353 inches. We are 95% confident that the mean average July precipitation is somewhere between these limits when the average July high temperature is 83 degrees.
(f) The predicted value of average July precipitation for a city with an average July high temperature of 83 degrees is 0.373 inches. This is our best estimate of average July precipitation for a city with an average July high temperature of 83 degrees.
(g) The 95% prediction interval for the average July precipitation of a city with an average high July temperature of 83 degrees is -2.314 to 7.517 inches. The lower prediction limit can be changed to 0.00 since negative precipitation amounts are not possible. We are 95% certain that the average July precipitation is somewhere between these limits for a city with an average July high temperature of 83 degrees.

Copyright © 2012 Pearson Education, Inc. Publishing as Addison-Wesley.

(h) The error in the estimate of the mean average July precipitation of cities with average July high temperature of 83 degrees is due only to the fact that the population regression line is being estimated by a sample regression line; whereas, the error in the prediction of the average July precipitation of a single city with an average July high temperature of 83 degrees is due to that fact plus the variation in average July precipitation for cities with average high July temperatures of 83 degrees.

(i) Choose **Stat ▶ Basic Statistics ▶ Correlation**, enter <u>HIGH</u> and <u>PRECIPITATION</u> in the **Variables** text box, check the box to **Display P-values**, and click **OK**. The result is
    Pearson correlation of HIGH and PRECIPITATION = 0.036
    P-Value = 0.809
Since the P-value is greater than 0.05, we do not reject the null hypothesis that $\rho = 0$.

(j) For the residual analysis, we look at the graph produced by the procedure at the beginning of this solution and also the results of a Ryan-Joiner normality test on the residuals, which were stored earlier.

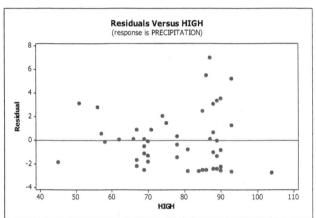

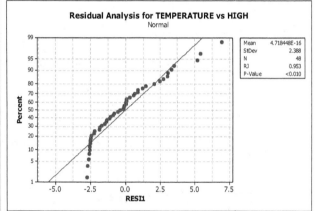

The first graph does not show a band of residuals centered on and symmetric about the horizontal axis. They may even be somewhat U-shaped. The normal probability plot shows several points well off the line and the Ryan-Joiner test has a P-value that is less than 0.01. Thus, Assumption 3 for making regression inferences appears to be violated. In addition, since a linear model accounts for only 0.1% of the variation in average July precipitation, it appears that Assumption 1 (of a linear model) is also violated. The regression inferences above are very unlikely to be valid and are pretty useless.

19. Using Minitab, with the data in columns named FAT CONSUMPTION and DEATH RATE, we choose **Stat ▶ Regression ▶ Regression...**, select `DEATH RATE` in the **Response:** text box, select `FAT CONSUMPTION` in the **Predictors:** text box, click on the **Graphs** button, click in the **Residuals versus the variables** text box and select `FAT CONSUMPTION`, and click **OK**. Looking ahead to parts (b-h) and (j), click on the **Options...** button, enter <u>92</u> in the **Prediction intervals for new observations:** text box, enter <u>95</u> in the **Confidence level:** text box, and click **OK**. Click on the **Storage...** button, check the **Residuals** box, click **OK**, and click **OK**. Then choose **Stat ▶ Basic statistics ▶ Normality test...**, select <u>RESI1</u> in the **Variable** text box, select the **Ryan-Joiner** option button from the **Tests for Normality** field, and click **OK**. We will first show all of the written output and then answer the questions.

Copyright © 2012 Pearson Education, Inc. Publishing as Addison-Wesley.

The graphical output will be shown in part (k).

```
The regression equation is
DEATH RATE = - 1.06 + 0.113 FAT CONSUMPTION

Predictor            Coef   SE Coef      T      P
Constant           -1.063    1.170   -0.91   0.371
FAT CONSUMPTION   0.11336   0.01126   10.06   0.000

S = 2.29453   R-Sq = 78.3%   R-Sq(adj) = 77.6%

Analysis of Variance

Source            DF      SS      MS      F       P
Regression         1  533.31  533.31  101.30   0.000
Residual Error    28  147.42    5.26
Total             29  680.73

Unusual Observations

Obs  FAT CONSUMPTION  DEATH RATE    Fit  SE Fit  Residual  St Resid
  5              96      4.800   9.820   0.419    -5.020    -2.23R
 30             132     18.400  13.901   0.575     4.499     2.03R

R denotes an observation with a large standardized residual.

Predicted Values for New Observations

New
Obs    Fit  SE Fit       95% CI           95% PI
  1  9.366   0.423  (8.500, 10.232)  (4.587, 14.145)

Values of Predictors for New Observations

New
Obs  FAT CONSUMPTION
  1             92.0
```

(a) The regression equation is DEATH RATE = - 1.06 + 0.113 FAT CONSUMPTION
(b) The standard error of the estimate is 2.29453. Roughly speaking, this indicates how much, on average, the predicted death rates differ from the observed values of death rates.
(c) The coefficient of determination is 0.783, indicating that 78.3% of the variation in death rates is explained by a linear relationship with the fat consumption. The t-test for the slope results in t = 10.06 with a P-value of 0.00. Thus, we would reject the null hypothesis that $\beta_1 = 0$. This is a fairly strong relationship and indicates that fat consumption is quite useful for predicting death rates if the assumptions for making such inferences are reasonable.
(d) The point estimate for the conditional mean death rate for a country with a fat consumption of 92 grams per day is 9.366 per 100,000 males. This is the point on the sample regression line that best estimates the corresponding point on the population regression line.
(e) The 95% confidence interval for the conditional mean of death rates for countries with a fat consumption of 92 grams per day is 8.500 to 10.232 per 100,000 males. We are 95% confident that the mean death rate is somewhere between these limits when the average fat consumption is 92 grams per day.
(f) The predicted death rate for a country with an average fat consumption of 92 grams per day is 9.366 per 100,000 males. This is our best estimate of the death rate for a country with an average fat consumption of 92 grams per day.

Copyright © 2012 Pearson Education, Inc. Publishing as Addison-Wesley.

(g) The 95% prediction interval for the death rate of a country with an average fat consumption of 92 grams per day is 4.587 to 14.145 per 100,000 males. We are 95% certain that the death rate is somewhere between these limits for a country with an average fat consumption of 92 grams per day.

(h) The error in the estimate of the mean death rate of countries with an average fat consumption of 92 grams per day is due only to the fact that the population regression line is being estimated by a sample regression line; whereas, the error in the prediction of the death rates of a single country with an average fat consumption of 92 grams per day is due to that fact plus the variation in death rates for countries with an average fat consumption of 92 grams per day.

(i) Choose **Stat ▶ Basic Statistics ▶ Correlation**, enter 'DEATH RATE' and 'FAT CONSUMPTION' in the **Variables** text box, check the box to **Display P-values**, and click **OK**. The result is
Pearson correlation of DEATH RATE and FAT CONSUMPTION = 0.885
P-Value = 0.000

The P-value for a right-tailed test is half of that shown (still 0.000) if r is positive. Since the P-value is less than 0.05, we reject the null hypothesis that $\rho = 0$.

(j) For the residual analysis, we look at the graph produced by the procedure at the beginning of this solution and also the results of a Ryan-Joiner normality test on the residuals, which were stored earlier.

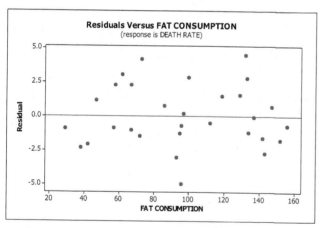

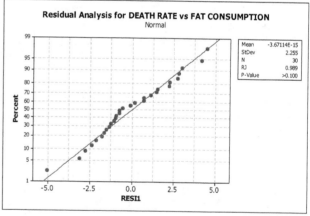

In the left-hand graph, the residuals are located in a horizontal band centered on and somewhat symmetrical about the horizontal axis. The P-value for the Ryan-Joiner test is > 0.100 indicating that normality of the residuals is a reasonable assumption. Since the linear model also appears to be appropriate, the regression inferences made earlier in this problem are likely to be reasonable.

20. Using Minitab, with the data in columns named FIRST ROUND and SECOND ROUND, we choose **Stat ▶ Regression ▶ Regression...**, select SECOND ROUND in the **Response:** text box, select FIRST ROUND in the **Predictors:** text box, click on the **Graphs** button, click in the **Residuals versus the variables** text box and select FIRST ROUND, and click **OK**. Looking ahead to parts (b-h) and (j), click on the **Options...** button, enter 72 in the **Prediction intervals for new observations:** text box, enter 95 in the **Confidence level:** text box, and click **OK**. Click on the **Storage...** button, check the **Residuals** box, click

Copyright © 2012 Pearson Education, Inc. Publishing as Addison-Wesley.

OK, and click **OK**. Then choose **Stat ▶ Basic statistics ▶ Normality test...**, select <u>RESI1</u> in the **Variable** text box, select the **Ryan-Joiner** option button from the **Tests for Normality** field, and click **OK**. We will first show all of the written output and then answer the questions. The graphical output will be shown in part (k).

```
The regression equation is
SECOND ROUND = 56.9 + 0.227 FIRST ROUND

Predictor        Coef   SE Coef      T      P
Constant       56.917     6.202   9.18  0.000
FIRST ROUND   0.22750   0.08265   2.75  0.007

S = 3.05993   R-Sq = 7.9%   R-Sq(adj) = 6.9%

Analysis of Variance

Source           DF       SS       MS      F      P
Regression        1   70.941   70.941   7.58  0.007
Residual Error   88  823.959    9.363
Total            89  894.900

Unusual Observations

       FIRST  SECOND
Obs    ROUND   ROUND     Fit  SE Fit  Residual  St Resid
  7     71.0  67.000  73.069   0.459    -6.069    -2.01R
 24     80.0  67.000  75.117   0.528    -8.117    -2.69R
 80     76.0  81.000  74.207   0.334     6.793     2.23R
 82     84.0  75.000  76.027   0.815    -1.027    -0.35 X
 84     78.0  81.000  74.662   0.410     6.338     2.09R
 85     79.0  81.000  74.889   0.465     6.111     2.02R
 86     79.0  81.000  74.889   0.465     6.111     2.02R
 89     89.0  74.000  77.164   1.206    -3.164    -1.13 X
 90     80.0  84.000  75.117   0.528     8.883     2.95R

R denotes an observation with a large standardized residual.
X denotes an observation whose X value gives it large leverage.

Predicted Values for New Observations

New Obs     Fit  SE Fit       95% CI              95% PI
      1  73.297   0.404  (72.494, 74.100)  (67.163, 79.431)

Values of Predictors for New Observations

          FIRST
New Obs   ROUND
      1    72.0
```

(a)   The regression equation is SECOND ROUND = 56.9 + 0.227 FIRST ROUND

(b)   The standard error of the estimate is 3.05993. Roughly speaking, this indicates how much, on average, the predicted values of the second round differ from the observed values of the second round.

(c)   The coefficient of determination is 0.079, indicating that 7.9% of the variation in second round scores is explained by a linear relationship with the first round scores. The t-test for the slope results in t = 2.75 with a P-value of 0.007. Thus, we would reject the null hypothesis that $\beta_1 = 0$. Even though the P-value is small, the low coefficient of determination indicates that this is a fairly weak relationship and that first round score is not very useful in making

Copyright © 2012 Pearson Education, Inc. Publishing as Addison-Wesley.

predictions of the second round score if the assumptions for making such inferences are valid.

(d) The point estimate for the conditional mean second round score for a player with a first round score of 72 is 73.297. This is the point on the sample regression line that best estimates the corresponding point on the population regression line.

(e) The 95% confidence interval for the conditional mean second round score for players with a first round score of 72 is 72.494 to 74.100.. We are 95% confident that the mean second round score is somewhere between these limits for players with a first round score of 72.

(f) The predicted second round score for a player with a first round score of 72 is 73.297. This is our best estimate of the second round score for a player with a first round score of 72.

(g) The 95% prediction interval for the second round score for a player with a first round score of 72 is 67.163 to 79.431. We are 95% certain that the second round score is somewhere between these limits for a player with a first round score of 72.

(h) The error in the estimate of the mean second round score for players with a first round score of 72 is due only to the fact that the population regression line is being estimated by a sample regression line; whereas, the error in the prediction of the second round score of a player with a first round score of 72 is due to that fact plus the variation in second round scores for players with a first round score of 72.

(i) Choose **Stat ▸ Basic Statistics ▸ Correlation**, enter <u>FIRST ROUND</u> and <u>SECOND ROUND</u> in the **Variables** text box, check the box to **Display P-values**, and click **OK**. The result is

    Pearson correlation of FIRST ROUND and SECOND ROUND = 0.282
    P-Value = 0.007

The P-value for a right-tailed test is half of that shown (0.007/2 = 0.0035) if r is positive. Since the P-value is less than 0.05, we reject the null hypothesis that $\rho = 0$.

(j) For the residual analysis, we look at the graph produced by the procedure at the beginning of this solution and also the results of a Ryan-Joiner normality test on the residuals, which were stored earlier.

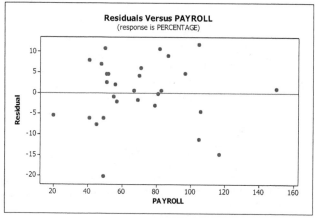
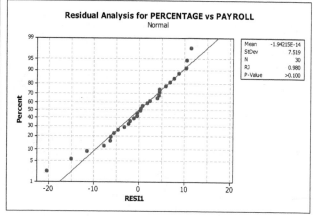

In the left-hand graph, the residuals are located in a horizontal band centered on and somewhat symmetrical about the horizontal axis. The P-value for the Ryan-Joiner test is > 0.100 indicating that normality of the residuals is a reasonable assumption. Since the linear model also appears to be appropriate, although not very useful, the regression inferences made earlier in this problem are likely to be reasonable.

Copyright © 2012 Pearson Education, Inc. Publishing as Addison-Wesley.